THE
INTERNATIONAL SERIES
OF
MONOGRAPHS ON PHYSICS

GENERAL EDITORS

J. BIRMAN   S.F. EDWARDS   R. FRIEND
C.H. LLEWELLYN SMITH   M. REES
D. SHERRINGTON   G. VENEZIANO

INTERNATIONAL SERIES OF MONOGRAPHS ON PHYSICS

105. Y. Kuramoto, Y. Kitaoka: *Dynamics of heavy electrons*
104. D. Bardin, G. Passarino: *The Standard Model in the making*
103. G. C. Branco, L. Lavoura, J. P. Silva: *CP violation*
102. T. C. Choy: *Effective medium theory*
101. H. Araki: *Mathematical theory of quantum fields*
100. L. M. Pismen: *Vortices in nonlinear fields*
 99. L. Mestel: *Stellar magnetism*
 98. K. H. Bennemann: *Nonlinear optics in metals*
 97. D. Salzmann: *Atomic physics in hot plasmas*
 96. M. Brambilla: *Kinetic theory of plasma waves*
 95. M. Wakatani: *Stellarator and heliotron devices*
 94. S. Chikazumi: *Physics of ferromagnetism*
 93. A. Aharoni: *Introduction to the theory of ferromagnetism*
 92. J. Zinn-Justin: *Quantum field theory and critical phenomena*
 91. R. A. Bertlmann: *Anomalies in quantum field theory*
 90. P. K. Gosh: *Ion traps*
 89. E. Simánek: *Inhomogeneous superconductors*
 88. S. L. Adler: *Quaternionic quantum mechanics and quantum fields*
 87. P. S. Joshi: *Global aspects in gravitation and cosmology*
 86. E. R. Pike, S. Sarkar: *The quantum theory of radiation*
 84. V. Z. Kresin, H. Morawitz, S. A. Wolf: *Mechanisms of conventional and high T_c superconductivity*
 83. P. G. de Gennes, J. Prost: *The physics of liquid crystals*
 82. B. H. Bransden, M. R. C. McDowell: *Charge exchange and the theory of ion–atom collision*
 81. J. Jensen, A. R. Mackintosh: *Rare earth magnetism*
 80. R. Gastmans, T. T. Wu: *The ubiquitous photon*
 79. P. Luchini, H. Motz: *Undulators and free-electron lasers*
 78. P. Weinberger: *Electron scattering theory*
 76. H. Aoki, H. Kamimura: *The physics of interacting electrons in disordered systems*
 75. J. D. Lawson: *The physics of charged particle beams*
 73. M. Doi, S. F. Edwards: *The theory of polymer dynamics*
 71. E. L. Wolf: *Principles of electron tunneling spectroscopy*
 70. H. K. Henisch: *Semiconductor contacts*
 69. S. Chandrasekhar: *The mathematical theory of black holes*
 68. G. R. Satchler: *Direct nuclear reactions*
 51. C. Møller: *The theory of relativity*
 46. H. E. Stanley: *Introduction of phase transitions and critical phenomena*
 32. A. Abragam: *Principles of nuclear magnetism*
 27. P. A. M. Dirac: *Principles of quantum mechanics*
 23. R. E. Peierls: *Quantum theory of solids*

Dynamics of Heavy Electrons

Y. KURAMOTO

Department of Physics
Tohoku University

Y. KITAOKA

Department of Physical Science
Osaka University

CLARENDON PRESS • OXFORD

2000

OXFORD

UNIVERSITY PRESS

Great Clarendon Street, Oxford OX2 6DP

Oxford University Press is a department of the University of Oxford.
It furthers the University's objective of excellence in research, scholarship,
and education by publishing worldwide in

Oxford   New York

Athens   Auckland   Bangkok   Bogotá   Buenos Aires   Calcutta
Cape Town   Chennai   Dar es Salaam   Delhi   Florence   Hong Kong   Istanbul
Karachi   Kuala Lumpur   Madrid   Melbourne   Mexico City   Mumbai
Nairobi   Paris   São Paulo   Singapore   Taipei   Tokyo   Toronto   Warsaw

and associated companies in   Berlin   Ibadan

Oxford is a registered trade mark of Oxford University Press
in the UK and in certain other countries

Published in the United States
by Oxford University Press Inc., New York

© Y. Kuramoto and Y. Kitaoka, 2000

A catalogue record for this book is available from the British Library

Library of Congress Cataloging in Publication Data

Kuramoto, Y. (Yoshio)
Dynamics of heavy electrons / Y. Kuramoto, Y. Kitaoka.
(The international series of monographs on physics; 105)
1. Mesons.   2. Energy-band theory of solids.   I. Kitaoka, Y. (Yoshio)
II. Title.   III. Series: International series of monographs on physics ; v. 105.
QC793.5.M42K87   2000   530.4'1–dc21   99–42821

ISBN 0 19 851767 X

Typeset by Newgen Imaging Systems (P) Ltd., Chennai, India

Printed in Great Britan
on acid-free paper by
TJ International Ltd., Padstow

PREFACE

The present book has two-fold purposes. First, the book is intended to be a monograph on heavy electrons, which have been the focus of very active experimental and theoretical studies in the last two decades. Heavy electrons are found in a number of lanthanide and actinide compounds, and are characterized by a large (comparable to the mass of a muon) effective mass. The heavy electrons exhibit rich phenomena, such as smooth crossover to local-moment behaviour with increasing temperature, unconventional superconductivity, weak antiferromagnetism and pseudo-metamagnetism. Although some mysteries still remain, the authors feel that, overall, a reasonably coherent understanding of heavy electrons is available as a whole. Therefore, a survey of the properties of heavy electrons from a global and unified point of view is in order.

The physics of heavy electrons is one of the typical examples where close interaction between theory and experiment has revealed systematically the origin of the seemingly incredible behaviour of electrons. One of the present authors (Kuramoto) has been engaged in formulating a theoretical scheme which is now applied widely to magnetic impurities and heavy electrons and, among other things, has derived the dynamical response functions. The other author (Kitaoka) has been involved in NMR experiments on heavy-electron systems and has observed the formation of metallic, insulating, super-conducting, and weakly magnetic phases of heavy electrons through the NMR. Thus, the authors think it appropriate for them to write a book which focuses on the dynamical aspects of heavy electrons. In view of the innumerable and diverse activities in the field, it is difficult to treat every aspect in detail, although the important thermodynamic and transport results have been covered reasonably well.

The second purpose of the book is to serve as an advanced textbook on the theoretical and experimental physics of strongly correlated electrons. The necessity to understand the fascinating experimental results on heavy electrons has stimulated intensive theoretical efforts to go beyond the existing schemes. It has turned out that many established ideas and techniques are insufficient for understanding heavy electrons. Theoretically, the extensive use of quantum fluctuation techniques has lowered the superiority of the celebrated mean-field theories. On the other hand, extreme experimental conditions, such as application of strong magnetic fields and pressures at ultralow temperatures, are required. Thus, heavy-electron systems have been a target of a case study for applying and testing almost all tools in theoretical and experimental condensed-matter physics. Those graduate students and researchers who want to work on strongly correlated condensed-matter systems will find in the book many examples on how the conventional concepts in general solid systems work or do not work in the heavy-electron systems. Although we try to make the book self-contained for those readers who have the knowledge of condensed-matter physics at the undergraduate level, they may sometimes find the description of the book too concise. In such a case, we advise consulting any of the available textbooks on elementary many-body physics.

The book tries to build a coherent picture of heavy electrons from a collection of apparently diverse experimental and theoretical results. However, the authors neither intend to be exhaustive in citing papers, nor are they excessively on the alert for including all of the recent developments. At any rate, a monograph cannot compete with a review in scientific journals for covering the latest findings. Instead, the book tries to provide a broad outlook which should be especially useful to newcomers to the field of strongly correlated electrons, experimental or theoretical, and tries to help them to initiate a new stage of investigation. Thus, the book emphasizes new theoretical methods and concepts which might be effective also in other related areas. It also emphasizes the utility and the limitation of the dynamical information brought about by NMR and neutron scattering. Special attention is paid to similarity with other strongly correlated systems, e.g. the copper oxide high-temperature superconductors. Since the research area of heavy electrons is vast, our treatment inevitably reflects our favourite choice of the topics, and pays unavoidable attention to our own contribution to the field. We would apologize to those authors whose contributions are not touched upon, or are not treated properly. Especially, we have omitted many topics on transport, thermodynamic properties, and dynamical measurements other than the NMR and neutron scattering.

As an existing reference on the Kondo effect and heavy electrons, we mention a book by A. C. Hewson (Cambridge 1993). The present book puts more emphasis on the periodic systems rather than magnetic impurities, and also on experimental aspects. Hence, we hope that the present monograph serves to be complementary to Hewson's book. In order to keep the size of the book reasonable, we omit some interesting theoretical topics like the Bethe Ansatz and the conformal field theory. These topics are extensively dealt with in other monographs.

The contents of the book are arranged as follows: Chapter 1 presents first the heavy-electron phenomena from thermodynamic measurements, and then introduces the two basic pictures of electrons in solids: itinerant and localized states. These pictures represent opposite limiting behaviours of electrons and do not reconcile with each other at the naive level. A brief account of the Fermi-liquid theory is given. Some basic aspects of NMR and neutron scattering experiments are explained.

Chapter 2 deals with the single-magnetic-impurity problem. The Kondo effect is discussed from various viewpoints; scaling, $1/n$ expansion, local Fermi liquid, and so on. We emphasize an effective Hamiltonian approach which includes the concept of the renormalization group. The local non-Fermi-liquid state is treated with emphasis on the unique excitation spectrum. Detailed experimental results are presented that can be interpreted without taking account of the intersite interactions among f electrons.

In Chapter 3, we turn to the periodic systems where heavy electrons are formed. Experimental facts on both metallic and insulating systems are presented. Although the metallic ground state is basically understood by the Fermi-liquid picture, the system shows strange behaviours in a magnetic field. We present intuitive ideas and a theoretical apparatus for understanding the basic features of heavy electrons. Practical methods for computing physical quantities such as the density of states and the dynamical susceptibility are explained.

These two chapters as a whole represent complementary approaches to heavy electrons from the local and itinerant limits. In Chapter 4, we discuss the anomalous magnetism which emerges as a result of a complicated interference between quasi-particles. The relevant phenomena include the weak antiferromagnetism with tiny ordered moments, and the orbital ordering accompanying the quadrupole moments. Alternatively, the anomalous magnetism is interpreted as originating from a competition between the Kondo screening and the intersite interaction, if one adopts the viewpoint from high temperatures. Although there is no established understanding of these phenomena, we present a possible view which emphasizes the dual nature of electrons with strong correlations: the simultaneous presence of itinerant and localized characters.

Chapter 5 deals with the superconductivity of heavy electrons. We begin with an introductory treatment of anisotropic pairing in general and symmetry classification of the singlet and triplet pairings. Then, a detailed review of the experimental situation for representative superconducting systems is presented. We emphasize the interplay between superconductivity and anomalous magnetism, and subtle coupling to the lattice degrees of freedom.

In Chapter 6, we compare heavy-electron systems with some cuprates which show high-temperature superconductivity. The cuprates also have strongly correlated electrons, but the parameters characterizing the electrons are rather different from those for heavy electrons. Interestingly, some dynamical properties, such as the NMR relaxation rate below the superconducting transition, show similar temperature dependence if one scales the temperature by the transition temperature of each system. We discuss similarities and differences between these two systems, and pursue a unified picture for understanding the strongly correlated electrons including the non-Fermi-liquid states.

In the appendices, we summarize the theoretical techniques frequently used in the many-body theory. They constitute a compact treatise which should be understandable, independently of the content of the main text.

It took us an unexpectedly long time to complete the writing of this book. During this period, we have benefited from discussions with our colleagues and friends. We are grateful to all of them for helping our understanding on the subject, and especially to Erwin Müller-Hartmann, who introduced one of the authors (Kuramoto) to the early stage of the field 20 years ago, and to Kunisuke Asayama and Hiroshi Yasuoka, who introduced the other author (Kitaoka) to the field of NMR in condensed-matter physics. We acknowledge fruitful cooperation with and valuable suggestions by Jacques Flouquet, Kenji Ishida, Yusuke Kato, Hiroaki Kusunose, Frank Steglich, Hideki Tou, and Hisatoshi Yokoyama. Special thanks are due to Frank Steglich, Yusuke Kato, and Stephen Julian, who read the first manuscript and made a number of constructive remarks to improve the book.

<table>
<tr><td>Sendai</td><td>Yoshio Kuramoto</td></tr>
<tr><td>Osaka</td><td>Yoshio Kitaoka</td></tr>
<tr><td>June 1999</td><td></td></tr>
</table>

CONTENTS

1

FUNDAMENTAL PROPERTIES OF ELECTRONS IN SOLIDS

1.1 What are heavy electrons?

Electrons in metals have an effective mass m^* which is in general different from the mass in vacuum, $m_0 = 9.1 \times 10^{-28}$ g. Experimentally, m^* can be measured by such methods as cyclotron resonance, de Haas–van Alphen (dHvA) effect, specific heat, and so on. The temperature dependence of the specific heat C much below the room temperature is usually fitted in the form

$$C = \gamma T + \beta T^3, \tag{1.1}$$

where the second term comes from lattice vibrations in ordinary metals. In the free electron model of metals with the spherical Fermi surface, γ is given by

$$\gamma = \frac{\pi^2}{3} k_{\mathrm{B}}^2 \rho^*(\mu), \tag{1.2}$$

where k_{B} denotes the Boltzmann constant and $\rho^*(\mu) = m^* k_{\mathrm{F}}/(\pi^2 \hbar^2)$ is the density of states at the Fermi level μ. For a general shape of the Fermi surface, the effective mass in the above formula should be interpreted as the average over the Fermi surface with the same volume enclosed as that of a sphere with radius k_{F}. The spin part χ of the magnetic susceptibility at low temperatures is also determined by $\rho^*(\mu)$:

$$\chi = (g\mu_{\mathrm{B}})^2 \rho^*(\mu), \tag{1.3}$$

where g is the g-factor close to 2 and $\mu_{\mathrm{B}} = e\hbar/(2m_0 c)$ is the Bohr magneton.

In simple metals such as K or Al, γ is of the order of mJ/(K^2 mol). This means that the effective mass is of the order of m_0. However, in some materials it becomes larger than m_0 by as much as three orders of magnitude. As an example, the temperature dependence of γ is shown in Fig. 1.1 for $\mathrm{Ce}_x\mathrm{La}_{1-x}\mathrm{Cu}_6$ with varying Ce concentrations [1,2]. We interpret the large γ as coming from the large effective mass, since k_{F} is determined by the density of conduction electrons, which is of the order of 10^{22} cm^{-3} independent of metals. Hence, these compounds are called heavy-electron (or heavy-fermion) systems. The validity of this interpretation depends on the Fermi-liquid theory, which is explained later. It is remarkable that these data fall on a single curve if divided by x, which means that each Ce ion contributes to the enhanced specific heat rather independently. In other words, possible intersite interaction among Ce ions does not play an important role in γ. Thus, the steep increase of γ below 10 K is essentially a single-site effect.

The magnetic susceptibility for various Ce concentrations is shown in Fig. 1.2. Like the specific heat, the susceptibility per mole of Ce is almost independent of x up to

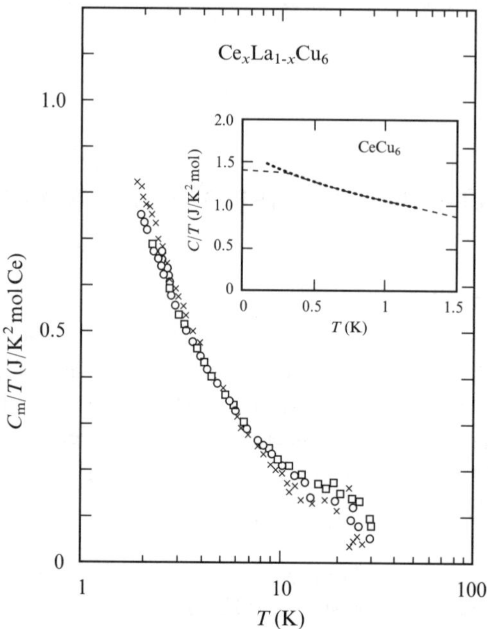

FIG. 1.1. Temperature dependence of 4f-derived specific heat per mole of Ce as C_m/T plotted against T for $Ce_xLa_{1-x}Cu_6$ with $x = 1$ (o), 0.8 ($\square$) and 0.5 ($\times$) [1]. Inset indicates low-T data. Dashed curve is the calculation for an $S = 1/2$ Kondo impurity with $T_K = 4.2$ K.

$x = 0.9$ [1]. In general, thermodynamic data point to the importance of the single-ion effect. Most of the apparent deviation from the single-ion behaviour can be understood by taking into account the change of the lattice parameter and other peripheral effects. As we shall discuss in detail later in this book, the origin of heavy effective masses is traced back to strong local correlations. As explained in detail later, the Kondo effect is the key to understand all the anomalous properties in a universal fashion.

Table 1.1 summarizes the values of γ and χ for representative compounds. The average valency of Ce and Yb in heavy-electron systems is very close to 3+. However, in some Yb materials which have smaller values of γ, of the order of 10^2 mJ/(K^2 mol), the average valency is between 2+ and 3+. These systems are called intermediate valence compounds. Since the valency fluctuates quantum mechanically between 2+ and 3+, the phenomenon is also called valence fluctuation, a term with emphasis on the dynamical aspect.

The transport properties of these materials also show an unusual temperature dependence. As an illustration, Fig. 1.3 shows the temperature dependence of the magnetic part of the resistivity for $Ce_xLa_{1-x}Cu_6$ [1]. The magnetic part is obtained by subtracting the resistivity of LaCu$_6$ with no f electrons. In the dilute limit, nearly logarithmic increase of the resistivity is observed up to over two orders of T. This is a typical signature of the Kondo effect. On the other hand, in the pure limit of CeCu$_6$, this logarithmic dependence disappears as $T \rightarrow 0$. Instead, the T^2 law of the resistivity holds. The latter behaviour is

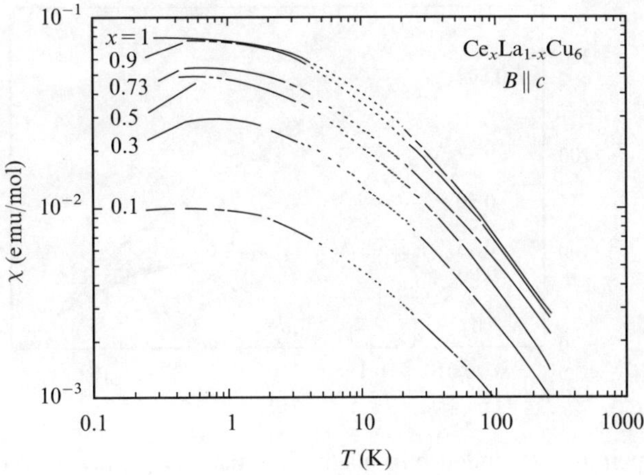

FIG. 1.2. Temperature dependence of the magnetic susceptibility per mole of formula unit for $Ce_xLa_{1-x}Cu_6$ [1].

Table 1.1 *Low-temperature susceptibility and specific heat coefficient of some heavy-electron systems. The susceptibilities along the c-axis, shown as (c) in the table, and in the ab-plane shown as (a) can be different in non-cubic systems.*

System	χ (10^{-3} emu/mol)	γ (mJ/mol K^2)
$CeCu_6$	80 (c)	1600 [1]
$CeCu_2Si_2$	8 (c), 4 (ab)	1100 [3]
$CeRu_2Si_2$	30 (c)	350 [4]
UBe_{13}	13	1100 [5]
UPt_3	4 (c), 8 (ab)	400 [6]

typical of the itinerant electrons with strong mutual interaction. Thus, the resistivity of $Ce_xLa_{1-x}Cu_6$ depends on x in a way that is completely different from the dependence of specific heat and susceptibility.

In order to study how the heavy electrons are realized in condensed-matter systems, and how one can understand their very rich properties, we begin in the following with the fundamental consideration about electrons in solids. Henceforth, we take units such that $\hbar = k_B = 1$, to simplify the notation, except for deriving dimensional numerical values.

1.2 Itinerant and localized states

Real solids are quantum many-body systems consisting of a large number of atomic nuclei and electrons. We simplify this complicated system by keeping only those degrees

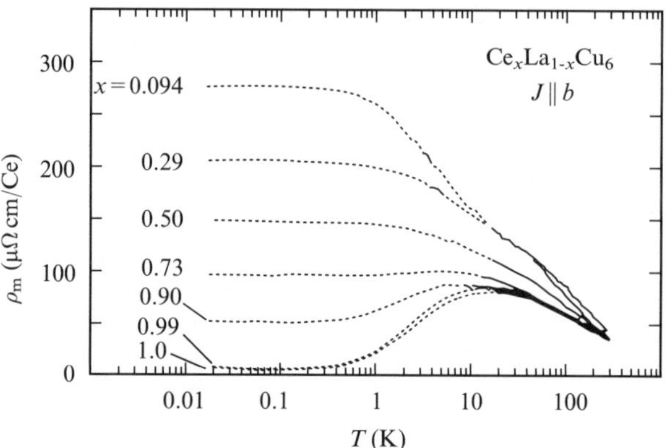

FIG. 1.3. Temperature dependence of magnetic resistivity [1] per mole of formula unit for $Ce_xLa_{1-x}Cu_6$.

of freedom that are essential to realize the most fundamental properties of electrons in solids. In order to clarify the point step by step, we begin with the simplest model that shows the essence of itinerant as well as the localized nature of electrons in solids. Consider the simple cubic lattice of hydrogen atoms. We assume, for simplicity, that the protons have no spin and are fixed at lattice points R_i with distance a from the nearest neighbours. By this simplification, we neglect the effect of lattice vibrations present in real solids. The hydrogen lattice does not have incomplete atomic shells with 3d, 4f, or 5f orbitals which are important for the occurrence of heavy electrons. These complications will be discussed in later chapters.

The second-quantization Hamiltonian is given by

$$H = \sum_\sigma \int d\boldsymbol{r}\, \Psi_\sigma^\dagger(\boldsymbol{r}) \left[-\frac{\Delta}{2m} - \sum_i v(\boldsymbol{r} - \boldsymbol{R}_i) \right] \Psi_\sigma(\boldsymbol{r}) + \frac{1}{2} \sum_{i \neq j} v(\boldsymbol{R}_i - \boldsymbol{R}_j) + V_C$$

(1.4)

where $\Psi_\sigma(\boldsymbol{r})$ is the field operator of electron with spin σ, $v(\boldsymbol{r}) = e^2/|\boldsymbol{r}|$, and V_C is the Coulomb interaction among the electrons:

$$V_C = \frac{1}{2} \int d\boldsymbol{r} \int d\boldsymbol{r}'\, v(\boldsymbol{r} - \boldsymbol{r}') \sum_{\sigma,\sigma'} \Psi_\sigma^\dagger(\boldsymbol{r})\Psi_{\sigma'}^\dagger(\boldsymbol{r}')\Psi_{\sigma'}(\boldsymbol{r}')\Psi_\sigma(\boldsymbol{r}). \qquad (1.5)$$

Two extreme cases are considered: (i) a is much smaller than the Bohr radius $a_B = 1/(me^2)$, and (ii) $a \gg a_B$. Figure 1.4 illustrates the charge density of electrons in each case.

1.2.1 *Formation of energy bands*

In the case (i), the wave functions of each 1s state overlap with each other substantially. As a result, mixing with 2s, 2p, and higher orbitals is so significant that the wave function is better constructed from plane waves. The kinetic energy of the electrons is the most

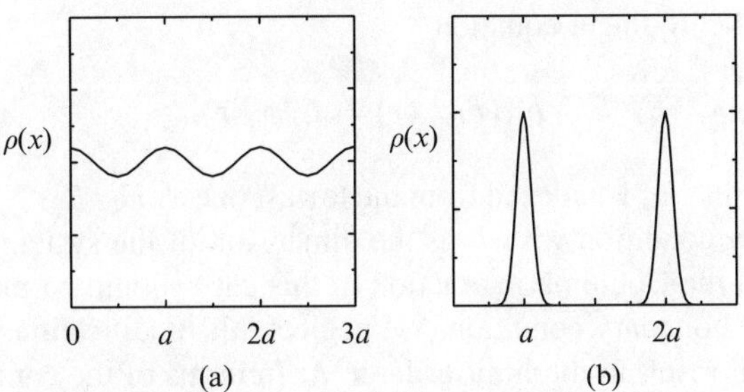

FIG. 1.4. Schematic view of the charge density of electrons: (a) the case of $a \ll a_{\mathrm{B}}$, and (b) $a \gg a_{\mathrm{B}}$.

important part and the fluctuation $\langle (n_\sigma(r) - \langle n_\sigma(r)\rangle)(n_{\sigma'}(r') - \langle n_{\sigma'}(r')\rangle)\rangle$ of the operator $n_\sigma(r) = \Psi_\sigma^\dagger(r)\Psi_\sigma(r)$ becomes small as compared with $\langle n_\sigma(r)\rangle \langle n_{\sigma'}(r')\rangle$. Then the simplest approximation is to replace the electron–electron interaction by a mean field acting on $n_\sigma(r)$:

$$\Psi_\sigma^\dagger(r)\Psi_{\sigma'}^\dagger(r')\Psi_{\sigma'}(r')\Psi_\sigma(r) \sim n_\sigma(r)\langle n_{\sigma'}(r')\rangle + \langle n_\sigma(r)\rangle n_{\sigma'}(r')$$
$$- \langle n_\sigma(r)\rangle \langle n_{\sigma'}(r')\rangle. \tag{1.6}$$

This is called the Hartree approximation. Note that eqn (1.6) does not exhaust factorization into fermion bilinear operators. The remaining ones are called exchange terms to be explained shortly. In the Hartree approximation, H is approximated by

$$H_{\mathrm{H}} = \int dr \sum_\sigma \Psi_\sigma^\dagger(r)h_{\mathrm{H}}(r)\Psi_\sigma(r) + E_{\mathrm{D}}, \tag{1.7}$$

where the constant E_{D} is given by

$$E_{\mathrm{D}} = \frac{1}{2}\sum_{i\neq j} v(R_i - R_j) - \frac{1}{2}\int dr \int dr'\, v(r - r') \sum_{\sigma,\sigma'} \langle n_\sigma(r)\rangle \langle n_{\sigma'}(r')\rangle. \tag{1.8}$$

The effective single-particle Hamiltonian $h_{\mathrm{H}}(r)$ is defined as

$$h_{\mathrm{H}}(r) = -\frac{\Delta}{2m} + v_{\mathrm{H}}(r), \tag{1.9}$$

where the one-body potential $v_{\mathrm{H}}(r)$ is given by

$$v_{\mathrm{H}}(r) = \int dr'\, v(r - r')\left[\sum_\sigma \langle n_\sigma(r')\rangle - \sum_j \delta(r' - R_j)\right]. \tag{1.10}$$

We notice that the electron–ion attraction alone would give divergent potential. Only with partial cancellation of it by the electron–electron repulsion does one obtain a finite potential as given by eqn (1.10).

Consider the Schrödinger equation

$$h_{\mathrm{H}}(r)\phi_n(r) = E_n\phi_n(r),\qquad(1.11)$$

where the eigenvalue E_n is indexed from the lowest one as E_1, E_2, ... Here we take the periodic boundary condition with L as the dimension of the system in each direction. Strictly speaking, the Coulomb interaction in this case should be modified in order to meet the periodic boundary condition. We neglect this modification since this does not influence the final result in the limit of large L. In terms of the complete set $\{\phi_n\}$, the field operator is expanded as

$$\Psi_\sigma(r) = \sum_n c_{n\sigma}\phi_n(r),\qquad(1.12)$$

where $c_{n\sigma}$ is the annihilation operator. The N-body ground state $|g\rangle$, with N even, is constructed from the vacuum $|\mathrm{vac}\rangle$ by filling $N/2$ states for each spin:

$$|g\rangle = \prod_\sigma \prod_{n=1}^{N/2} c_{n\sigma}^\dagger |\mathrm{vac}\rangle.\qquad(1.13)$$

In the first quantization, the state $|g\rangle$ is represented by a single Slater determinant. The electron density is given by

$$\langle n_\sigma(r')\rangle = \sum_{n=1}^{N/2} |\phi_n(r')|^2,\qquad(1.14)$$

which in turn determines the Hartree potential $v_{\mathrm{H}}(r)$ self-consistently by eqn (1.10). In the ground state the N electrons fill the lower half of the first band. The set of highest occupied k states constitutes the Fermi surface. The volume V_{F} enclosed by the Fermi surface is related to the electron density $n = N/L^3$ by $V_{\mathrm{F}}/(2\pi)^3 = n/2$.

The Hartree approximation has a weakness that v_{H} gives nonzero electron–electron interaction even when there is only a single electron. This is due to the neglect of the quantum nature of $n_\sigma(r)$, the proper account of which should exclude the self-interaction of an electron. To remedy this, one should take account of other contributions to the mean field from

$$\Psi_\sigma^\dagger(r)\Psi_{\sigma'}^\dagger(r')\Psi_{\sigma'}(r')\Psi_\sigma(r)\qquad(1.15)$$
$$\rightarrow\quad -\langle\Psi_\sigma^\dagger(r)\Psi_{\sigma'}(r')\rangle\Psi_{\sigma'}^\dagger(r')\Psi_\sigma(r) - \Psi_\sigma^\dagger(r)\Psi_{\sigma'}(r')\langle\Psi_{\sigma'}^\dagger(r')\Psi_\sigma(r)\rangle$$
$$+\langle\Psi_\sigma^\dagger(r)\Psi_{\sigma'}(r')\rangle\langle\Psi_{\sigma'}^\dagger(r')\Psi_\sigma(r)\rangle.\qquad(1.16)$$

The approximation which includes this contribution, in addition to $v_{\mathrm{H}}(r)$, is called the Hartree–Fock approximation. In this case the mean field acts not only on $n_\sigma(r)$ but on a nonlocal operator $\Psi_{\sigma'}^\dagger(r')\Psi_\sigma(r)$. If a magnetic order is present, $\langle\Psi_{\sigma'}^\dagger(r')\Psi_\sigma(r)\rangle$ may not be diagonal in spin indices. The scheme to allow for this possibility is often called the unrestricted Hartree–Fock theory.

One obtains the Hamiltonian which is bilinear in the electron fields as

$$H_{\text{HF}} = \int d\boldsymbol{r} \int d\boldsymbol{r}' \sum_{\sigma,\sigma'} \Psi_\sigma^\dagger(\boldsymbol{r}) h_{\sigma\sigma'}(\boldsymbol{r},\boldsymbol{r}') \Psi_{\sigma'}(\boldsymbol{r}') + E_{\text{DX}}, \qquad (1.17)$$

where

$$h_{\sigma\sigma'}(\boldsymbol{r},\boldsymbol{r}') = h_{\text{H}}(\boldsymbol{r}) \delta_{\sigma,\sigma'} \delta(\boldsymbol{r}-\boldsymbol{r}') - v(\boldsymbol{r}-\boldsymbol{r}') \langle \Psi_{\sigma'}^\dagger(\boldsymbol{r}') \Psi_\sigma(\boldsymbol{r}) \rangle, \qquad (1.18)$$

$$E_{\text{DX}} = E_{\text{D}} + \frac{1}{2} \int d\boldsymbol{r} \int d\boldsymbol{r}' \, v(\boldsymbol{r}-\boldsymbol{r}') \sum_{\sigma,\sigma'} \langle \Psi_{\sigma'}^\dagger(\boldsymbol{r}') \Psi_\sigma(\boldsymbol{r}) \rangle \langle \Psi_\sigma^\dagger(\boldsymbol{r}) \Psi_{\sigma'}(\boldsymbol{r}') \rangle. \qquad (1.19)$$

The one-body Schrödinger equation reads

$$\int d\boldsymbol{r}' \sum_{\sigma'} h_{\sigma\sigma'}(\boldsymbol{r},\boldsymbol{r}') \psi_n(\boldsymbol{r}',\sigma') = E_{n\sigma} \psi_n(\boldsymbol{r},\sigma), \qquad (1.20)$$

which takes into account a possible spin polarization in the ground state.

The Hartree–Fock Hamiltonian H_{HF} can have Bloch states as a self-consistent solution for the paramagnetic state. Then the ground-state wave function is given by a single Slater determinant as in eqn (1.13). It is noted that the self-consistent solution is not necessarily unique. Among the states given by a single Slater determinant, each self-consistent solution satisfies the variational property that the ground-state energy is stationary against small variation around the solution. The state that gives a local minimum is either the ground state or a metastable state.

The average magnitude of the electron–electron interaction is much smaller than the width of the energy band in the case (i). Thus, in the zeroth approximation the Bloch picture of nearly free electrons is valid. One can naturally examine the importance of fluctuations which have been neglected so far. Because the excitation has no gap in the band state, this may lead to singularities in the perturbation theory. In the one-dimensional system, it is known that virtual transitions across the Fermi surface make the Hartree–Fock state unstable, however weak the two-body interaction is. This is characteristic in one dimension because the phase space available to the particle–hole excitation is particularly large in one dimension. In three dimensions, a small two-body interaction only perturbs the Hartree–Fock ground state. Thus, the metallic character with the Fermi surface remains. In the presence of strong two-body interactions, the paramagnetic state may become unstable against the magnetic or superconducting states. If the state remains paramagnetic, a theorem based on the Fermi-liquid theory shows that the volume enclosed by the Fermi surface is the same as the one determined by the mean-field theory [7]. However, the excitation spectrum may be drastically affected by the correlation. In the case of high electron density, the difference from the Hartree approximation is not important qualitatively. The common feature is that the ground state is a metal with a Fermi surface.

We will return to the discussion of the correlation problem in detail; for the moment, we continue on using the rather elementary picture of electrons.

1.2.2 *Localized states*

In the case (ii) with $a \gg a_B$, one starts from the atomic picture of the ground state in which each proton has an electron in its 1s orbital, and the influence from other hydrogen atoms is a small perturbation. In this case the ground state should be insulating. According to Mott [8], such an insulating state can be found in actual materials, although the electronic state is much more complicated than the hydrogen lattice. In the energy band picture, the insulating state needs a long-range order of spins so that the unit cell contains an even number of electrons. In reality, some materials, such as NiS, remain insulating even above the critical temperature for transition to an antiferromagnetic state. If one applies pressure, or replaces some constituent atoms to cause an equivalent chemical effect, systems such as V_2O_3 [8] could be changed into a metal. Such systems are often called Mott insulators.

We inspect whether the mean-field theory can deal with the case $a \gg a_B$. If we start from the paramagnetic ground state, the many-electron wave function given by the Slater determinant of Bloch states gives substantial probability of having both up- and down-spin electrons in a 1s orbital. However, with $a \gg a_B$, there should be either up or down spin but not both in each 1s orbital.

Therefore, we consider artificially the situation where all electrons have spin up. Also, in this case the Hartree approximation cannot deal with the situation because $v_H(r)$ cannot reproduce the potential for the hydrogen atom. The exchange potential, on the other hand, cancels most of the direct Coulomb interaction from the same cell. As a result, the Hartree–Fock Hamiltonian reproduces the limit of weakly coupled hydrogen atoms as long as the spin is completely polarized.

This observation permits us to define the basis functions that are necessary to establish the localized picture [9,10]. Suppose we have solved the Hartree–Fock equation assuming a fully polarized ground state. Then the lowest energy band, which consists dominantly of the 1s state at each site, is full. The next lowest bands separated by the energy gap consist of 2s and 2p states. In the case (ii), the gap should be about 0.75 Ryd. From the Bloch functions $\psi_k(r)$ of the lowest band we construct the Wannier function by

$$w_i(r) = \frac{1}{\sqrt{N}} \sum_k \psi_k(r) \exp(-i\,k \cdot R_i), \qquad (1.21)$$

which is similar to the 1s wave function at site R_i. In contrast to the atomic wave function, the Wannier functions constitute an orthogonal set. The annihilation operator $c_{i\sigma}$ corresponding to this Wannier state is defined by

$$c_{i\sigma} = \int dr\, w_i^*(r)\Psi_\sigma(r). \qquad (1.22)$$

Furthermore, the eigenenergy E_k of the first band is Fourier transformed to give

$$t_{ij} = \frac{1}{\sqrt{N}} \sum_k E_k \exp[i\,k \cdot (R_i - R_j)]. \qquad (1.23)$$

Here, $t_{ii} \equiv \epsilon_a$ represents the 1s level with a possible shift, and t_{ij} with $i \neq j$ is the hopping energy between the sites i and j. Explicitly, t_{ij} is given by

$$t_{ij} = \int d\mathbf{r} \int d\mathbf{r}'\, w_i^*(\mathbf{r}) h_{\uparrow\uparrow}(\mathbf{r}, \mathbf{r}') w_j(\mathbf{r}'). \tag{1.24}$$

We are now ready to represent the original Hamiltonian given by eqn (1.4) in terms of the Wannier basis. Since the complete set has infinitely many energy bands, the number of Wannier states is also infinite. If we are interested in the ground state and the low-lying excitations, the subspace consisting of 1s states, but with arbitrary spin configurations, is the most important one. We introduce a projection operator P_{1s} to this subspace. Then we get

$$P_{1s} H P_{1s} = H_1 + H_2 + E_{DX}, \tag{1.25}$$

$$H_1 = \sum_{i\sigma} \epsilon_a c_{i\sigma}^\dagger c_{i\sigma} + \sum_{i \neq j,\sigma} t_{ij} c_{i\sigma}^\dagger c_{j\sigma}, \tag{1.26}$$

$$H_2 = \frac{1}{2} \sum_{ijlm} \sum_{\sigma\sigma'} \langle ij|v|ml\rangle (c_{i\sigma}^\dagger c_{j\sigma'}^\dagger c_{m\sigma'} c_{l\sigma} - c_{i\sigma}^\dagger c_{l\sigma} \delta_{jm}$$

$$+ c_{i\sigma}^\dagger c_{m\sigma'} \delta_{\sigma\sigma'} \delta_{jl}), \tag{1.27}$$

where

$$\langle ij|v|ml\rangle = \int d\mathbf{r} \int d\mathbf{r}'\, v(\mathbf{r} - \mathbf{r}') w_i^*(\mathbf{r}) w_j^*(\mathbf{r}') w_m(\mathbf{r}') w_l(\mathbf{r}). \tag{1.28}$$

Note that, throughout the book, we use the ordering of basis in the bra as defined by $\langle ij| = |ji\rangle^\dagger$. The bilinear terms in H_2 compensate the mean-field contribution to t_{ij}. Without the mean-field contribution, t_{ij} would become infinite because of the long range of the Coulomb interaction. Thus, the mean field is by no means a small perturbation.

The nature of $w_i(\mathbf{r})$ localized around $\mathbf{R}_i$ makes most of the $\langle ij|v|ml\rangle$ terms small, leaving two kinds of dominant terms in H_2: (i) terms of the type $\langle ij|v|ji\rangle$ that remain significant even for a pair of distant sites i and j, and (ii) the exchange terms of the type $\langle ij|v|ij\rangle$ for neighbouring sites. The first type is called the Coulomb integral and is responsible for the charge fluctuations like plasma oscillation and exciton-type correlation. On the other hand, the second type is called the exchange integral and favours the parallel arrangement of spins at neighbouring sites.

The simplest way to see the effect of each term on the spin configuration is to take $N = 2$ with the free boundary condition, which makes possible the analogy with the hydrogen molecule. In H_2 we keep $U \equiv \langle 11|v|11\rangle$, $K \equiv \langle 12|v|21\rangle$, $J_d \equiv \langle 12|v|12\rangle$ and equivalent ones, all of which are positive. In this case, an exact solution of $H_1 + H_2$ is obtained easily. Of the six states in total, three are spin-triplet with energy $E_t = 2\epsilon_a$. Of the three singlet states, the lowest one has the energy

$$E_s = 2\epsilon_a + J_d + (U - K)/2 - [(U - K)^2/4 + 4t_{12}^2]^{1/2}, \tag{1.29}$$

and the double occupation of a site tends to zero in the limit $a \gg a_{\mathrm{B}}$. In this limit, U is much larger than $|t_{12}|$ and K. Then we may expand the square root to first order. It is convenient to use the projection operators P_{s} and P_{t} to the singlet and triplet pairs as given by

$$P_{\mathrm{s}} = \tfrac{1}{4} - \boldsymbol{S}_1 \cdot \boldsymbol{S}_2, \qquad P_{\mathrm{t}} = \tfrac{3}{4} + \boldsymbol{S}_1 \cdot \boldsymbol{S}_2, \tag{1.30}$$

in terms of spin operators $\boldsymbol{S}_1$ and $\boldsymbol{S}_2$. Then we obtain the effective Hamiltonian to describe the singlet and triplet states:

$$H_{\mathrm{spin}} = E_{\mathrm{s}} P_{\mathrm{s}} + E_{\mathrm{t}} P_{\mathrm{t}} \sim 2\epsilon_a + \frac{1}{4} J_{\mathrm{d}} - \frac{t_{12}^2}{(U-K)} + \left[\frac{4 t_{12}^2}{(U-K)} - J_{\mathrm{d}} \right] \boldsymbol{S}_1 \cdot \boldsymbol{S}_2. \tag{1.31}$$

The coefficient of the spin-dependent term is called the exchange interaction. Two competing effects are present here. The part with the transfer favours the singlet pair, which is because the doubly occupied site can occur as the intermediate state by perturbation in terms of t_{12}. On the other hand, J_{d} favours the triplet pair. This gain of energy originates from the Pauli principle, which discourages the overlap of wave functions with the same spin.

In the Heitler–London theory for the bonding of hydrogen molecule, the stability of the singlet state is interpreted in terms of the Coulomb attraction due to the electrons accumulated in between the protons. The stabilization by the kinetic exchange represents the same accumulation effect in different terms. Note that the use of orthogonal states is especially convenient in dealing with large systems. For arbitrary N, the competing exchange mechanisms described above are also present. Since J_{d} decays faster than t_{12}^2 as a/a_{B} increases, the kinetic exchange dominates in the case (ii). In addition, more complicated exchanges involving more than two sites are possible. In general, for arbitrary N we cannot expect the exact solution even in the 1s subspace. Instead, for a systematic treatment, we introduce below a version of perturbation theory.

Let us describe in general the effective Hamiltonian which acts on the restricted Hilbert space M called the model space [10,11]. We introduce a projection operator P to the model space M by

$$P = \sum_{\alpha \in M} |\alpha\rangle \langle \alpha|, \tag{1.32}$$

where $|\alpha\rangle$ denotes a state in M. From a state Ψ_i, which is an eigenstate of H with eigenvalue E_i, we construct a projected wave function $P\Psi_i$. We define the effective Hamiltonian H_{eff} such that the same eigenvalues are reproduced within the model space, i.e.

$$H_{\mathrm{eff}} P\Psi_i = E_i P\Psi_i. \tag{1.33}$$

Since the dimension of M is less than that of the original Hilbert space, some $P\Psi_i$ are actually zero. The effective Hamiltonian is useful only if the states of interest have nonzero $P\Psi_i$. If Ψ_i has a substantial weight in M, the projection can be performed accurately by lower-order perturbation theory. It should be noted, however, that in strongly correlated systems one often has to work with a model space with very small weight. This is typically the case when one deals with the Kondo effect.

The unperturbed Hamiltonian is defined as $H_0 = PHP + QHQ$ with $Q = 1 - P$. The perturbation part is then given by $V = H - H_0 = PHQ + QHP$. It is obvious that P and Q commute with H_0. The Schrödinger equation $H\Psi_i = E_i\Psi_i$ is written in the form

$$(E_i - H_0)Q\Psi_i = QV\Psi_i. \tag{1.34}$$

In terms of the operator $R(E) = (E - H_0)^{-1}$, called the 'resolvent', we get

$$Q\Psi_i = R(E_i)QV\Psi_i. \tag{1.35}$$

This form is convenient for iteration; namely, we obtain

$$\Psi_i = P\Psi_i + Q\Psi_i = \sum_{n=0}^{\infty} [R(E_i)QV]^n P\Psi_i \equiv \Omega(E_i)P\Psi_i, \tag{1.36}$$

where the wave operator $\Omega(E)$ is introduced in the last term. Thus, H_{eff} which satisfies eqn (1.33) is given by

$$H_{\text{eff}}(E_i) = PH\Omega(E_i)P, \tag{1.37}$$

where the last P could equally be removed by using the property that $P^2 = P$. The notable feature here is that $H_{\text{eff}}(E_i)$ depends on the eigenvalue to be derived. This kind of formalism with self-consistency condition for E_i is called the Brillouin–Wigner perturbation theory. It is possible to eliminate the energy dependence in H_{eff} by a modification of the above expansion. The resultant formalism is called the Rayleigh–Schrödinger perturbation theory, which is more commonly used in quantum mechanics. We explain the latter scheme in Appendix C in connection with the renormalization of the Kondo model. Up to second order in V, there is no difference between the two perturbation theories.

In the case of the hydrogen lattice with $a/a_{\text{B}} \gg 1$, it is appropriate to define the model space M as the singly occupied set of 1s Wannier orbitals. The perturbation part V consists of the hopping and the two-body interactions. In the lowest order in V, we replace E_i by the unperturbed one $N\epsilon_a$. Then we get

$$H_{\text{eff}} = H_0 + PV(N\epsilon_a - H_0)^{-1}QVP. \tag{1.38}$$

Of the two competing exchanges, J_{d} arises from H_0 and the kinetic exchange from the second term. If one uses the atomic 1s orbital, U is calculated to be 17 eV. However, for higher V, the parameters M_{d} and U are renormalized by the presence of 2s and other orbitals. In actual solids, this polarization effect reduces U drastically (to about half). It is therefore very difficult to obtain the exchange parameter from first principles. It should be emphasized that the magnitude of the exchange and other parameters in the effective Hamiltonian depends on the model space chosen. This feature is often overlooked in the literature and causes confusion in interpretation of experimental data.

Another higher-order effect of V is to produce many-spin exchange terms and estimating their magnitude is even more difficult than the two-spin exchange. We only note here that in some cases the experimentally observed spin ordering requires significant three-spin or four-spin interactions [12,13].

The most important feature in the case (ii) is that the energy gap of the order of 0.75 Ryd survives for any spin configuration, and that the system needs a finite threshold energy to conduct the electronic current. Thus, the system behaves as an insulator at absolute zero. However, spin excitations from the ground state are gapless. In the case of a magnetically ordered ground state, the gapless excitation is related to the Goldstone theorem on the spontaneous breakdown of the continuous symmetry [14]. We emphasize that this insulating state is different from the band insulator where there are an even number of electrons per cell. In the latter case, there is energy gap for both spin and charge excitations as exemplified by solid He with $1s^2$ configuration. The band insulator is connected smoothly to weakly interacting atoms as the lattice parameter increases in this case.

Let us summarize the two fundamental characters of electrons in the hydrogen lattice. In the case where the kinetic energy is the most important quantity as in high densities, the ground state is a metal and the electrons form energy bands. In the opposite case where the Coulomb repulsion among the electrons is the most important one, the ground state is insulating and the electrons are localized. Except in one dimension, the ground state has a certain kind of magnetic order.

The fundamental question to ask is what happens in the case of $a \sim a_B$. The question was raised by Mott [8], and the effort to answer this question has opened remarkably rich outcomes in both experiment and theory. We will explore the actual examples in this book. In closing this section, we introduce the simplest model that is still capable of describing both the itinerant and the localized features of electrons. Namely, in H_2 given by eqn (1.28) we keep only the largest term in $\langle ij|v|lm \rangle$ that corresponds to the case of $i = j = l = m$. Then H_2 is replaced by

$$H_2 \rightarrow U \sum_i (n_{i\uparrow}n_{i\downarrow} - n_{i\uparrow} - n_{i\downarrow}) \equiv H_U - U N_e. \tag{1.39}$$

where $U = \langle ii|v|ii \rangle$ and $n_{i\sigma} = a_{i\sigma}^\dagger a_{i\sigma}$, and N_e is the total number of electrons which may not be equal to N. The resultant model $H_1 + H_U$ is called the Hubbard model [15–17], the Hubbard Hamiltonian H_{Hub} being given by

$$H_{\text{Hub}} = \sum_{i\sigma} \epsilon_a c_{i\sigma}^\dagger c_{i\sigma} + \sum_{i \neq j,\sigma} t_{ij} c_{i\sigma}^\dagger c_{j\sigma} + U \sum_i n_{i\uparrow}n_{i\downarrow}. \tag{1.40}$$

We note that the fully polarized Hartree–Fock state becomes an eigenstate of the Hubbard model. However, this ferromagnetic state accompanies more energy in H_1 given by eqn (1.26) than the paramagnetic and antiferromagnetic states. According to the available exact solution in one dimension, the ferromagnetic state is not actually the ground state [18].

In heavy-electron systems, the Hubbard model with a single energy band is actually too simplistic. The minimum elements to be included are the conduction band and the nearly localized electrons. The simplest model with these two kinds of states is called the Anderson lattice, which will be discussed in Chapter 3.

1.3 Fundamentals of spin dynamics

1.3.1 *Itinerant magnetic moments*

In the case of a single conduction band, the magnetization M_i at site i is given by

$$M_i = -\tfrac{1}{2} g \mu_B c_{i\alpha}^\dagger \sigma_{\alpha\gamma} c_{i\gamma}, \tag{1.41}$$

where σ denotes the vector composed of the Pauli matrices, α and γ are spin components, and g is the g-factor. The Fourier transform $M(q)$ of M_i is given by

$$M(q) = -\frac{1}{2} g \mu_B \sum_{k\alpha\gamma} c_{k\alpha}^\dagger \sigma_{\alpha\gamma} c_{k+q\gamma}. \tag{1.42}$$

If the spin–orbit interaction is strong in the band, one should interpret the spin indices α and γ as representing a state in a Kramers doublet for each momentum. In this case the g-factor should also be replaced by a tensor. We shall encounter an important example of this case in the superconductivity of heavy electrons in Chapter 5.

The dynamical property of the magnetic moment is most conveniently characterized by the linear response against external perturbation. The linear response theory provides the dynamical magnetic susceptibility tensor $\mu_B^2 \chi_{\alpha\gamma}(q, \omega)$ in terms of the statistical average of the commutator

$$\mu_B^2 \chi_{\alpha\gamma}(q, \omega) = \frac{i}{V} \int_0^\infty dt \, \exp(i\omega t) \langle [M_\alpha(q, t), M_\gamma(-q)] \rangle, \tag{1.43}$$

where V is the volume of the system and $M_\alpha(q, t) = \exp(iHt) M_\alpha(q) \exp(-iHt)$ is the magnetization in the direction α. Here we take the Heisenberg picture with Hamiltonian H. The statistical average is taken with respect to the Boltzmann weight factor $\exp(-\beta H)/Z$, with Z being the partition function $\mathrm{Tr}\exp(-\beta H)$. We refer to Appendix A for a brief review of linear response theory.

We begin with the simplest case of a free electron gas for which the dynamical magnetic susceptibility can be obtained exactly. Note that

$$M_+(q) = M_x(q) + i M_y(q) = -\mu_B \sum_k c_{k\uparrow}^\dagger c_{k+q\downarrow}, \tag{1.44}$$

with $g = 2$, and that the time dependence is given simply by $c_{k\sigma}(t) = \exp(-i\epsilon_k t) c_{k\sigma}$. Then we use the commutation rule

$$[c_1^\dagger c_2, c_3^\dagger c_4] = \delta_{23} c_1^\dagger c_4 - \delta_{14} c_3^\dagger c_2, \tag{1.45}$$

where the obvious set of quantum numbers have been abbreviated symbolically. The magnetic susceptibility is then given by

$$\mu_B^2 \chi_0(q, \omega) = \frac{2\mu_B^2}{V} \sum_k \frac{f(\epsilon_{k+q}) - f(\epsilon_k)}{\omega - \epsilon_{k+q} + \epsilon_k + i\delta}, \tag{1.46}$$

which in this case is a scalar. Here $f(\epsilon) = [\exp(\beta\epsilon) + 1]^{-1}$ is the Fermi distribution function with the chemical potential being the origin of energy.

1.3.2 *Localized magnetic moments*

Heavy electrons are observed mainly in rare-earth and actinide compounds. These compounds have incomplete f shells which have finite magnetic moments. We summarize the elementary facts about the magnetic moments associated with f shells [19].

Let us consider a case of a free trivalent Ce ion with the electron configuration $(4f)^1$ on top of the Xe configuration. The spin–orbit interaction splits the 14-fold degenerate level with the orbital angular momentum $L = 3$ and the spin $S = 1/2$ into two levels with $J = L + S = 7/2$ and $J = L - S = 5/2$, where J is the total angular momentum. The six-fold degenerate level with $J = 5/2$ lies lower than the $J = 7/2$ level by about $0.3\,\mathrm{eV}$.

The magnitude of the magnetic moment M associated with the $J = 5/2$ level is given by $M = -g_J \mu_\mathrm{B} J$, where μ_B is the Bohr magneton and

$$g_J = \frac{3}{2} + \frac{S(S+1) - L(L+1)}{2J(J+1)} \tag{1.47}$$

is the Landé g-factor.

In the case of a free rare-earth ion with more f electrons, the energy levels are described by the LS-coupling (or Russel–Saunders) scheme. Here, one first neglects the spin–orbit interaction and takes a multiplet with definite angular momenta S and L. The effect of spin–orbit interaction is approximated by the term $H_{LS} = \lambda L \cdot S$, where λ gives the strength of the interaction. According to an empirical rule, namely the Hund's rule, the ground state of the atom takes the largest possible S and the largest L allowed for the S. The effective Hamiltonian H_{LS} should work if the magnitude of λ is much smaller than the splitting of multiplets with different L and S. The sign of λ is positive for a less-than-half filled shell, i.e. less than seven $4f$ electrons, and favours antiparallel L and S. On the other hand, λ is negative for a more-than-half filled shell. Thus, in the case of trivalent Yb with 13 electrons in the 4f shell, the spin–orbit interaction favours parallel L and S, leading to the ground level with $J = 7/2$. The spin–orbit splitting is larger than the Ce case, and amounts to $1.2\,\mathrm{eV}$.

The actinides have even larger spin–orbit interaction, and the LS-coupling scheme for describing levels within a multiplet is not so accurate as in rare-earths. However, at least for light actinides like U, the scheme is still practical for a discussion of energy levels.

A rare-earth or an actinide ion in a crystal feels the potential of the environment. The relevant point group symmetry is lower than the spherical one. As a result, the degeneracy $2J + 1$ for a spin–orbit level with the angular momentum J is split into sublevels. This splitting as well as associated physical effects are called crystalline electric field (CEF) effects. In the case of Ce^{3+} in a cubic CEF, for example, the $J = 5/2$ level is split into a doublet called Γ_7 and a quartet called Γ_8. The CEF eigenstates are given for the Γ_7 by

$$\psi_{7,\pm} = \sqrt{\frac{1}{6}}\left|\pm\frac{5}{2}\right\rangle - \sqrt{\frac{5}{6}}\left|\mp\frac{3}{2}\right\rangle, \tag{1.48}$$

in terms of the eigenstates of J_z. The eigenstates for Γ_8 are given by

$$\psi_{8A,\pm} = \sqrt{\frac{5}{6}}\left|\pm\frac{5}{2}\right\rangle + \sqrt{\frac{1}{6}}\left|\mp\frac{3}{2}\right\rangle, \qquad \psi_{8B,\pm} = \left|\pm\frac{1}{2}\right\rangle. \tag{1.49}$$

The magnetic moment operator M_z has nonvanishing matrix elements only within a CEF level. However, the M_x and M_y components have finite elements both within the CEF levels and between different levels.

In order to describe the localized levels and transitions between them, it is convenient to introduce an operator $X_{\mu\nu} \equiv |\mu\rangle\langle\nu|$ which describes a transition from a localized multi-electron state $|\nu\rangle$ to another state $|\mu\rangle$. In the case of $\nu = \mu$, $X_{\mu\mu} \equiv |\mu\rangle\langle\mu|$ becomes a projection operator onto $|\mu\rangle$. The set of localized levels $\{E_\mu\}$ with the inclusion of the spin–orbit interaction and the CEF effects is summarized by the Hamiltonian H_1 as

$$H_1 = \sum_\mu E_\mu X_{\mu\mu}. \tag{1.50}$$

The magnetic moment in the z-direction is expressed as

$$M_z = \sum_{\mu\nu} \langle\mu|M_z|\nu\rangle X_{\mu\nu}. \tag{1.51}$$

The Landé g-factor is included in the matrix element $\langle\mu|M_z|\nu\rangle$.

It is sometimes convenient to associate a pseudo-spin-1/2 with the doublet CEF. We derive the effective g-factor taking as an example the tetragonal CEF. The basis functions of the lowest doublet $|\pm\rangle$ are given by

$$|\pm\rangle = a|\pm 5/2\rangle - b|\mp 3/2\rangle, \tag{1.52}$$

where a and b are real parameters with the constraint $a^2 + b^2 = 1$. We calculate a matrix element

$$\frac{1}{\mu_B}\langle+|M_z|+\rangle = \left(\frac{5}{2}a^2 - \frac{3}{2}b^2\right)g_J \equiv -\frac{1}{2}g_z, \tag{1.53}$$

where we introduce the effective g-factor g_z along the z-direction. Similarly, we introduce the perpendicular component $g_\perp = g_x = g_y$ by

$$\langle+|M_x + iM_y|-\rangle = -g_\perp\mu_B. \tag{1.54}$$

Then a simple calculation gives

$$g_z = -\frac{6}{7}(1 + 4\cos 2\theta), \qquad g_\perp = \frac{6\sqrt{5}}{7}\sin 2\theta, \tag{1.55}$$

with $a = \cos\theta$, and we have put $g_J = -6/7$. In the special case of cubic symmetry, we get $g_z = g_\perp = 10/7$ with $\cos\theta = 1/\sqrt{6}$. Figure 1.5 shows each component as

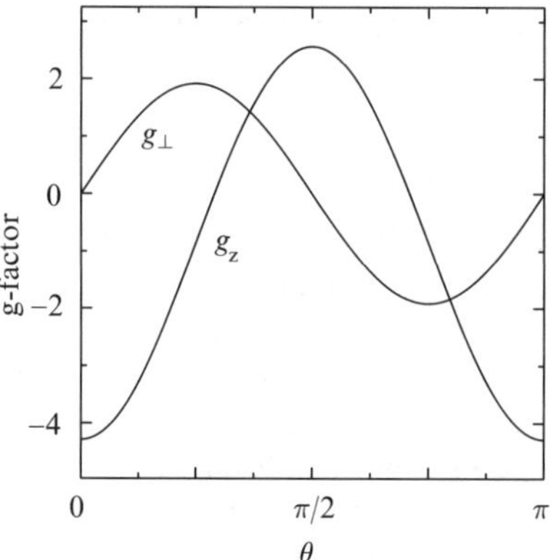

FIG. 1.5. The effective g-tensor as a function of the parameter θ. See text for details.

functions of θ. It should be noted that the effective g-factor can be very anisotropic depending on the CEF; it is Ising-like with $\theta = 0$, π and XY-like with $\tan^2 \theta = 5/3$. In reality, θ is determined as a function of the CEF parameters as explained, for example, in ref. [19]. The implication of the anisotropic g-factor for superconductivity seems to be an important problem, but has not yet been discussed in the literature.

The origin of the CEF is not only the Coulomb potential of the environment but also includes the overlap of wave function of the ligands. The latter is called hybridization. If the perturbation energy due to hybridization is smaller than the splitting of spin–orbit split levels, one can still retain the localized picture of f electrons. In the opposite limit, the 4f electrons can be better treated as itinerant electrons. The degree of hybridization depends on the material, but the general tendency is that the 4f electrons are more strongly localized than the 5f electrons. In particular, the U ions often have hybridization strength for which neither the localized picture nor the itinerant picture applies well.

In a hypothetical case of localized electrons with no interaction with the environment, one can obtain the dynamical susceptibility exactly. Using the X-operators, the linear response formula can be evaluated directly. There is no dependence on the wavenumber. The result for the susceptibility $\chi_{zz}(\omega)$ at a site of localized electrons is

$$\mu_B^2 \chi_{zz}(\omega) = \sum_{\mu\nu} |\langle\mu|M_z|\nu\rangle|^2 \frac{\langle X_{\nu\nu}\rangle - \langle X_{\mu\mu}\rangle}{\omega - E_\nu + E_\mu + i\delta}. \tag{1.56}$$

The small factor δ is required to make the integral converge and is related to causality.

If the external field is static, the susceptibility is obtained from the formula

$$\mu_B^2 \chi_{\alpha\gamma} = \int_0^\beta d\lambda \, \langle e^{\lambda H} M_\gamma e^{-\lambda H} M_\alpha \rangle = \mu_B^2 \chi_{\gamma\alpha}, \tag{1.57}$$

which corresponds to the isothermal susceptibility. On the other hand, the static limit of eqn (1.56) corresponds to the adiabatic susceptibility. In the system without hybridization, the result is given by

$$\chi_{zz} = \chi_{zz}(\omega \to 0) + \frac{1}{\mu_B T} \sum_{\mu\nu} |\langle\mu|M_z|\nu\rangle|^2 \langle X_{\mu\mu}\rangle, \tag{1.58}$$

where the sum in the second term is only over degenerate states μ, ν which do not contribute to the first term. The second term gives the Curie law, while the first term is called the Van Vleck term. Thus, if the ground state of the system is degenerate, the adiabatic and isothermal static susceptibilities differ by the second term. In reality, the localized electrons interact with the conduction electrons by hybridization, or with other localized electrons at different sites by higher-order interactions. These effects remove the degeneracy of the ground state. Then the time average becomes equivalent to the statistical average, which is referred to as ergodicity. Accordingly, the adiabatic and isothermal susceptibilities become the same.

1.3.3 *Random phase approximation*

In realistic systems, the electrons are neither free nor strictly localized. The difficulty to take reliable account of the interaction effects stimulated various theories for the dynamical response. Let us explain the simplest description of the interaction effect. To do this, we write the interaction H_U given by eqn (1.39) as

$$H_U = \frac{U}{4} \sum_i (n_i^2 - m_i^2 - 4n_i), \tag{1.59}$$

where $n_i = n_{i\uparrow} + n_{i\downarrow}$ and $m_i = n_{i\uparrow} - n_{i\downarrow}$. We focus attention on the term $-m_i^2$, and decompose it as

$$m_i^2 = (\langle m_i\rangle + \delta m_i)^2 = \langle m_i\rangle^2 + 2\langle m_i\rangle\delta m_i + (\delta m_i)^2, \tag{1.60}$$

where $\delta m_i = m_i - \langle m_i\rangle$, and $\langle m_i\rangle$ means a certain average to be determined self-consistently. The approximation we make consists in neglecting the last term in eqn (1.60). This resembles a mean-field approximation, and is often called the random phase approximation (RPA). This is equivalent to replacing H_U in the Heisenberg picture by

$$H_{\text{eff}} = -\sum_i \phi_i(t)m_i(t)$$

$$= -\sum_q \int_{-\infty}^{\infty} d\omega \, \exp(i\boldsymbol{q}\cdot\boldsymbol{R}_i)\phi(\boldsymbol{q},\omega)m(-\boldsymbol{q},\omega), \tag{1.61}$$

with $\phi_i(t) = U\langle m_i(t)\rangle/2$ being the dynamic effective field, and

$$m(-\boldsymbol{q},\omega) = \sum_i \int_{-\infty}^{\infty} dt \, \exp(-i\boldsymbol{q}\cdot\boldsymbol{R}_i + i\omega t)m_i(t), \tag{1.62}$$

$$\phi(\boldsymbol{q},\omega) = \frac{U}{2}\langle m(\boldsymbol{q},\omega)\rangle. \tag{1.63}$$

In the presence of a space- and time-dependent external magnetic field $H \exp(i\boldsymbol{q}\cdot\boldsymbol{R}_i - i\omega t)$ along the z-direction, the linear response of the system with

H_{eff} is given by

$$\langle m(q, \omega) \rangle \equiv \chi(q, \omega)h = \chi_0(q, \omega)[h + \phi(q, \omega)], \tag{1.64}$$

where $h = \mu_B H$ and $\chi_0(q, \omega)$ is given by eqn (1.46). By substituting eqn (1.63) for $\phi(q, \omega)$, the dynamical spin susceptibility $\chi(q, \omega)$ of the system is derived as

$$\chi(q, \omega)^{-1} = \chi_0(q, \omega)^{-1} - \frac{U}{2}. \tag{1.65}$$

We make two remarks about eqn (1.65). First, in the static and homogeneous limit we obtain, from eqn (1.46),

$$\chi_0(\mathbf{0}, 0) = \frac{2}{V} \sum_k \delta(\epsilon_k) \equiv \rho(\mu). \tag{1.66}$$

The RPA is a good approximation if $\alpha \equiv \rho(\mu)U/2 \ll 1$. As α increases toward unity, the contribution of fluctuations becomes important. Thus, the divergence of $\chi(\mathbf{0}, 0)$ at $\alpha = 1$ should not be taken literally. In fact, one of the most important problems in heavy electrons is to explain the stability of the paramagnetic state although the opposite limit, $\alpha \gg 1$, is realized. Secondly, since the system has rotational invariance with respect to the spin direction, $\chi(q, \omega)$ is a scalar, i.e. it is independent of the direction of the magnetic field.

1.3.4 *Fermi-liquid theory*

If the state near absolute zero of temperature T is a paramagnetic metal, the Fermi-liquid theory of Landau provides a very successful description. Let us review briefly the basic idea of the Fermi-liquid theory [20,21]. The starting point is the assumption that the low-lying excitations of interacting fermions are in one-to-one correspondence with those in a noninteracting counterpart. The assumption is equivalent to adiabatic continuity when turning on the interaction. Provided the perturbation expansion with respect to the interaction converges, the continuity should be satisfied no matter how strong the interaction is. Then any excited state can be characterized by the distribution $\{n_{p\sigma}\}$ of quasi-particles. A quasi-particle characterizes the one-to-one correspondence, and reduces to the original particle in the noninteracting limit. We consider a change of the distribution function of a quasi-particle from that of the ground state. For momentum p and spin s_z ($=\sigma/2 = \pm 1/2$), the change is written as $\delta n_{p\sigma}$. Then the excitation energy δE is expanded as

$$\delta E = \sum_{p\sigma} \epsilon_p \delta n_{p\sigma} + \frac{1}{2} \sum_{p\sigma} \sum_{p'\sigma'} f(p\sigma, p'\sigma') \delta n_{p\sigma} \delta n_{p'\sigma'} + O(\delta n_{p\sigma}^3), \tag{1.67}$$

where ϵ_p is the quasi-particle energy measured from the Fermi level, and $f(p\sigma, p'\sigma')$ describes the interaction between quasi-particles. Since ϵ_p is a small quantity, the first and second terms on the right-hand side are of the same order of magnitude. It is possible to neglect terms of higher order with respect to $\delta n_{p\sigma}$.

Let us assume spherical symmetry in the system. Then we can regard the absolute value of the momenta p and p' of quasi-particles equal to the Fermi momentum p_F. In terms of the angle $\theta_{pp'}$ between p and p', We write

$$\rho^*(\mu) f(p\sigma, p'\sigma') = \sum_{l=0}^{\infty} (F_l + \sigma\sigma' Z_l) P_l(\cos \theta_{pp'}), \qquad (1.68)$$

where $\rho^*(\mu)$ is the density of states of quasi-particles at the Fermi level. It is given by

$$\rho^*(\mu) = \frac{1}{V} \sum_{p\sigma} \delta(\epsilon_p) = \frac{m^* p_F}{\pi^2} \qquad (1.69)$$

with m^* being the effective mass of a quasi-particle. The dimensionless quantities F_l and Z_l are called Landau parameters. A similar decomposition for $\delta n_{p\sigma}$ is given in terms of spherical harmonics as

$$\delta n_{p\sigma} = \sum_{lm} \delta n_{lm\sigma} Y_{lm}(\hat{p}), \qquad (1.70)$$

where $\hat{p}$ denotes the solid angle of p. Then the second term of eqn (1.67) can be written as

$$\frac{1}{2\rho^*(\mu)} \sum_{lm} \sum_{\sigma\sigma'} (F_l + \sigma\sigma' Z_l) \delta n_{lm\sigma} \delta n_{lm\sigma'}. \qquad (1.71)$$

The observables at low excitation energies are described quantitatively in terms of a small number of Landau parameters. Thus, by comparison with measurement, these parameters can be determined experimentally. This is the most advantageous aspect of the Fermi-liquid theory. The interaction between quasi-particles plays no role for the specific heat C at low T; namely, we obtain

$$C = \frac{\pi^2}{3} \rho^*(\mu) T, \qquad (1.72)$$

which takes the same form as for the free electron gas. The interaction effect appears only in the effective mass.

Next we derive the static spin susceptibility χ. If the Landau parameters are all zero, the susceptibility is given by $\chi_0^* = \rho^*(\mu)$. This is regarded as the static and small $-q$ limit of the free quasi-particle susceptibility analogous to eqn (1.46). As the next simplest case, we assume the presence of homogeneous polarization density $M = V^{-1} \sum_p (n_{p\uparrow} - n_{p\downarrow})$. Then eqn (1.67) is rewritten as

$$\frac{\delta E}{V} = \frac{1}{2\chi_0^*} M^2 + \frac{1}{2\rho^*(\mu)} Z_0 M^2 \equiv \frac{1}{2\chi} M^2. \qquad (1.73)$$

Taking the second derivative of eqn (1.73) with respect to M we obtain

$$\chi = \chi_0^*/(1 + Z_0), \qquad (1.74)$$

where we have used the relation $\chi_0^* = \rho^*(\mu)$.

One can regard eqn (1.67) as an effective Hamiltonian for the quasi-particles. Then it should be possible to derive the dynamical properties of the system as well. This was done by Landau in the form of the transport equation. Here we follow an alternative approach which is easier to compare with the perturbation theory. Let us consider the situation where an external magnetic field H along the z-axis has a single Fourier component $\boldsymbol{q}$. Namely, we assume that

$$H_{\text{ex}}(t) = -m(-\boldsymbol{q}, t)h \exp(-i\omega t), \tag{1.75}$$

with $h = \mu_{\text{B}} H$. The dynamical susceptibility of noninteracting quasi-particles is written as $\chi_0^*(\boldsymbol{q}, \omega)$ and can be calculated as in eqn (1.46). From the static result, we anticipate that the dynamical response is given by

$$\langle m(\boldsymbol{q}, \omega)\rangle = \chi_0^*(\boldsymbol{q}, \omega)[h + h_{\text{eff}}(\boldsymbol{q}, \omega)], \tag{1.76}$$

where $h_{\text{eff}}(\boldsymbol{q}, \omega)$ is the self-consistent field. If we neglect the Landau parameters other than Z_0, we obtain

$$h_{\text{eff}}(\boldsymbol{q}, \omega) = -Z_0 \chi_0^*(\boldsymbol{q}, \omega)\langle m(\boldsymbol{q}, \omega)\rangle. \tag{1.77}$$

Thus, the full susceptibility $\chi(\boldsymbol{q}, \omega) = \langle m(\boldsymbol{q}, \omega)\rangle / h$ is given by

$$\chi(\boldsymbol{q}, \omega)^{-1} = \chi_0^*(\boldsymbol{q}, \omega)^{-1} + Z_0/\rho^*(\mu). \tag{1.78}$$

Since there are many nonzero Landau parameters in general, eqn (1.78) constitutes an approximation which may be called the quasi-particle RPA [22]. In the original RPA for bare particles in the Hubbard model, $\chi_0(\boldsymbol{q}, \omega)$ includes the bare spectrum and $Z_0/\rho^*(\mu)$ is replaced by $-U/2$. We emphasize that eqn (1.78) is meaningful only for excitations with small ω and $\boldsymbol{q}$.

In concluding this subsection, we note the complication in real materials. In the presence of multiple energy bands, the magnetic susceptibility consists of both the spin susceptibility as discussed above and the interband contribution which corresponds to the Van Vleck term in the local-moment case. It is not at all trivial to separate the two contributions from the measured susceptibility in heavy electrons. In fact, the interband term is often large and requires careful analysis in NMR and neutron scattering [23].

1.3.5 *Parametrization of the dynamical susceptibility*

For practical purposes, it is useful to parametrize the dynamical susceptibility approximately. Assuming that $q/k_{\text{F}} \ll 1$ and $\omega/v_{\text{F}}q < 1$, with v_{F} being the Fermi velocity, we obtain from eqn (1.46)

$$\chi_0^*(\boldsymbol{q}, \omega) = \rho^*(\mu)\left[1 - b^2 q^2 + i\frac{\pi\omega}{2v_{\text{F}}q}\right], \tag{1.79}$$

where b is a constant of $O(k_{\text{F}}^{-1})$. In the case of the parabolic spectrum, we have $b^{-2} = 12k_{\text{F}}^2$. To extrapolate to larger q and ω, we modify eqn (1.79) as follows:

$$\chi_0^*(\boldsymbol{q}, \omega) = \rho^*(\mu) \left[1 + b^2 q^2 - i \frac{\pi \omega}{2 v_F q} \right]^{-1}. \tag{1.80}$$

By this approximation, the original cut singularity in $\chi_0^*(\boldsymbol{q}, \omega)$ as a function of ω is replaced by the pole singularity. Then, using eqn (1.78) we obtain

$$\chi(\boldsymbol{q}, \omega) = \frac{\rho^*(\mu)}{1 + Z_0 + b^2 q^2 - i\pi\omega/(2 v_F q)}. \tag{1.81}$$

The imaginary part can be written in the form

$$\mathrm{Im}\, \chi(\boldsymbol{q}, \omega) = \frac{\chi}{1 + \xi^2 q^2} \frac{\omega \Gamma_q}{\omega^2 + \Gamma_q}, \tag{1.82}$$

where $\xi^2 = b^2/(1 + Z_0)$ and

$$\Gamma_q = 2 v_F (1 + Z_0) q (1 + \xi^2 q^2). \tag{1.83}$$

This 'double Lorentzian' form is often used for fitting the experimental results of neutron scattering. The correlation length ξ of the magnetic fluctuation becomes divergent at the ferromagnetic instability $Z_0 = -1$. The relaxation rate Γ_q is proportional to q for small wavenumber, which is a feature specific to the itinerant magnetism. The characteristic energy of the ferromagnetic spin fluctuation is given by $(1 + Z_0) v_F k_F$, and becomes much smaller than the Fermi energy near the ferromagnetic instability $1 + Z_0 = 0$.

Let us now consider the case where the antiferromagnetic (AF) correlation is dominant. The relevant wavenumber is near the ordering vector $\boldsymbol{Q}$. The relaxation rate $\Gamma(\boldsymbol{q})$ in the AF case remains finite at $\boldsymbol{q} = \boldsymbol{Q}$. The difference from eqn (1.83) exists because of the absence of a conservation law for the staggered magnetization. Let us assume that the dissipation in the system occurs dominantly through the local processes. Then the dynamical susceptibility can be parametrized in the form

$$\chi(\boldsymbol{q}, \omega) = \frac{\chi_L(\omega)}{1 - J(\boldsymbol{q}) \chi_L(\omega)}, \tag{1.84}$$

where $J(\boldsymbol{q})$ is the Fourier transform of the intersite exchange interaction. Note that the local susceptibility $\chi_L(\omega)$ describes the dissipation process. Antiferromagnetic order takes place when the denominator of eqn (1.84) becomes zero at $\boldsymbol{q} = \boldsymbol{Q}$. We postulate the Lorentzian form

$$\chi_L(\omega) = \frac{\chi_L \Gamma}{\Gamma - i\omega}, \tag{1.85}$$

where Γ is the local relaxation rate. Then we may write eqn (1.84) as

$$\chi(\boldsymbol{q}, \omega) = \frac{\chi(\boldsymbol{q}) \Gamma(\boldsymbol{q})}{\Gamma(\boldsymbol{q}) - i\omega}, \tag{1.86}$$

where $\Gamma(\boldsymbol{q}) = [1 - J(\boldsymbol{q}) \chi_L] \Gamma$ and $\chi(\boldsymbol{q}) = \chi(\boldsymbol{q}, 0)$. From this equation we learn that the $\boldsymbol{q}$-dependent relaxation rate $\Gamma(\boldsymbol{q})$ becomes small near the AF instability. In fact, we have the relation [24]

$$\chi(q)\Gamma(q) = \chi_{\mathrm{L}}\Gamma. \tag{1.87}$$

In order to study the q dependence of the relaxation, we expand $J(q)$ around $q = Q$ as

$$J(Q + q) = J(Q) - Aq^2 + \cdots, \tag{1.88}$$

where we assume isotropic q dependence of $J(Q + q)$ and $A = -J''(Q)/2$. Then we get

$$\chi(Q + q) = \frac{\chi(Q)}{1 + q^2\xi^2}, \tag{1.89}$$

$$\Gamma(Q + q) = \Gamma(Q)(1 + q^2\xi^2), \tag{1.90}$$

where $\xi^2 = A\chi(Q)$. Note that ξ represents the correlation length of the AF fluctuation. In the itinerant AF system, one can use the same parametrization as given by eqns (1.86), (1.89), and (1.90) even though one does not rely on the Ruderman–Kittel–Kasuya–Yosida (RKKY) interaction and the local nature of relaxation.

As noted already, the relaxation of magnetic moments depends strongly on whether or not the magnetization is a conserved quantity at the relevant wavenumber. If it is conserved, the magnetic relaxation is suppressed for small wavenumbers. We now turn to another phenomenological description which is useful in the presence of disorder in itinerant systems or in local-moment systems. We follow Kadanoff and Martin in deriving $\chi(q, \omega)$ in the hydrodynamic regime [25]. One starts with the continuity equation for the magnetization density $M(r, t)$ along the z-direction:

$$\frac{\partial}{\partial t}M(r, t) + \nabla \cdot j(r, t) = 0, \tag{1.91}$$

where $j(r, t)$ is the magnetization current. It is understood that the direction of magnetization is always along the z-axis, and that $M(r, t)$ and $j(r, t)$ are c-numbers. In terms of a parameter D, called the spin diffusion coefficient, the current is given by

$$j(r, t) = -D\nabla M(r, t). \tag{1.92}$$

Substituting this into the continuity equation, we get a diffusion equation for the magnetization,

$$\frac{\partial}{\partial t}M(r, t) = D\nabla^2 M(r, t). \tag{1.93}$$

We note that this equation is valid only when the magnetic field is absent. In order to make connection with the linear response, we take the Laplace transform of eqn (1.93). Then we obtain

$$M(q, z) = \frac{M(q, t = 0)}{-iz + Dq^2}, \tag{1.94}$$

where

$$M(\boldsymbol{q}, z) = \int_0^\infty dt\, M(\boldsymbol{q}, t) \exp(izt). \tag{1.95}$$

Here the complex variable z has a positive imaginary part. Equation (1.94) gives the Laplace transform of the relaxation function, since $M(\boldsymbol{q},\, t = 0)$ is determined by the static magnetic field present up to $t = 0$. Then, using the relation to the retarded Green function as given by eqn (B.10) in Appendix B, we obtain the dynamical susceptibility as

$$\chi(\boldsymbol{q}, \omega) = \frac{\chi(\boldsymbol{q}, 0) Dq^2}{-i\omega + Dq^2}. \tag{1.96}$$

The form (1.96) is often used in analysing the experimental results at small $\boldsymbol{q}$ for systems with conserved local moments. In the case of pure itinerant ferromagnetism, the diffusion equation (1.93) does not hold since there is no hydrodynamic regime.

1.3.6 *Mode-coupling picture of spin fluctuations*

To account for the spin fluctuations a step beyond the RPA, one can use the Hartree approximation for the coupling of spin fluctuations [26–28]. In this section we critically survey the fundamental ideas of the mode-coupling theory. For simplicity, we take the Hubbard model with the paramagnetic ground state, and consider the Helmholtz free energy F per site under the condition that homogeneous spin polarization $m = \langle n_{i\uparrow} - n_{i\downarrow}\rangle$ is present. The magnetic susceptibility χ in units of μ_{B}^2, which is the homogeneous and static limit of the dynamical susceptibility $\chi(\boldsymbol{q}, \omega)$, is given by the thermodynamic derivative

$$\chi^{-1} = \frac{\partial^2 F}{\partial m^2}. \tag{1.97}$$

The task is to account for the interaction effect in F. We rewrite H_U as

$$H_U = -\frac{U}{2} \sum_i \{S_i^+, S_i^-\} + \frac{U}{2} N \tag{1.98}$$

in terms of spin flip operators at each site i. The free energy of the system with the coupling strength gU $(0 < g < 1)$ is given by

$$F(g) = -T \ln \mathrm{Tr} \exp\left[-(\beta H_1 + g H_U)\right], \tag{1.99}$$

where H_1 is given by eqn (1.26) and the trace is over all states with fixed number of particles. Then we obtain

$$g\frac{\partial}{\partial g} F(g) = \langle H_{gU}\rangle_g, \tag{1.100}$$

where the average is taken for the system with the interaction gU. In the noninteracting state, the average of the anticommutator can be derived easily as

$$\langle\{S_i^+, S_i^-\}\rangle_0 = n + \tfrac{1}{2}(n^2 - m^2). \tag{1.101}$$

Thus, we obtain

$$F = F_0 + \frac{1}{2\chi_0} m^2 - \frac{1}{4} U m^2 + \Delta F, \tag{1.102}$$

where F_0 is the free energy of the noninteracting system with $m = 0$, and ΔF accounts for the correlation effect. The latter is given by

$$\Delta F = \int_0^1 dg \, [\langle \{ S_i^+, S_i^- \} \rangle_g - \langle \{ S_i^+, S_i^- \} \rangle_0]. \tag{1.103}$$

If ΔF is neglected in eqn (1.97), it leads to the Hartree–Fock susceptibility, which corresponds to the static and homogeneous limit of the dynamical susceptibility in the RPA. In computing ΔF, we notice that the average in eqn (1.103) is given by

$$\langle \{ S_i^+, S_i^- \} \rangle_g = \int_{-\infty}^{\infty} \frac{d\omega}{2\pi} \coth \left(\frac{1}{2} \beta \omega \right) \operatorname{Im} \chi_{m,gU}^{\perp}(\omega)$$
$$= T \sum_n \chi_{m,gU}^{\perp}(i\nu_n), \tag{1.104}$$

where $\chi_{m,gU}^{\perp}(\omega) = -\langle [S_i^+, S_i^-] \rangle_g (\omega)$ is the local transverse susceptibility in the presence of magnetization m. This is an example of the fluctuation–dissipation theorem explained in Appendix B. In the second equality, $i\nu_n$ denotes the Matsubara frequency (see Appendix B).

In the RPA, the dynamical susceptibility is given by

$$\chi_{m,gU}^{\perp}(\boldsymbol{q}, \omega)^{-1} = \chi_{m,0}^{\perp}(\boldsymbol{q}, \omega)^{-1} - gU. \tag{1.105}$$

Then one can easily derive the correlation part in the RPA as

$$\Delta F_{\mathrm{RPA}} = T \sum_n \sum_{\boldsymbol{q}} \left\{ \ln[1 - \frac{U}{2} \chi_{m,0}^{\perp}(\boldsymbol{q}, i\nu_n)] + \frac{U}{2} \chi_{m,0}^{\perp}(\boldsymbol{q}, i\nu_n) \right\}. \tag{1.106}$$

The RPA theory has an obvious inconsistency [27]: the static and homogeneous limit of eqn (1.105) with $g = 1$ does not agree with the one derived using eqns (1.106) and (1.97). This discrepancy becomes serious if one wants to calculate the Curie temperature T_c by using the condition $\chi^{-1} = 0$, since the RPA susceptibility diverges at T_c^{HF}, which is different from T_c calculated from eqn (1.97).

The simplest way to remedy this drawback is to introduce a parameter λ by

$$\chi^{-1} = (1 + \lambda) \chi_0^{-1} - \tfrac{1}{2} U. \tag{1.107}$$

The correlation effect represented by λ is related to ΔF as

$$\lambda \chi_0^{-1} = \frac{\partial^2 \Delta F}{\partial m^2} \bigg|_{m=0}. \tag{1.108}$$

One then modifies eqn (1.106) in the following manner [27]:

$$\Delta F_{\text{SCR}} = T \sum_n \sum_q \left\{ \ln[1 + \lambda - \frac{U}{2}\chi^{\perp}_{m,0}(q, i\nu_n)] + \frac{U}{2}\chi^{\perp}_{m,0}(q, i\nu_n) \right\}. \quad (1.109)$$

Provided that $\chi^{\perp}_{m,0}(q, 0)$ has a maximum at $q = 0$ and $m = 0$, a singularity in ΔF_{SCR} occurs at T_c as the ferromagnetic instability. This instability at T_c is consistent with the divergence of χ given by eqn (1.107). Thus, the inconsistency of the RPA has been removed.

Equation (1.109) is of central importance in the so-called self-consistent renormalization (SCR) theory of Moriya and Kawabata [27]. Since $\chi^{\perp}_{m,0}(q, i\nu_n)$ can be derived explicitly for a given spectrum of electrons, it is possible to set up a self-consistency equation for λ at each temperature. One obtains $\lambda(T) - \lambda(0) \propto T^{4/3}$, since the local susceptibility with $m = 0$ in eqn (1.104) behaves as $\omega^{1/3}$ for small ω. This power-law behaviour will be derived shortly. The second derivative of $\chi^{\perp}_{m,0}(q, i\nu_n)$ with respect to m does not change this leading behaviour at $m = 0$. We emphasize that the T dependence of $\lambda(T)$ is controlled by the spectrum of the ferromagnetic spin fluctuation and is independent of the details of the band structure. In the case of antiferromagnetic metals, the different q dependence of spin fluctuations leads to Im $\chi(\omega) \propto \omega^{1/2}$ and $\lambda(T) - \lambda(0) \propto T^{3/2}$.

The temperature dependence of λ leads to an approximate Curie–Weiss-like behaviour of the susceptibility, which is the most remarkable effect of the mode coupling. The SCR theory has been used widely in explaining the temperature dependence of physical quantities in transition metals and some of their compounds. The temperature dependence deviates from the RPA theory significantly near the magnetic transition. The most appealing feature of the SCR theory is that the mode-coupling effect is taken into account in a simple and practical manner. However, eqn (1.109) cannot be justified microscopically near the magnetic transition. Since the SCR theory relies on $\chi^{\perp}_{m,0}(q, \omega)$ as the basic ingredient like the RPA, the parameter λ should be small to justify the SCR as a weak-coupling theory. However, λ becomes actually of the order of unity by the self-consistency requirement. This is simply because ΔF_{SCR} does not vanish even with $\lambda = 0$ and, therefore, $\lambda \ll 1$ cannot be a self-consistent solution of eqn (1.108) near the magnetic instability. For this reason, one should regard the SCR theory as a semi-phenomenological scheme to deal with the mode coupling among spin fluctuations, rather than a microscopic theory for the Hubbard model. The original Moriya–Kawabata theory as presented above was later modified in a few different ways. We refer to Moriya's book [29] for a comprehensive account of the SCR theory.

An alternative approach to take account of the mode-coupling effect has been proposed by Dzyaloshinskii and Kondratenko [28] by the use of the Fermi-liquid theory. Let us assume again the paramagnetic Fermi-liquid ground state. We consider the leading finite-temperature correction to the dynamical susceptibility $\chi(q, \omega; T)$ at low temperature T. According to the Matsubara Green function formalism the leading correction is obtained by the following procedure: from a Feynman diagram for $\chi(q, i\omega; T)$ we pick up a part $\Gamma(i\nu; T)$ with low frequency ν flowing from and to the rest of the diagram.

FIG. 1.6. The dominant fluctuation contribution to the susceptibility.

Then we replace the frequency summation in this part by

$$\left(T \sum_\nu - \int_{-\infty}^{\infty} \frac{d\omega}{2\pi} \right) \Gamma(i\nu; T), \tag{1.110}$$

and integration is performed over all the remaining frequencies.

In the case of a weak ferromagnet, the dominant contribution to $\Gamma(i\nu; T)$ comes from a spin fluctuation with small wavenumber and frequency. Particle–hole excitations have the energy scale of the Fermi energy and can safely be neglected. Thus, the relevant diagram is represented by Fig. 1.6 and constitutes the self-energy-like correction to the susceptibility:

$$\chi(0, 0; 0)^{-1} - \chi(0, 0; T)^{-1}$$
$$= -\gamma \int_0^{\infty} \frac{d\omega}{2\pi} \int \frac{d\boldsymbol{q}}{(2\pi)^3} \frac{1}{\exp(\omega/T) - 1} \mathrm{Im}\, \chi(\boldsymbol{q}, \omega; T)$$
$$\equiv -\gamma \beta(T), \tag{1.111}$$

where γ is a parameter for the mode coupling, and the Bose distribution function emerges from the analytic continuation of eqn (1.110). One can regard γ as a constant since the dependence on wavenumber, frequency, and temperature is on the much larger scale of one-particle excitations.

The temperature-dependent correction given by eqn (1.111) corresponds to $\lambda(T) - \lambda(0)$ in the SCR theory. The magnitude of γ cannot be evaluated by perturbation theory since higher-order terms are equally important in general. In the nearly ferromagnetic ground state, $1 + Z_0 \equiv \alpha$ is much smaller than unity. Then the characteristic temperature $T_{\mathrm{SF}} = \alpha v_{\mathrm{F}} k_{\mathrm{F}}$ of spin fluctuations is much lower than the Fermi temperature $T_{\mathrm{F}} = v_{\mathrm{F}} k_{\mathrm{F}}/2$. In the process of integration in eqn (1.111) for the temperature range $T_{\mathrm{SF}} \ll T \ll T_{\mathrm{F}}$, one may use the parametrized form, eqn (1.81), for $\chi(\boldsymbol{q}, \omega; T)$. This is firstly because only the small q/k_{F} values are important, and secondly the detailed analytic structure with respect to ω is not important for small T_{SF}/T. We scale the integration variables as $q = \omega^{1/3}\bar{q}$, and then $\omega = T\bar{\omega}$. Integrations with respect to $\bar{q}$ and $\bar{\omega}$ in eqn (1.111) give rise to the factor

$$\beta(T) \sim T^{4/3}. \tag{1.112}$$

With the nearly ferromagnetic ground state, the magnetic susceptibility follows $\chi(T) \sim T^{-4/3}$. Thus, the same temperature dependence as that of the SCR theory [27] has been obtained. This agreement is not accidental, as is clear from the above derivation.

1.4 Nuclear magnetic resonance

For understanding the specific properties of heavy-electron systems, it is important to investigate and describe the low-energy excitations. In this context, both nuclear magnetic resonance (NMR) and neutron scattering experiments play central roles in obtaining information about the dynamical response of heavy-electron systems. Whereas the NMR probes the local environment of one particular nucleus and, therefore, a wave-vector average of the dynamical response function with a small energy transfer comparable to the nuclear Zeeman energy (10^{-6}–10^{-8} meV), the neutron scattering experiments can scan wider energies and wave vectors corresponding to the whole Brillouin zone. In real situations, the energy transfer is usually limited below about 100 meV with the use of thermal neutrons. Thus, both experiments are complementary. The advantage of NMR is that it can extract the lowest-energy excitation and detect a magnetic instability with a tiny moment if it appears. Furthermore, in the superconducting state, NMR can provide a detailed structure of the response function reliably. In this section, we begin with the description of the experimental aspects of NMR. For more detailed descriptions on NMR, we refer to standard textbooks [30].

1.4.1 *Phenomenology*

A nucleus with spin I has a magnetic moment $\boldsymbol{\mu}_n = \gamma_n \boldsymbol{I}$, where γ_n is the nuclear gyromagnetic ratio. If an isolated nucleus is placed in an external magnetic field $\boldsymbol{H}_0$, the nuclear spin levels are split into $2I + 1$ levels with an equal energy separation:

$$\Delta E = \gamma_n H_0. \tag{1.113}$$

According to the quantum mechanical selection rule, the magnetic dipole transitions are induced only between the adjacent levels. Thus, the nuclear spin system undergoes resonance absorption if the perturbing oscillatory magnetic field $\boldsymbol{H}_1$ is applied perpendicularly to $\boldsymbol{H}_0$ with the condition $\omega_n = \gamma_n H_0$. Actually, a nucleus in a nonmagnetic solid experiences a magnetic field which is different from the external field; the correction ΔH contains contributions from dipolar fields of other nuclei, atomic diamagnetism and the chemical shift. The correction is proportional to the induced electronic moment, and gives rise to the frequency shift and linewidth of the NMR absorption spectrum. On the other hand, in a magnetically ordered solid a nuclear spin experiences a dominant internal magnetic field associated with the spontaneous magnetization of the electrons.

Let $\boldsymbol{H}_0$ be applied along the z-axis, and $\boldsymbol{M}(t)$ be the nuclear magnetization averaged over the sample at time t. In most cases, the z-component $M_z(t)$ will recover to the equilibrium value M_0 exponentially. The equation of motion for $\boldsymbol{M}(t)$ is then described by

$$\frac{dM_z(t)}{dt} = \gamma_n (\boldsymbol{M}(t) \times \boldsymbol{H})_z + \frac{1}{T_1}(M_0 - M_z(t)), \tag{1.114}$$

where T_1 is the nuclear spin–lattice relaxation time and characterizes the rate at which energy is transferred from the nuclear spin system to the electron spin system. In heavy-electron systems, the spin fluctuations resulting from exchange couplings among the localized electron spins and/or with conduction electron spins are responsible for the nuclear spin–lattice relaxation process. The magnetic field $\boldsymbol{H}$ is in general different from $\boldsymbol{H}_0$ because of polarization of the medium.

In addition to T_1, there is another relaxation time T_2 which is related to the component $\boldsymbol{M}_\perp(t)$ of the magnetization perpendicular to the external field. The transverse magnetization is associated with the forced precession of the nuclear moment caused by the perturbing oscillation field H_1. The equation of motion for $\boldsymbol{M}_\perp(t)$ is described by

$$\frac{d\boldsymbol{M}_\perp(t)}{dt} = \gamma_\mathrm{n}(\boldsymbol{M}(t) \times \boldsymbol{H}) - \frac{\boldsymbol{M}_\perp(t)}{T_2^*}. \tag{1.115}$$

Note that $\boldsymbol{M}_\perp$ is zero in thermal equilibrium. The characteristic time T_2^* does not directly reflect the relaxation of the nuclear spin, since a possible inhomogeneity of the external field over the sample and/or a static distribution of hyperfine fields at nuclear sites in the magnetic substance contribute to T_2^*.

In order to determine the intrinsic transverse relaxation time T_2, one typically uses the spin-echo method illustrated in Fig. 1.7. The experiment proceeds as follows. Under the external field H_0 along the z-axis, one applies, along the x-axis, an alternating field H_1 with the resonance frequency ω_n for a short time $\delta t = \pi/(2\gamma_\mathrm{n} H_1)$ ($\ll T_1$). This pulse causes $\boldsymbol{M}$ to rotate toward the y-axis by an angle $\pi/2$, and hence is called a $\pi/2$-pulse. After the pulse is turned off, each nuclear spin contributing to $M_\perp(t)$ begins to precess in the x–y plane. Because of the nuclear spin–spin interaction and the inhomogeneous field, the precession frequency of each spin is different. Then the free induction signal, which arises from the sum total of $M_\perp(t)$ from all portions of the sample, decays exponentially with a time constant T_2^*. Waiting for a time τ ($>T_2^*$) after the first $\pi/2$-pulse, one applies the second pulse for a time duration $\delta t' = \pi/(\gamma_\mathrm{n} H_1)$, which generates a rotation of each nuclear spin by an angle π around the x-axis. The nuclear spins, which are pointing in various directions in the x–y plane under the inhomogeneous field, resume precession and finally refocus along the $-y$-direction after a time 2τ from the first $\pi/2$-pulse. This refocusing results in the recovery of $M_\perp(t)$, which is probed as a peak in the free induction signal like an echo. Hence the phenomenon is called spin-echo.

The amplitude $M_\mathrm{s}(2\tau)$ of the spin-echo usually decays exponentially as a function of 2τ as

$$M_\mathrm{s}(2\tau) = M_\perp(0) \exp\left(-\frac{2\tau}{T_2}\right). \tag{1.116}$$

The refocusing occurs only when the spins precessing under the inhomogeneous field keep the memory of their phase. Then the condition $\tau \ll T_1, T_2$ is required for spin-echo. If $T_2 \ll T_1$, the decay of spin-echo is caused by the nuclear spin–spin interaction without any influence of the inhomogeneous field distribution. Hence, T_2 provides information about the intrinsic properties of the system. We will discuss T_1 and T_2 in greater detail later for nuclei embedded in a magnetic solid.

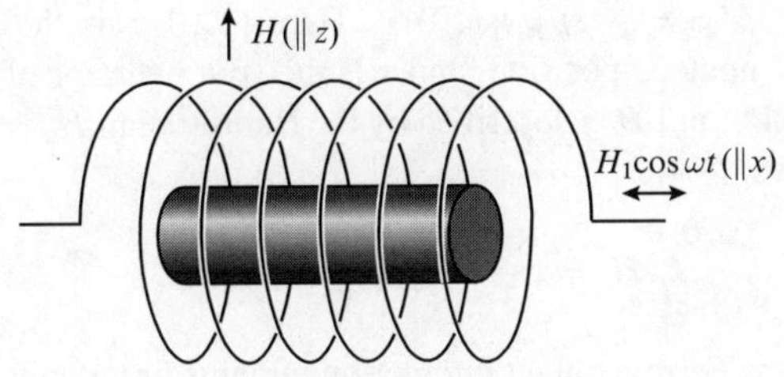

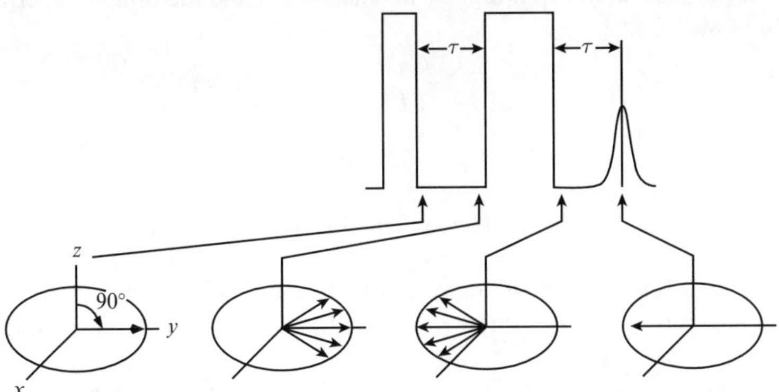

FIG. 1.7. Illustration of the spin-echo NMR method.

1.4.2 *Magnetic hyperfine interaction*

Most NMR experiments in heavy-electron systems have been carried out for nuclei of non-lanthanide constituents. As an exception, Yb NMR has been observed in the valence fluctuating compounds $YbAl_2$ and $YbAl_3$ [31], where the average valency of Yb is between 2+ and 3+ and the valency fluctuates quantum mechanically. One can discuss directly the magnetic relaxation rate of f electrons, which is related to the linewidth of the quasi-elastic neutron scattering intensity. This discussion, of course, requires a reliable knowledge of the hyperfine field. On the other hand, the hyperfine interaction between the non-lanthanide nuclei and f electrons is caused by a mixing between the f electrons and conduction electrons, which, in turn, interact with the nuclei. This transferred hyperfine interaction dominates the magnetic hyperfine interaction between non-lanthanide nuclei and f electrons in most of the heavy-electron compounds. The interaction is mediated by the Fermi contact interaction, the spin-dipolar, and the orbital hyperfine interactions with nuclei at non-lanthanide sites. The Fermi contact interaction acts only for s electrons and the latter two act only for non-s electrons, as explained below.

Let us first discuss the Fermi contact interaction between a nucleus and an electron in an s-orbital. Provided the magnetization density $M_n = \gamma_n I / v_n$ of the nucleus is uniform inside the nuclear volume v_n, the magnetic field B produced by M_n is given by $B = (8\pi/3)\,M_n$. On the other hand, the electronic magnetization M_e at the nucleus position is given by $M_e = -g\mu_B s |\phi_s(0)|^2$. Here $|\phi_s(0)|^2$ is the probability density of the electron at the nucleus position, and s is the spin operator of the s electron. The interaction between M_e and B is described by the Hamiltonian $H_c = -B \cdot M_e v_n$, which is rewritten as follows:

$$H_c = \frac{8\pi}{3} \gamma_n g \mu_B |\phi_s(0)|^2 I \cdot s. \tag{1.117}$$

We emphasize that this Fermi contact interaction remains finite only for s electrons with nonvanishing values of $|\phi_s(0)|^2$.

On the other hand, an electron in a non-s orbital has zero amplitude at the nuclear position. It has instead the spin-dipolar interaction H_d and the orbital hyperfine interaction H_l of the form

$$H_d = -\gamma_n g \mu_B \frac{1}{r^3} \left[I \cdot s - \frac{3(I \cdot r)(s \cdot r)}{r^2} \right], \tag{1.118}$$

$$H_l = \gamma_n \mu_B \frac{2}{r^3} I \cdot l, \tag{1.119}$$

where l is the orbital angular momentum of the non-s electron. In actual metals the orbital motion appears as the orbital Knight shift, which is usually independent of the temperature.

Furthermore, we mention the inner core polarization, which is due to unpaired non-s electrons. These electrons have vanishing amplitude at the nucleus, but can induce an unbalanced spin density of paired inner s electrons. In contrast with the Fermi contact interaction, the sign of the hyperfine field due to the core polarization is negative.

We now proceed to the many-body description of the hyperfine interaction. In heavy-electron systems like the Ce and U compounds, the spin–orbit interaction is so strong that the electronic state of an ion is characterized by the total angular momentum as $J = L + S$ and the magnetic moment $-g_J \mu_B J$. The lowest J manifold is split into several sublevels by the CEF. For simplicity, we consider a nucleus at the origin. The nuclear spin I and the f electrons at site R_j interact through the hyperfine interaction as

$$H_{\text{hyper}} = \sum_j I \cdot A(R_j) \cdot J(R_j), \tag{1.120}$$

where the sum is taken over all nearest-neighbour ions of lanthanides or actinides. The tensor $A(R_j)$ consists of the orbital and the spin parts projected onto the total angular momentum $J(R_j)$. This is in contrast to transition metals where spin and orbital parts survive separately because of much weaker spin–orbit interaction. In the hyperfine coupling tensor $A(R_j)$ in heavy-electron systems, the transferred (indirect) hyperfine interaction plays a main role for the magnetic interaction for non-lanthanide nuclei. The dipole–dipole interaction between f-electron spins and nuclear spins is unimportant and is neglected in the following.

The hyperfine interaction causes the magnetic field at the nucleus to deviate from the external one, H_0, by an amount ΔH. In order to avoid complexity, here we take the direction α of H to be one of the principal axes of the crystal. The shift satisfies the relation

$$\gamma_{\mathrm{n}}\Delta H = A_\alpha \langle J_\alpha \rangle, \tag{1.121}$$

where $\langle J_\alpha \rangle$ is the average of $\boldsymbol{J}(\boldsymbol{R}_j)$ in the α-direction, and

$$A_\alpha = \mathrm{Av}_j[\boldsymbol{A}(\boldsymbol{R}_j)]_{\alpha\alpha}$$

is the spatial average of the diagonal element of the hyperfine interaction. Let the number of f electrons per unit volume be n. By the definition of the static susceptibility χ_α, we have

$$ng_J\mu_{\mathrm{B}}\langle J_\alpha \rangle = \chi_\alpha H_0.$$

Then the Knight shift K_α defined by $K = \Delta H/H_0$ is expressed as

$$K_\alpha = \frac{A_\alpha}{n\gamma_{\mathrm{n}}g_J\mu_{\mathrm{B}}}\chi_\alpha. \tag{1.122}$$

One can extract the transferred hyperfine field by plotting the Knight shift vs the susceptibility, with temperature as an implicit parameter, i.e.

$$A_\alpha = n\gamma_{\mathrm{n}}g_J\mu_{\mathrm{B}}\left(dK_\alpha/d\chi_\alpha\right).$$

It often happens that the K vs χ plot is not linear in the whole temperature range. A straight line in the high-T range goes over to another one below a certain crossover temperature. This means that A_α in the low-T range is no longer the same as that in the high-T range. This change occurs possibly because the conduction electron spin polarized by the c–f mixing varies with temperature, which reflects the change of the character of the f electrons from localized to itinerant. In the NMR investigation for non-lanthanide nuclei, the knowledge of the transferred hyperfine tensor enables us to extract a dynamical response function from the nuclear spin–lattice relaxation measurements. This leads to a quantitative comparison with that derived by neutron scattering experiments.

1.4.3   *Electric quadrupolar interaction*

We now describe the electric quadrupolar interaction between electrons and nuclei with $I > 1/2$. These nuclei have an aspherical charge distribution $\rho(\boldsymbol{r})$, and, therefore, have an electric quadrupole moment $Q'_{jk} = \int \rho(\boldsymbol{r})x_jx_k\,d\boldsymbol{r}$. Let $V(\boldsymbol{r})$ denote the electrostatic potential due to the surrounding charges around the nucleus, and $V_{jk} = \partial^2 V(\boldsymbol{r})/\left(\partial x_j\,\partial x_k\right)_{r=0}$ be the field gradient tensor at the nucleus. Then the interaction Hamiltonian H_Q of the quadrupole moment of a nucleus with the field gradient is given by

$$H_Q = \frac{1}{2}\sum_{j,k} Q'_{jk}V_{jk} + \cdots. \tag{1.123}$$

We can express the traceless tensor $Q_{jk} = 3Q'_{jk} - \delta_{jk}\sum_i Q'_{ii}$ in terms of the nuclear spin I as

$$Q_{jk} = \frac{eQ}{I(2I-1)}\left[\frac{3}{2}(I_j I_k + I_k I_j) - \delta_{jk} I^2\right].$$ (1.124)

For an axially symmetric electric field gradient we define eq by

$$eq = \left(\frac{\partial^2 V}{\partial z^2}\right)_{r=0} = -e\int d\boldsymbol{r}\,\frac{3z^2-r^2}{r^5}n(\boldsymbol{r}),$$ (1.125)

where $n(\boldsymbol{r})$ denotes the electronic density. Then H_Q is given by

$$H_Q = \frac{e^2 q Q}{4I(2I-1)}\left[3I_z^2 - I(I+1)\right],$$ (1.126)

with

$$eQ = Q_{zz} = 3Q'_{zz} - \sum_i Q'_{ii} = \int \rho(\boldsymbol{r})(3z^2 - r^2)\,d\boldsymbol{r}.$$ (1.127)

If $V(\boldsymbol{r})$ has no uniaxial symmetry, an additional term $\eta(I_x^2 - I_y^2)$ enters into H_Q with an asymmetry parameter $\eta = (V_{xx} - V_{yy})/V_{zz}$. Here x, y, z are taken to be the principal axes of the field gradient. This electric interaction splits the nuclear spin levels into several multiplets, and the transition between them causes a resonant absorption of radio frequency. This is called the nuclear quadrupole resonance (NQR). The important feature of NQR is that the experiment does not need an external magnetic field. For example, the NQR of ^{63}Cu was observed at 3.91 and 3.43 MHz for CeCu$_6$ and CeCu$_2$Si$_2$, respectively. The NQR experiment is particularly advantageous in investigating low-energy excitations in the superconducting state at zero magnetic field. We note that the coupling of the nuclear quadrupole moment with the lattice vibration opens another relaxation channel mediated by the charge fluctuation in ionic or metallic crystals including nuclei with a large quadrupole moment. However, in heavy-electron systems the magnetic relaxation channel prevails in most cases.

1.4.4 *Nuclear spin–lattice relaxation time T_1*

Nuclear spin–lattice relaxation in heavy-electron systems is principally caused by f-electron spin fluctuations through the transferred hyperfine interaction given by eqn (1.120). The relaxation rate $1/T_1$ defined by eqn (1.114) corresponds to that of the nuclear magnetic susceptibility. In Appendix D we have given the formulae necessary to derive the relaxation rate by using perturbation theory. The dynamics of I_z, the z-component of the nuclear spin, is described by

$$\frac{\partial I_z}{\partial t} = i[H_{\text{hyper}}, I_z] = \sum_j [-A_x(\boldsymbol{R}_j)J_x(\boldsymbol{R}_j)I_y + A_y(\boldsymbol{R}_j)J_y(\boldsymbol{R}_j)I_x].$$ (1.128)

In the case where the hyperfine interaction can be treated in the lowest order, we obtain, from eqn (D.12),

$$\frac{1}{T_1} = \frac{T}{2} \sum_q \Big[|A_x(\boldsymbol{q})|^2 \langle J_x(\boldsymbol{q})|\delta(\omega_{\mathrm{n}} - \mathcal{L}_0)|J_x(\boldsymbol{q})\rangle$$

$$+ |A_y(\boldsymbol{q})|^2 \langle J_y(\boldsymbol{q})|\delta(\omega_{\mathrm{n}} - \mathcal{L}_0)|J_y(\boldsymbol{q})\rangle \Big], \tag{1.129}$$

where $\mathcal{L}_0$ is the unperturbed Liouville operator, and ω_{n} is the frequency at the resonance which is much smaller than the characteristic frequency of the electrons. The factor T on the right-hand side of eqn (1.129) comes from $\langle I_z|I_z\rangle = \beta I(I+1)/3$ in the denominator. The quantity $\langle \cdots \rangle$ is related to the dynamical susceptibility of the f electrons; namely, using eqn (B.10) we have

$$\langle J_x(\boldsymbol{q})|\delta(\omega - \mathcal{L}_0)|J_x(\boldsymbol{q})\rangle = (g_J\mu_B)^2 \, \mathrm{Im}\, \chi_{xx}(\boldsymbol{q}, \omega)/\omega. \tag{1.130}$$

From eqn (1.130) we see that $1/T_1$ reflects sensitively the low-energy magnetic excitations. If the dynamical response is dominated around a special wave vector $\boldsymbol{Q}$, $1/T_1 T$ is directly connected to the slope at the low energy tail of $\mathrm{Im}\,\chi(\boldsymbol{Q}, \omega)$ around $\omega \sim 0$. Below we take typical examples of local moment and itinerant electron systems and discuss the temperature dependence of $1/T_1$ for each case.

Local-moment system We first consider the case of localized f electrons, the dynamics of which is described by the Heisenberg model with the exchange interaction J_{ex}. According to the formalism in Appendix D, the spectral shape becomes Gaussian type if the short-time relaxation behaviour dominates the spectrum. This is the case in the paramagnetic state of the system. Then we have, from eqns (D.18)–(D.21),

$$\langle M(t)|M\rangle = (g_J\mu_B)^2 \frac{J(J+1)}{3} \exp\left[-\frac{1}{2}\langle\omega^2\rangle_{\mathrm{c}} t^2\right], \tag{1.131}$$

where $\langle\omega^2\rangle_{\mathrm{c}} = J_{\mathrm{ex}}^2 z_{\mathrm{n}} J(J+1)$, with z_{n} being the number of nearest neighbours of interacting local moments. From eqn (1.129), we get

$$\frac{1}{T_1} = \sqrt{2\pi} \frac{C|A_{\mathrm{hf}}|^2}{\sqrt{\langle\omega^2\rangle_{\mathrm{c}}}}, \tag{1.132}$$

where the $\boldsymbol{q}$-dependent hyperfine coupling, eqn (1.129), is replaced by the average A_{hf} over $\boldsymbol{q}$. We note that $1/T_1$ is independent of the temperature in the present case.

On the other hand, if the relaxation of local moments is caused by the exchange interaction J_{cf} with conduction electrons, the relaxation function is dominated by the long-time behaviour. Then the spectrum undergoes motional narrowing with the linewidth of the order of J_{cf}^2/D, where D is the bandwidth of the conduction electrons. The relaxation function is given by

$$\langle M(t)|M\rangle = (\mu_B g_J)^2 \frac{J(J+1)}{3} \exp\left[-\Gamma_{\mathrm{cf}} t\right] \tag{1.133}$$

for large time t. Since the long-time behaviour given above determines the spectrum, we can parametrize the dynamical susceptibility as

$$\chi(\omega) = \frac{C\Gamma_{\mathrm{cf}}}{T(\Gamma_{\mathrm{cf}} - i\omega)}. \tag{1.134}$$

From eqns (1.129) and (1.130) we obtain

$$\frac{1}{T_1} = \frac{C|A_{\mathrm{hf}}|^2}{\Gamma_{\mathrm{cf}}}. \tag{1.135}$$

As will be discussed in the next section, Γ_{cf} has a linear T dependence if f spins decay through coupling with conduction electrons. Thus, $1/T_1$ decreases as $1/T$ upon heating.

Itinerant electron system   For itinerant electrons at low temperatures, the Fermi-liquid theory provides a reliable description as presented in the previous section. We discuss $1/T_1$ on the basis of the quasi-particle RPA given by eqn (1.78). In terms of the quasi-particle susceptibility $\chi_0^*(\boldsymbol{q}, \omega)$, we obtain

$$\mathrm{Im}\,\chi(\boldsymbol{q}, \omega) = \left| \frac{\chi(\boldsymbol{q}, \omega)}{\chi_0^*(\boldsymbol{q}, \omega)} \right|^2 \mathrm{Im}\,\chi_0^*(\boldsymbol{q}, \omega). \tag{1.136}$$

Since $\mathrm{Im}\,\chi(\boldsymbol{q}, \omega)$ is much smaller than $\mathrm{Re}\,\chi(\boldsymbol{q}, \omega)$ in the relevant frequency range, we can approximate

$$\left| \frac{\chi(\boldsymbol{q}, \omega)}{\chi_0^*(\boldsymbol{q}, \omega)} \right|^2 \sim \left(\frac{1}{1 + \chi_0^*(\boldsymbol{q}, 0)Z_0/\rho^*(\mu)} \right)^2, \tag{1.137}$$

which acts as an enhancement factor in $1/T_1$.

We get the imaginary part of $\chi_0^*(\boldsymbol{q}, \omega)$ near the Fermi energy μ as

$$\mathrm{Im}\,\chi_0^*(\boldsymbol{q}, \omega) = \pi\omega \sum_k \left(-\frac{\partial f}{\partial \epsilon_k} \right) \delta(\omega - \epsilon_{k+q} + \epsilon_k). \tag{1.138}$$

Furthermore, we replace the enhancement factor by its average. Then we get

$$\frac{1}{T_1} = \pi A_{\mathrm{hf}}^2 \rho^*(\mu)^2 T \left\langle \left(\frac{1}{1 + \chi_0^*(\boldsymbol{q}, 0)Z_0/\rho^*(\mu)} \right)^2 \right\rangle_{\mathrm{FS}}, \tag{1.139}$$

where $\langle \cdots \rangle_{\mathrm{FS}}$ means the average over the Fermi surface. If the enhancement factor is the same as that for the $\boldsymbol{q} = 0$ case, we have the so-called Korringa relation:

$$\frac{1}{T_1 T} = \pi \left(\frac{\gamma_{\mathrm{n}}}{\mu_{\mathrm{B}}} \right)^2 K^2, \tag{1.140}$$

where K is the Knight shift. Note that the result obtained above neglects the presence of anisotropy. In reality, although magnetic properties in most heavy-electron systems are very anisotropic, there are cases where the Korringa relation is valid for the actinide heavy-electron compounds. Since the Landau parameters provide a quantitative description only in the isotropic case, due care is necessary in interpreting the experimental data.

When the temperature is much smaller than the Fermi energy, the quantity $1/(T_1 T)$ is independent of T. This property holds as long as the Fermi liquid is realized. We often call the constancy of $1/(T_1 T)$ the Korringa law even though the stronger relation given by eqn (1.140) is not confirmed.

1.4.5 *Nuclear spin–spin relaxation time T_2*

In heavy-electron systems, the indirect nuclear spin–spin interaction plays a primary role in the decay of the transverse nuclear magnetization [32]. A nuclear spin at site R interacts with other nuclear spins indirectly. First, it interacts with the spin of a conduction electron at r through the hyperfine interaction tensor $A(r - R)$. The conduction electron propagates to another nuclear site and interacts with the second nucleus via the hyperfine interaction. This indirect interaction is called the Ruderman–Kittel–Kasuya–Yosida or just RKKY interaction. The interaction Hamiltonian H_{ij} between two nuclear spins I_i and I_j is expressed as

$$H_{ij} = -I_i \hat{\Phi}(R_{ij}) I_j \tag{1.141}$$

where

$$\hat{\Phi}(R_{ij}) = \int dr \int dr'\, A(R_i - r)\hat{\chi}(r - r')A(r' - R_j), \tag{1.142}$$

with $\hat{\chi}(r - r')$ being the susceptibility tensor of the conduction electrons. Consequently, the calculation of the indirect nuclear spin coupling is reduced to that of magnetic susceptibility $\hat{\chi}(r)$ or its Fourier transform $\hat{\chi}(q)$. If we assume a contact hyperfine interaction given by $A(R_i - r) = A_s\delta(R_i - r)$, the range of the interaction is determined simply by $\hat{\chi}(r)$. In this case, the indirect interaction is written as

$$\hat{\Phi}(R_{ij}) = \frac{1}{N} \sum_q A_s \hat{\chi}(q) A_s \exp(i q \cdot R_{ij}). \tag{1.143}$$

For the free electron gas model, the above formula leads to the Ruderman–Kittel form for the indirect nuclear spin–spin interaction. On the other hand, in magnetically ordered states at temperatures well below the transition temperature, $\hat{\chi}(q)$ can be calculated by a spin wave approximation.

In the case where $\Phi(R_{ij})_{zz}$ dominates over all the other components, the transverse relaxation spectrum has a Gaussian shape. Thus, in the spin-echo experiment one has

$$M_\perp(2\tau) = M_\perp(0) \exp\left[-\frac{1}{2}\left(\frac{2\tau}{T_{2G}}\right)^2\right], \tag{1.144}$$

where the relaxation rate $1/T_{2G}$ is given by $(1/T_{2G})^2 = \langle\omega^2\rangle$, with $\langle\omega^2\rangle$ defined by eqn (D.21) in Appendix D. For simplicity, in the lowest-order perturbation theory, we assume that the anisotropy of $\hat{\Phi}(R_{ij})$ arises due to the large A_{zz} and that $\hat{\chi}(q)$ is isotropic. Then we have

$$\left(\frac{1}{T_{2G}}\right)^2 = \frac{3I(I+1)}{4} \sum_{i \neq j} |\Phi_{zz}(R_{ij})|^2$$
$$= A_{zz}^4 \frac{3I(I+1)}{4}\left[\mathrm{Av}_q\, \chi(q)^2 - \left(\mathrm{Av}_q\, \chi(q)\right)^2\right], \tag{1.145}$$

where Av_q means the average over q. On the other hand, a possible exchange interaction causes the correlation function of the hyperfine field to decay with a time constant τ_C.

If the exchange interaction is strong, one has $1/\tau_C^2 \gg \langle \omega^2 \rangle$, where $\langle \omega^2 \rangle$ is the second moment of the local field. In this case, the nuclear spin cannot feel the full magnitude of the local field and the NMR spectrum is narrowed. This is called exchange narrowing. As a result, spin-echo has the simple exponential law with the decay rate $1/T_{2L}$, which is roughly given by $1/T_{2L} \sim \tau_C \langle \omega^2 \rangle$.

In heavy-electron systems such as $CeCu_2Si_2$, the AF spin correlation develops at low temperatures even though no magnetic order takes place. Then the enhancement of the susceptibility at the corresponding $\boldsymbol{Q}$ vector should be reflected in the temperature dependence of T_2. To show the characteristic behaviour we use the parametrization of the q-dependent susceptibility as given by eqn (1.89). Then we obtain

$$\left(\frac{1}{T_2}\right)^2 \sim \int \left(\frac{\chi(\boldsymbol{Q})}{1+q^2\xi^2}\right)^2 d\boldsymbol{q} \sim \frac{\chi(\boldsymbol{Q})^2}{\xi^3}. \tag{1.146}$$

In the typical case of $\chi(\boldsymbol{Q}) \sim \xi^2(T)$, we infer that $(1/T_2)^2 \sim \xi(T)$. Therefore, $1/T_2$ is strongly enhanced as the magnetic correlation length $\xi(T)$ increases.

To summarize, important physical quantities to be obtained by the NMR in magnetic solids are the Knight shift, the nuclear spin–lattice relaxation time (T_1), the nuclear spin–spin relaxation time (T_2). For the nuclear electric quadrupole resonance, which does not need a magnetic field, the NQR frequency and the relaxation time are relevant quantities. The internal field probed by the NMR in the ordered state also provides valuable information on the dispersion relation of the spin-wave excitations.

1.5 Neutron scattering

1.5.1 *Characteristics and utility*

Low-energy neutrons with typical energy of 30–500 K and with wavelength in a range of 1–5 Å interact with nuclei and magnetic moments of electrons in solids. The corresponding scattering length is of the order of 10^{-13} cm for the nuclear scattering, and of $e^2/mc^2 = 2.8 \times 10^{-13}$ cm for the magnetic scattering. For magnetic materials, both cross sections are comparable. The magnetic Bragg scattering, i.e. elastic scattering, provides information about the orientation and length of the magnetic moments in the ordered state, and the shape of the spin density of magnetic ions. Inelastic scattering in the ordered state may determine the spectrum of elementary excitations like spin-waves. In the paramagnetic state, on the other hand, a study of wavenumber dependence and the energy distribution of scattering neutrons provides information about the magnetic correlation length and a characteristic energy scale in the spin system.

Neutron scattering experiments are a powerful tool to investigate the spin dynamics in the heavy-electron system, since the magnetic neutron scattering cross section is proportional to the dynamical structure factor $S(\boldsymbol{q}, \omega)$ and to Im $\chi(\boldsymbol{q}, \omega)$. For magnetically ordered heavy-electron systems with tiny moments, the elastic scattering profile often shows diffuse scattering instead of the true Bragg one. This shows that the correlation length is finite, in contrast to the case of true long-range order. This raises a fundamental problem about the nature of heavy-electron magnetism, which is not yet resolved. In the following, we describe very briefly the basics of neutron scattering by magnetic moments. For more details of neutron scattering, we refer to standard monographs [33,34].

1.5.2 *Magnetic scattering*

The interaction between a neutron (n) at the origin and an electron (e) at r is given by

$$V_{\rm e} = -\boldsymbol{\mu}_{\rm e} \cdot \boldsymbol{H}_{\rm n}(\boldsymbol{r}), \tag{1.147}$$

where $\boldsymbol{\mu}_{\rm e}$ is the magnetic moment of the electron, and $\boldsymbol{H}_{\rm n}(\boldsymbol{r})$ is the magnetic field due to the neutron,

$$\boldsymbol{H}_{\rm n} = \nabla \times \left[\nabla \times \left(\frac{\boldsymbol{\mu}_{\rm n}}{r} \right) \right], \tag{1.148}$$

with $\boldsymbol{\mu}_{\rm n}$ as the magnetic moment of the neutron. In the momentum space, the interaction is written simply as

$$\int d\boldsymbol{r} \, \exp\left(-i\boldsymbol{q} \cdot \boldsymbol{r}\right) V_{\rm e}(\boldsymbol{r}) = -4\pi \left[\boldsymbol{\mu}_{\rm n} \cdot \boldsymbol{\mu}_{\rm e} - (\boldsymbol{\mu}_{\rm n} \cdot \boldsymbol{e}_q)(\boldsymbol{\mu}_{\rm e} \cdot \boldsymbol{e}_q) \right]$$

$$= -4\pi \left(\boldsymbol{\mu}_{\rm n} \cdot \boldsymbol{\mu}_{\rm e\perp} \right), \tag{1.149}$$

where $\boldsymbol{e}_q$ is the unit vector along $\boldsymbol{q}$, and $\boldsymbol{\mu}_{\rm e\perp}$ denotes the moment of an electron perpendicular to $\boldsymbol{q}$. In deriving this result, we use the fact that the Fourier transform of $1/r$ is given by $4\pi/q^2$ and that of ∇ by $i\boldsymbol{q}$.

We consider a collision where the incoming neutron with momentum $\boldsymbol{k}_0$ and energy $k_0^2/2M$ is scattered by the potential $\boldsymbol{H}_{\rm s} = \sum_j V_{\rm e}(\boldsymbol{r}_j)$ due to many electrons in the solid. The outgoing neutron loses momentum $\boldsymbol{q}$ and energy ω. The resultant momentum $\boldsymbol{k}_1$ and energy $k_1^2/2M$ should then satisfy

$$\boldsymbol{k}_0 = \boldsymbol{k}_1 - \boldsymbol{q}, \qquad \frac{k_1^2}{2M} = \frac{k_0^2}{2M} - \omega. \tag{1.150}$$

The differential cross section $d^2\sigma/d\Omega\,d\omega$, which represents the number of outgoing neutrons in the solid angle $d\Omega$ and in the energy range $d\omega$ around the average value defined by $\boldsymbol{k}_1$, is the central quantity. We specify the spin state of the neutron by σ and the state of the scattering system by n. Then we obtain

$$\frac{d^2\sigma}{d\Omega\,d\omega} = \left(\frac{M}{2\pi} \right)^2 \left| \frac{k_1}{k_0} \right| \sum_{\sigma,\sigma'} \sum_{n,n'} P_\sigma P_n \mid \langle \boldsymbol{k}_0 \sigma n \mid H_{\rm s} \mid \boldsymbol{k}_1 \sigma' n' \rangle \mid^2$$

$$\times \delta(E_n - E_{n'} + \omega), \tag{1.151}$$

in the Born approximation. The Born approximation is justified since, with R the ionic radius, the magnitude of the interaction is of order $\mu_{\rm e}\mu_{\rm n}/R^3$, which is much smaller than the kinetic energy of neutrons, $1/(2MR^2)$. In eqn (1.151), $P_n = (1/Z)\exp\left(-E_n/T\right)$ is the probability of finding the scattering system in the initial state n, and P_σ is the probability of finding the neutron spin in the initial state σ.

In order to derive the matrix element in eqn (1.151) we assume that only the spin part of the electronic wave function is relevant to the collision, and that the spatial part remains the same before and after the collision. In such a case, we have

$$H_{\rm s} = -4\pi \sum_{\boldsymbol{q}} F(\boldsymbol{q}) \boldsymbol{M}_\perp(\boldsymbol{q}) \cdot \boldsymbol{\mu}_{\rm n}, \tag{1.152}$$

where we introduce the form factor $F(q)$, which represents the spin density in momentum space, and the transverse magnetization $M_\perp(q)$, which is defined only inside the Brillouin zone. We have

$$F(q)M_\perp(q) = \sum_j \mu_{e\perp} \exp(-i q \cdot r_j).$$

The matrix element is then written as

$$\left(\frac{M}{2\pi}\right) \langle k_0 \sigma n \mid H_s \mid k_1 \sigma' n' \rangle = \frac{A}{2\mu_B} F(q) \langle \sigma n \mid \sigma \cdot M_\perp(q) \mid \sigma' n' \rangle, \qquad (1.153)$$

where $A = -g_n e/(2Mc\hbar) = 0.537 \times 10^{-12}$ cm with $g_n = -1.91$ being the g-factor of the neutron.

With these notations, we obtain the compact formula

$$\frac{d^2\sigma}{d\Omega\,d\omega} = \left(\frac{A}{2\mu_B}\right)^2 \left|\frac{k_1}{k_0}\right| \mid F(q) \mid^2 \sum_{\alpha,\beta} \langle \sigma_\alpha \sigma_\beta \rangle$$
$$\times \int \frac{dt}{2\pi} e^{-i\omega t} \langle M_\perp^\alpha(q,0) M_\perp^\beta(-q,t)\rangle, \qquad (1.154)$$

where the symbol $\langle \cdots \rangle$ denotes the statistical average. In the case where neutrons are unpolarized, i.e. $\langle \sigma_\alpha \sigma_\beta \rangle = \delta_{\alpha\beta}$, the cross section is reduced to

$$\frac{d^2\sigma}{d\Omega\,d\omega} = A^2 \left|\frac{k_1}{k_0}\right| \mid F(q) \mid^2 \sum_{\alpha,\beta} (\delta_{\alpha\beta} - e_q^\alpha e_q^\beta) S^{\alpha\beta}(q,\omega), \qquad (1.155)$$

where the tensor $S^{\alpha\beta}(q,\omega)$ is called the dynamical structure factor. This quantity is related to the imaginary part of the dynamical susceptibility tensor $\chi^{\alpha\beta}(q,\omega)$ via the fluctuation–dissipation theorem:

$$S^{\alpha\beta}(q,\omega) = \frac{1}{4\pi} \frac{\mathrm{Im}\,\chi^{\alpha\beta}(q,\omega)}{1 - \exp(-\omega/T)}. \qquad (1.156)$$

Here, the factor $1/4$ comes from the definition of $\chi^{\alpha\beta}(q,\omega)$ according to eqn (1.56). The fluctuation–dissipation theorem and its background are explained in Appendix B.

Since the correlation between $M_\perp^\alpha(q,0)$ and $M_\perp^\beta(q,t)$ disappears for $t \to \infty$,

$$\lim_{t\to\infty} \langle M_\perp^\alpha(q,0) M_\perp^\beta(-q,t)\rangle = \langle M_\perp^\alpha(q)\rangle \langle M_\perp^\beta(-q)\rangle \qquad (1.157)$$

is valid. Accordingly, the result of eqn (1.154) is divided into two parts,

$$\frac{d^2\sigma}{d\Omega\,d\omega} = \left(\frac{d^2\sigma}{d\Omega\,d\omega}\right)_{\text{elastic}} + \left(\frac{d^2\sigma}{d\Omega\,d\omega}\right)_{\text{inelastic}} \qquad (1.158)$$

$$\left(\frac{d^2\sigma}{d\Omega d\omega}\right)_{\text{elastic}} = \left(\frac{A}{2\mu_B}\right)^2 \left|\frac{k_1}{k_0}\right| \mid F(\boldsymbol{q}) \mid^2 \sum_{\alpha,\beta} \langle \sigma_\alpha \sigma_\beta \rangle \langle M_\perp^\alpha(\boldsymbol{q}) \rangle \langle M_\perp^\beta(-\boldsymbol{q}) \rangle \delta(\omega),$$

$$\left(\frac{d^2\sigma}{d\Omega d\omega}\right)_{\text{inelastic}} = \left(\frac{A}{2\mu_B}\right)^2 \left|\frac{k_1}{k_0}\right| \mid F(\boldsymbol{q}) \mid^2 \sum_{\alpha,\beta} \langle \sigma_\alpha \sigma_\beta \rangle$$

$$\times \int \frac{dt}{2\pi} e^{-i\omega t} \left[\langle M_\perp^\alpha(\boldsymbol{q},0) M_\perp^\beta(-\boldsymbol{q},t) \rangle \right.$$

$$\left. - \langle M_\perp^\alpha(\boldsymbol{q}) \rangle \langle M_\perp^\beta(-\boldsymbol{q}) \rangle \right] \tag{1.159}$$

The first elastic part is referred to as the magnetic Bragg scattering and is nonzero only for a discrete set of $\boldsymbol{q}$ values. If the magnetization density is periodic with some well-defined magnetic unit cell, the average value of $\langle M(\boldsymbol{q}) \rangle$ remains nonzero only for $\boldsymbol{q} = \pm \boldsymbol{Q} + \boldsymbol{K}_n$ where $\boldsymbol{Q}$ is associated with the magnetic order and $\boldsymbol{K}_n$ is the reciprocal lattice vector. Due to this selection rule of the Bragg scattering, the detailed spin structure is determined even for complex spin arrangement such as helixes or spin density waves.

The second inelastic part has no strong selection rules on $\boldsymbol{q}$ and ω, and gives the diffuse scattering. The magnetic inelastic cross section in the paramagnetic state where $\langle M(\boldsymbol{q}) \rangle = 0$ is proportional to the Fourier component of the spin correlation function of $S(\boldsymbol{q})$. The integration of eqn (1.159) over ω is derived with eqns (1.155) and (1.156) as

$$\frac{d\sigma}{d\Omega} = A^2 \left|\frac{k_1}{k_0}\right| \mid F(\boldsymbol{q}) \mid^2 T \sum_{\alpha,\beta} (\delta_{\alpha\beta} - e_q^\alpha e_q^\beta) \chi^{\alpha\beta}(\boldsymbol{q}), \tag{1.160}$$

which is proportional to the static susceptibility, $\chi(\boldsymbol{q})$. Therefore, the scattering intensity is increased significantly upon cooling towards the ordering temperature T_Q, since χ_Q diverges at T_Q. This is the so-called critical scattering. In the ordered state, the magnetic inelastic scattering can probe spin-wave excitations, leading to a direct measure of the dispersion relation.

The essential feature of the magnetic scattering profile in heavy-electron compounds is that it consists of a superposition of two contributions at low temperatures: (i) a $\boldsymbol{q}$-independent (single-site) quasi-elastic contribution which is of the Lorentzian shape, and (ii) a strongly peaked inelastic contribution at finite wave vector $\boldsymbol{Q}$ associated with intersite spin correlations. Furthermore, there is a case where the most relevant wave vector $\boldsymbol{Q}$ for the low-energy response changes as the temperature decreases. The same happens also as a function of the energy transfer at low temperature. Thus, the neutron scattering experiment provides a direct way to obtain detailed information on the wave number and energy dependence of the dynamical response function in heavy-electron systems.

By combining both NMR and neutron scattering experiments, we can study dynamical properties of heavy-electron systems with different types of ground states. This is a central issue to be addressed in the sequel in detail from both experimental and theoretical points of view.

Bibliography

[1] Y. Onuki and T. Komatsubara, *J. Magn. Magn. Mater.* **63–64**, 281 (1987).

[2] F. Steglich, U. Rauchschwalbe, U. Gottwick, H. M. Mayer, G. Sparn, N. Grewe, and U. Poppe, *J. Appl. Phys.* **57**, 3054 (1985).

[3] W. Assmus, M. Herrman, U. Rauchschwalbe, S. Riegel, W. Lieke, H. Spille, S. Horn, G. Weber, and F. Steglich, *Phys. Rev. Lett.* **52**, 469 (1984).

[4] J.-M. Mignot, J. Flouquet, P. Haen, F. Lapierre, L. Puech, and J. Voiron, *J. Magn. Magn. Mater.* **76–77**, 97 (1988).

[5] H. R. Ott, H. Rudigier, T. M. Rice, K. Ueda, Z. Fisk, and J. L. Smith, *Phys. Rev. Lett.* **52**, 1915 (1984).

[6] L. Taillefer, J. Flouquet, and G. G. Lonzarich, *Physica B* **169**, 257 (1991).

[7] J. M. Luttinger, *Phys. Rev.* **121**, 942 (1961).

[8] N. F. Mott, *Metal insulator transitions* (Taylor and Francis, London 1990).

[9] P. W. Anderson, *Solid state physics* Vol. 14 (Academic Press, New York, 1963), p. 14.

[10] B. Brandow, *Adv. Phys*, **26**, 651 (1977).

[11] I. Lindgren and J. Morrison, *Atomic many-body theory* (Springer Verlag, Berlin, 1986).

[12] S. Inagaki and K. Yosida, *J. Phys. Soc. Jpn.* **50**, 3268 (1981).

[13] M. Roger, J. H. Hetherington, and J. M. Delrieu, *Rev. Mod. Phys.* **55**, 1 (1983).

[14] J. Zinn-Justin, *Quantum field theory and critical phenomena* (Oxford University Press, Oxford, 1993).

[15] J. Hubbard, *Proc. Roy. Soc. A* **276**, 238 (1963).

[16] J. Kanamori, *Prog. Theor. Phys.* **30**, 275 (1963).

[17] M. Gutzwiller, *Phys. Rev. Lett.* **10**, 159 (1963).

[18] E. H. Lieb and F. Y. Wu, *Phys. Rev. Lett.* **20**, 1445 (1968).

[19] P. Fulde, In *Handbook on the physics and chemistry of rare earths, Vol. II* (Eds, K. A. Gschneider, Jr and L. Eyring, North-Holland, Amsterdam, 1978), Ch. 17.

[20] L. D. Landau, *Sov. Phys. JETP* **3**, 920; ibid., **5**, 101 (1957).

[21] D. Pines and P. Nozieres, *The theory of quantum liquids* (W. A. Benjamin, New York, 1966).

[22] J. W. Serene and D. Rainer, *Phys. Reports* **101**, 222 (1983).

[23] N. R. Bernhoeft and G. C. Lonzarich, *J. Phys. C* **7**, 7325 (1995).

[24] Y. Kuramoto, *Solid State Commun.* **63**, 467 (1987).

[25] L. P. Kadanoff and P. C. Martin, *Ann. Phys.* **24**, 419 (1963).

[26] K. K. Murata and Doniach, *Phys. Rev. Lett.* **29**, 85 (1972).

[27] T. Moriya and A. Kawabata, *J. Phys. Soc. Jpn.* **34**, 639 (1973).

[28] I. E. Dzyaloshinskii and P. S. Kondratenko, *Sov. Phys. JETP* **43**, 1036 (1976).

[29] T. Moriya, *Spin fluctuations in itinerant electron magnetism* (Springer Verlag, Berlin, 1985).

[30] A. Abragam, *The principles of nuclear magnetism* (Oxford University Press, London, 1961); C. P. Slichter, *Principles of magnetic resonance*, 2nd edition (Springer Series in Solid State Sciences, New York, 1978); J. Winter, *Magnetic resonance in metals* (Oxford University Press, New York, 1967).

[31] T. Shimizu, M. Takigawa, H. Yasuoka, and J. H. Wernick, *J. Magn. Magn. Mater* **52**, 187 (1985).

[32] T. Moriya, *Prog. Theor. Phys.* **28** (1962) 371.

[33] S. W. Lovesey, *Theory of neutron scattering from condensed matter Vols. 1 & 2* (Clarendon Press, Oxford 1984).

[34] G. L. Squires, *Introduction to the theory of thermal neutron scattering* (Cambridge University Press, Cambridge, 1978).

2

CROSSOVER FROM LOCALIZED MOMENT TO LOCAL FERMI OR NON-FERMI LIQUID

2.1 Description of singlet formation

2.1.1 Renormalization of the exchange interaction

A fundamental theoretical model that can deal with the Kondo effect is called the Anderson model [1]. In the simplest version of the model, one considers an impurity with a nondegenerate localized orbital in a metallic environment which has a single conduction band with the spectrum ϵ_k. The localized orbital has the on-site Coulomb repulsion U, and the electron with spin σ is annihilated by the operator f_σ. The model is given by

$$H_{\mathrm{A}} = \sum_{k\sigma} \left[\epsilon_k c_{k\sigma}^{\dagger} c_{k\sigma} + \frac{1}{\sqrt{N}} V_k \left(c_{k\sigma}^{\dagger} f_\sigma + f_\sigma^{\dagger} c_{k\sigma} \right) \right] + \sum_\sigma \epsilon_{\mathrm{f}} f_\sigma^{\dagger} f_\sigma + \frac{1}{2} U \sum_{\sigma \neq \sigma'} n_{\mathrm{f}\sigma} n_{\mathrm{f}\sigma'},$$

$$(2.1)$$

where V_k is the strength of the hybridization and N the number of lattice sites. It is straightforward to generalize the Anderson model so that more realistic structures of the f shell and the conduction bands are taken into account.

In spite of its simple appearance, the Anderson model has a remarkably rich physics including the Kondo effect. There are already an enormous number of treatises on the Kondo effect in general [2–4], and on the exact solution with use of the Bethe Ansatz [4–6] in particular. In view of this situation, we avoid repetition of such a treatment in this book. Instead, we emphasize the effective Hamiltonian approach which naturally includes the concepts of scaling and the renormalization group.

As the first step to construct the effective Hamiltonian, we consider the situation where the average occupation of the f states is unity, and their charge fluctuation can be neglected. This is realized when the f level ϵ_{f} is deep below the Fermi level (taken to be zero), and $\epsilon_{\mathrm{f}} + U$ is far above the Fermi level. Thus, we work with the model space where f states are neither vacant nor doubly occupied. The hybridization part H_{hyb} in H_{A} connects the model space and the other states to be projected out. According to eqn (1.38), the effective interaction in the lowest order is given by

$$H_{\mathrm{int}} = P H_{\mathrm{hyb}} (E_i - H_{\mathrm{c}} - H_{\mathrm{f}})^{-1} Q H_{\mathrm{hyb}} P. \qquad (2.2)$$

Here the energy E_i can be taken as the zeroth-order value. The intermediate f states orthogonal to P are either vacant or doubly occupied. Only a singlet pair consisting of an incoming conduction electron and an f electron change into the doubly occupied state, which decays again by hybridization. In the latter process, we have two possibilities: either the spin of the conduction electron remains the same as the incoming one or not. The resultant effective model is given by

$$H_{\text{eff}} = H_{\text{c}} + \frac{1}{N} \sum_{k\alpha} \sum_{k'\beta} \left[\frac{1}{2} J_{kk'} \, \boldsymbol{S} \cdot \boldsymbol{\sigma}_{\beta\alpha} c^{\dagger}_{k'\beta} c_{k\alpha} + K_{kk'} \delta_{\alpha\beta} c^{\dagger}_{k'\beta} c_{k\alpha} \right], \qquad (2.3)$$

where $\boldsymbol{S}$ is the impurity spin operator, and

$$J_{kk'} = 2 V_k V_{k'} \left[\frac{1}{\epsilon_{k'} - \epsilon_{\text{f}}} - \frac{1}{\epsilon_k - \epsilon_{\text{f}} - U} \right], \qquad (2.4)$$

$$K_{kk'} = \frac{1}{2} V_k V_{k'} \left[\frac{1}{\epsilon_{k'} - \epsilon_{\text{f}}} + \frac{1}{\epsilon_k - \epsilon_{\text{f}} - U} \right]. \qquad (2.5)$$

The model H_{eff} can be simplified further by replacing V_k by a constant V, and by assuming that ϵ_k and $\epsilon_{k'}$ are negligible as compared to ϵ_{f} or $\epsilon_{\text{f}} + U$. In the particular case where $\epsilon_{\text{f}} + U = |\epsilon_{\text{f}}|$, called the symmetric case, the potential scattering given by $K_{kk'}$ vanishes, and $J_{kk'}$ becomes $J = 2V^2/|\epsilon_{\text{f}}|$. Then we obtain the so-called Kondo model (or the s–d model):

$$H_K = H_{\text{c}} + J \boldsymbol{S} \cdot \boldsymbol{s}_{\text{c}} = H_{\text{c}} + H_{\text{ex}}, \qquad (2.6)$$

where $\boldsymbol{s}_{\text{c}}$ is the spin operator of conduction electrons at the impurity site, and is given by

$$\boldsymbol{s}_{\text{c}} = \frac{1}{2N} \sum_{kk'} \sum_{\mu\nu} c^{\dagger}_{k\mu} \boldsymbol{\sigma}_{\mu\nu} c_{k'\nu}. \qquad (2.7)$$

The Kondo model, which may appear to be even simpler than the Anderson model, in fact generates unmanageable singularities in the perturbation theory in J. It is now understood that the singularities mean that the starting point of the perturbation theory is inappropriate at zero temperature, however small J is. The ground state is understood more easily if one uses the Anderson model for all values of the parameters. On the other hand, at finite temperature the perturbation theory becomes more and more accurate as one includes higher orders in J. The best theoretical machinery to understand these contrasting properties of the Kondo model is called the renormalization group, which will be discussed shortly. Before analysing the Kondo model, we introduce another related model, namely, the Coqblin–Schrieffer model H_{CS} [7] defined by

$$H_{\text{CS}} = H_{\text{c}} + \frac{J}{2} \mathcal{P}_{\text{spin}}, \qquad (2.8)$$

where $\mathcal{P}_{\text{spin}}$ is the spin permutation operator written explicitly as

$$\mathcal{P}_{\text{spin}} = \frac{1}{N} \sum_{kk'} \sum_{\nu\mu} f^{\dagger}_{\nu} f_{\mu} c^{\dagger}_{k\mu} c_{k'\nu} = 2 \boldsymbol{S} \cdot \boldsymbol{s}_{\text{c}} + \frac{1}{2} \hat{n}_{\text{f}} \hat{n}_{\text{c}}. \qquad (2.9)$$

where $\hat{n}_{\text{f}}$ ($= 1$) is the number operator of f electrons, and $\hat{n}_{\text{c}}$ is that of conduction electrons at the impurity site. The Coqblin–Schrieffer model is used frequently in the case of an orbitally degenerate magnetic impurity because of its simplicity to generalize to arbitrary degeneracy.

Let us now turn to the renormalization treatment of the Kondo model. We use the effective Hamiltonian method explained in Chapter 1 as a variant of the perturbation theory. Suppose that the conduction band extends from $-D$ to D, and the Fermi level is at the centre of the band $\mu = 0$. Then we choose the model space as the one which does not involve the conduction states near the band edges. With this restriction, the exchange interaction should be modified into the effective one. This change of interaction, which occurs in the process of reducing the model space, is called renormalization. A very powerful way to accomplish the renormalization is to eliminate the high-energy states by an infinitesimal amount in each step, and continue the process successively [8]. Let us illustrate the process in the following. In the lowest order of the Brillouin–Wigner perturbation theory, we obtain the effective Hamiltonian as

$$H_{\text{eff}} = P(H_c + H_{\text{ex}})P + P H_{\text{ex}}(E_i - H_c)^{-1} Q H_{\text{ex}} P, \tag{2.10}$$

where E_i is the energy of the state to be derived. We are interested in low-energy states where $|E_i|$ is close to the ground-state energy E_g of H_c. Choosing Q to project onto a space with conduction states $[D + \delta D, D]$ and $[-D, -D - \delta D]$ with $\delta D < 0$ infinitesimal, we can safely replace E_i by E_g in the denominator. Figure 2.1 shows the perturbation process.

Let us assign the spins σ, ξ, and σ' for the incoming conduction electron, the intermediate state with high energy, and the outgoing electron, respectively. Then we look at scattering with a spin component $J S^\alpha s_c^\alpha$ for the right part, and $J S^\beta s_c^\beta$ for the left part. In Fig. 2.1(a), the matrix elements of s_c are given by

$$\langle \sigma' | s_c^\beta | \xi \rangle \langle \xi | s_c^\alpha | \sigma \rangle, \tag{2.11}$$

which is called the direct scattering. For the scattering shown in Fig. 2.1(b), we similarly obtain

$$\langle \sigma' | s_c^\alpha | \xi \rangle \langle \xi | s_c^\beta | \sigma \rangle. \tag{2.12}$$

Since the energy of the intermediate state does not depend on ξ, the summation over ξ can be carried out independently of the energy denominator. Note that the process (b)

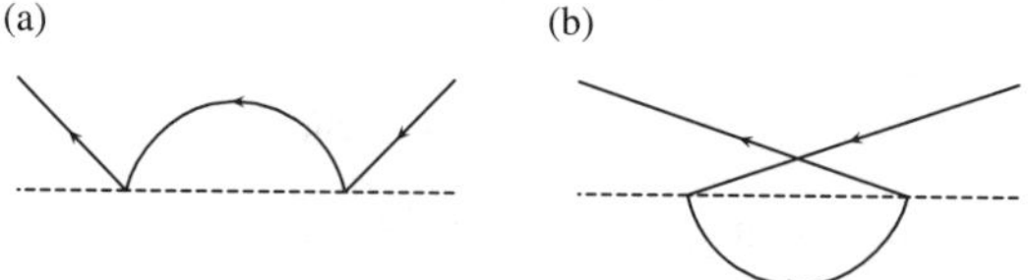

FIG. 2.1. Exchange scattering processes in second order. The solid line shows a conduction-electron state, while the dashed line the impurity spin. The projection operator Q requires the intermediate conduction-electron states to have energies near the band edges, which are opposite to each other in the cases (a) and (b).

acquires an extra minus sign because an interchange of fermion operators is involved in evaluating the product of operators. Then we obtain, summing (a) and (b),

$$\frac{J^2}{-D} \sum_{\alpha\beta} S^\beta S^\alpha \left[s_c^\beta, s_c^\alpha \right] |\delta D| \rho_c, \tag{2.13}$$

as the second-order effective Hamiltonian. Here $\rho_c = (2D)^{-1}$ is the density of conduction states per spin per site and is assumed to be uniform. By using the commutation rule

$$\left[s_c^\beta, s_c^\alpha \right] = -i \epsilon_{\alpha\beta\gamma} s_c^\gamma, \tag{2.14}$$

where $\epsilon_{\alpha\beta\gamma}$ is the completely antisymmetric unit tensor, we find that the effective interaction takes the same form as the original H_{ex}, except that the strength is modified. Thus, we obtain the change δJ of the exchange interaction as

$$\delta J = -\frac{\delta D}{D} J^2 \rho_c. \tag{2.15}$$

This differential relationship is called the scaling equation [8], or the renormalization-group equation. The latter name stems from the fact that a succession of two renormalization processes can be regarded as a different single process, and that each process of renormalization constitutes a kind of group element. In contrast to the usual definition of a group, the renormalization group does not necessarily have the inverse of an element. More precisely, the renormalization group is a semigroup.

One can continue to eliminate the conduction states near the new band edge as long as the final effective bandwidth $2D_{\text{eff}}$ is much larger than the energy scale we are interested in. This condition is also necessary to justify the lowest-order perturbation theory in each renormalization step. Under this restriction, one can integrate the scaling equation with the result

$$J_{\text{eff}} = \frac{J}{1 - J\rho_c \ln(D/D_{\text{eff}})}. \tag{2.16}$$

Here the boundary condition is such that $J_{\text{eff}} = J$ for $D_{\text{eff}} = D$.

The important feature of J_{eff} is that it increases as one decreases D_{eff}, and finally diverges at $D_{\text{eff}} = D \exp[-1/(J\rho_c)]$. This characteristic energy $D \exp[-1/(J\rho_c)]$ in temperature units is called the Kondo temperature T_K. Note that T_K is a nonanalytic function of the exchange interaction. Of course, there is no guarantee for the renormalization to be valid down to $D_{\text{eff}} = T_K$. It simply says that the perturbation theory breaks down as one goes lower and lower in energy scales. As we shall see later, T_K also enters into physical quantities such as the susceptibility and specific heat near zero temperature.

If the temperature T is sufficiently higher than T_K, it is necessary to take explicit account of thermally excited conduction states. Then the renormalization should stop at D_{eff}, which is of the order of T. The proportionality constant can be set to unity within the logarithmic accuracy. We define the effective interaction in this case as

$$J_{\text{eff}}(T) = \frac{J}{1 - J\rho_c \ln(D/T)}. \tag{2.17}$$

If one replaces J by $J_{\text{eff}}(T)$ in the lowest-order perturbative calculation of a physical quantity, say, the resistivity or the magnetic susceptibility, the result reproduces an infinite summation of leading logarithmic terms [9].

Let us first explain the case of the resistivity ρ. In the simplest theory of electronic conduction in metals, the conductivity $\sigma = 1/\rho$ is given by

$$\sigma = \frac{ne^2\tau}{m^*}, \tag{2.18}$$

where n is the density of conduction electrons with effective mass m^*. The lifetime τ of conduction electrons near the Fermi surface is given in the Born approximation by

$$\frac{1}{\tau} = 2\pi c_{\text{imp}} J^2 \sum_\alpha S_\alpha^2 \, \text{Tr}_c \, s_\alpha^2 = \frac{3\pi}{4} c_{\text{imp}} \frac{J^2}{\rho_c}, \tag{2.19}$$

where c_{imp} is the concentration of magnetic impurities. We write the resultant resistivity as ρ_0. In the Born approximation, one does not take account of intermediate states with high excitation energies. However, in higher-order perturbation in J, these states contribute to the leading logarithmic singularity. If one uses $J_{\text{eff}}(T)$ in place of J, the intermediate states with high energies are correctly taken into account. As a result of this replacement, we obtain

$$\rho(T) = \frac{\rho_0}{[1 - J\rho_c \ln(D/T)]^2} = \frac{\rho_0}{[J\rho_c \ln(T/T_{\text{K}})]^2}. \tag{2.20}$$

This result is the same as the one obtained by infinite summation of perturbation terms [9], and reduces to the original result of Kondo in $O(J^3)$. The resistivity thus increases logarithmically as the temperature is decreased. The divergence at $T = T_{\text{K}}$ lies beyond the valid region of the effective Hamiltonian theory and is merely an artefact of the approximation. The determination of the range of validity requires more accurate analysis, which will be discussed later.

As the next example, we consider the relaxation rate Γ of the local moment. In the Born approximation, Γ is given by

$$\Gamma = \frac{3\pi}{4N} \sum_{k,p} J^2 f(\epsilon_k)\big[1 - f(\epsilon_p)\big]\delta(\epsilon_k - \epsilon_p) = \frac{3}{4}\pi(J\rho_c)^2 T, \tag{2.21}$$

which is usually referred to as the Korringa relaxation. Replacing J here by J_{eff} leads to

$$\Gamma(T) = \pi \left(\frac{J\rho_c}{1 - J\rho_c \ln(D/T)} \right)^2 T. \tag{2.22}$$

Thus, the relaxation rate is enhanced by the Kondo effect. In contrast to the resistivity to which ordinary scattering also contributes, magnetic relaxation is dominantly given by the exchange interaction. Hence, the enhancement by the Kondo effect is conspicuous and the temperature dependence deviates drastically from the linear behaviour. Again the divergence at $T = T_K$ should not be trusted.

2.1.2 *Numerical renormalization group method*

As the temperature approaches T_K from above, the dimensionless coupling $J_{\text{eff}}\rho_c$ becomes of the order of unity. Then the simplest renormalization described above breaks down. In order to proceed further, Wilson developed a numerical renormalization group (NRG) approach [10]. In this approach, one can deal with the newly generated terms in the effective Hamiltonian very easily. We first pick out the s-wave part of the whole conduction states, since other parts do not couple to the magnetic impurity in the Kondo model. Then we approximate the spectrum ϵ_k by the following form:

$$\epsilon_k = v_F k, \tag{2.23}$$

where k is measured relative to k_F. The range of momentum k is set to be $k \in [-k_F, k_F]$, which means that the bandwidth is $2D = 2k_F v_F$. In the following, we choose units so that the Fermi momentum is unity, i.e. $k_F = 1$. We introduce a set of discretized spherical orbitals ϕ_n around the impurity by averaging the s-wave states over the k-space interval $[\Lambda^{-n+1}, \Lambda^{-n}]$, where the parameter Λ (>1) controls the discretization. For negative momentum, we form the corresponding states by averaging over $[-\Lambda^{-n+1}, -\Lambda^{-n}]$. Then we make a unitary transformation from ϕ_n to a new Wannier-type basis, in terms of which the Hamiltonian matrix for the conduction band is given in the tridiagonal form. We refer to the original paper [10] for the fairly complicated procedure to construct the basis set explicitly. The final form of the Hamiltonian in the new basis is given by

$$H_\Lambda = D \sum_{n=0}^{\infty} \Lambda^{-n/2} \sum_{\sigma} (c_{n\sigma}^{\dagger} c_{n+1\sigma} + c_{n+1\sigma}^{\dagger} c_{n\sigma}) + \frac{1}{2}\tilde{J} \sum_{\alpha\beta} c_{0\alpha}^{\dagger} \boldsymbol{\sigma}_{\alpha\beta} c_{0\beta} \cdot \boldsymbol{S}, \tag{2.24}$$

where $\tilde{J}$ is the exchange interaction in the new basis set. It is evident that the new Hamiltonian has the tight-binding form where the transfer $\Lambda^{-n/2}$ decreases exponentially as the orbital number n increases. Only the innermost orbital 0 is coupled to the impurity spin $\boldsymbol{S}$. We note that eqn (2.24) with $\tilde{J} = 0$ describes the free-electron band by definition.

In order to construct the effective Hamiltonian, which is nothing but the renormalization of the model, we perform the projection Q successively. Namely, we first take certain manageable number ν of Wannier-type orbitals near the origin and diagonalize the Hamiltonian matrix with this truncated subspace. By explicit diagonalization, we can arrange the resultant many-body states involving the impurity spin according to their energies in the subspace. Then we shall keep only the lowest N_P states and neglect the others. This neglect amounts to the first step of projection Q in constructing the model space. We newly add the orbital next to the νth one and construct the many-body states by the direct product with the N_P states obtained by the first step of the renormalization. Then we diagonalize the Hamiltonian within the newly constructed many-body subspace, and keep only the lowest N_P states again. By repeating the same procedure we can include states more and more remote from the origin in the effective Hamiltonian. In this way we can see how the effective Hamiltonian converges in the limit of $n \to \infty$.

The convergent form of the effective Hamiltonian, or the ground and low-lying states it describes, is called a fixed point of the renormalization group. In the special case of $J = 0$ the fixed point must be that of free electrons. The energy scale $\Lambda^{-n/2}$ introduced in the nth iteration corresponds to the separation $2\pi v_{\mathrm{F}}/L_n$ of the free-electron spectrum with L_n the size of the system corresponding to the renormalization step. In other words, the minimum momentum in the logarithmic discretization is proportional to the inverse of the system size. The spectrum of a free Fermi gas is given by

$$E = \frac{2\pi v_{\mathrm{F}}}{L_n}\left[\frac{Q^2}{8} + \frac{S(S+1)}{4} + n_Q + n_S\right], \qquad (2.25)$$

where $Q = \delta n_\uparrow + \delta n_\downarrow$ describes the number of electrons relative to the pure Fermi sea, and $n_Q(n_S)$ are the number of spin conserving (spin flipping) particle–hole excitations. These numbers are analogous to the number of phonon excitations. Equation (2.25) is just another way of writing the energy of conduction bands. For an even number of free conduction electrons, the ground state is nondegenerate and the first excited level is separated by $2\pi v_{\mathrm{F}}/L_n$ which has excess charge $Q = \pm 1$ and spin $1/2$. For an odd number of free conduction electrons, the ground state is doubly degenerate with spin $1/2$ since the one-electron level nearest to the Fermi level is singly occupied.

In the case of the Kondo model, the fixed point Hamiltonian turns out to have the same spectrum as the free conduction band. However, the spectrum with an even number of conduction electrons corresponds to that of an odd number of free conduction electrons. This represents the fact that the magnetic impurity couples tightly with one conduction electron spin. The resultant spectrum with an even number of conduction electrons has spin $1/2$ and that with odd number has spin 0. We discuss this aspect of the spectrum in more detail later in § 2.3.1.

One can do the same thing for the Anderson model by just replacing the exchange interaction term in eqn (2.24) by the hybridization term. It was demonstrated by Krishnamurthy *et al.* [11] that, as the iteration proceeds, the spectrum shows two alternating sequences as in the case of the Kondo model.

The most important conclusion of the numerical renormalization group is to elucidate the way the ground state is approached as temperature is decreased. In the numerical renormalization-group method, one can calculate the susceptibility for all temperatures [4,10–12]. Below T_{K}, the increase of the impurity susceptibility becomes slow and saturates to a value of the order of C/T_{K}, with C the Curie constant.

2.1.3   *Local Fermi-liquid theory*

The fixed point of the Kondo model is described by a local Fermi-liquid theory of Nozieres, who revealed that the phase shift of conduction electrons incorporates the many-body effect [13]. The term 'local Fermi-liquid' means that low-lying eigenstates of the system are in one-to-one correspondence with those of the Anderson model with $U = 0$. Hence, the Fermi-liquid fixed point is accessible by perturbation theory with respect to U if one works in the Anderson model. By analysing the perturbation series up to an infinite order, Yamada derived a nontrivial relationship between the susceptibility and the specific heat [14].

Here we present an alternative phenomenological approach which is in parallel to the original Fermi-liquid theory of Landau. We first diagonalize the Anderson model without the Coulomb repulsion. If one assumes, for simplicity, spherical symmetry for hybridization, only the s-wave part of conduction states is mixed. In the following, therefore, we neglect other conduction states with finite angular momentum, since these states are not affected by the magnetic impurity. Each eigenstate is characterized by the radial momentum p of the scattered s wave and the spin. The Coulomb interaction among the f electrons now appears as an interaction among hybridized electrons. The quasi-particles in this case are characterized by p and σ. Then the Landau interaction function is given by $f(p\sigma, k\sigma')$. The quasi-particle distribution function is identical to the noninteracting one in the ground state. Namely, it is unity if the energy of the state is below the Fermi level, and zero otherwise. The deviation $\delta n_{p\sigma}$ from the ground state determines the energy of the excited state. The latter is expanded as

$$E = E_g + \sum_{p\sigma} \epsilon_{p\sigma} \delta n_{p\sigma} + \frac{1}{2} \sum_{p\sigma} \sum_{k\sigma'} f(p\sigma, k\sigma') \delta n_{p\sigma} \delta n_{k\sigma'} + O\left(\delta n_{p\sigma}^3\right). \quad (2.26)$$

As in the translationally invariant case, the expansion up to second-order terms is sufficient to describe the low-energy excitations. This includes the magnetic susceptibility and the low-temperature limit of the specific heat. The validity of this assumption can be confirmed by analysing the infinite-order perturbation series [14,15].

For low-energy excitations, $\delta n_{p\sigma}$ is strongly peaked around the Fermi level. Provided the interaction $f(p\sigma, k\sigma')$ is a smooth function of the momentum, one may replace p and k here by the Fermi momentum p_F. By rotational invariance, we are then left with only two dimensionless parameters F and Z given by

$$F = \tfrac{1}{2}[f(p_F\sigma, p_F\sigma) + f(p_F\sigma, p_F\bar{\sigma})]\rho^*, \quad Z = \tfrac{1}{2}[f(p_F\sigma, p_F\sigma) - f(p_F\sigma, p_F\bar{\sigma})]\rho^*, \quad (2.27)$$

where $\bar{\sigma} = -\sigma$ and ρ^* is the density of states of quasi-particles (for both spins) at the Fermi level. Since we are dealing with a single impurity in a macroscopic number N of s-wave states, the parameters F and Z are of order $1/N$.

In terms of the effective Hamiltonian picture, $f(p\sigma, k\sigma')$ is regarded as the effective interaction for zero momentum transfer in a fictitious one-dimensional system. The momentum in this system must be positive since it is originally the radial part. In order to obtain the scattering amplitude between the quasi-particles with small momentum transfer q and small energy transfer ω, we have to consider the repeated scattering as shown in Fig. 2.2. Each bubble in the figure corresponds to the zeroth-order susceptibility and is given by

$$\chi_0(q, \omega) = \frac{2}{N} \sum_k \frac{kq/m^*}{\omega - kq/m^* + i\delta} \left(-\frac{\partial f(\epsilon_k)}{\partial \epsilon_k}\right), \quad (2.28)$$

where m^* is the effective mass of the quasi-particles. The susceptibility is a function of the ratio $v_F q/\omega$, with v_F being the Fermi velocity. In the case of $v_F q \gg \omega$, called the q-limit, $\chi_0(q, \omega)$ tends to ρ^*. In the opposite case of $v_F q \ll \omega$, called the ω-limit, it tends to zero. Note that eqn (2.28) is the small-momentum limit of the form given by eqn (1.46).

The Fermi-liquid parameters F and Z correspond to the forward scattering amplitude of quasi-particles in the ω-limit. On the other hand, the forward scattering amplitude

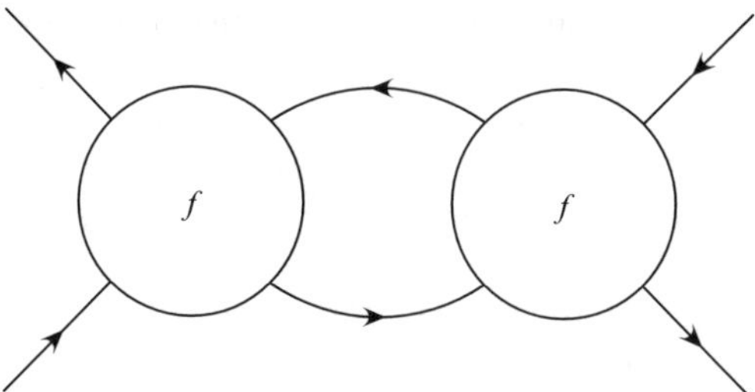

FIG. 2.2. Quasi-particle scattering in the second order in the effective interaction $f(p\sigma, k\sigma')$. The bubble in the centre corresponds to $\chi_0(q, \omega)$.

$a(p\sigma, k\sigma')$ in the q-limit is given by

$$a(p\sigma, k\sigma') = f(p\sigma, k\sigma') - \rho^* \sum_\tau f(p\sigma, p\tau)a(p\tau, k\sigma'),\qquad(2.29)$$

which is obtained by a repetition of the correction shown in Fig. 2.2. From combinations of $a(p\sigma, k\sigma')$ as in eqn (2.27) we introduce the parameters A (spin-symmetric part) and B (spin-antisymmetric part) which are related to F and Z by

$$A = F/(1 + F), \qquad B = Z/(1 + Z).\qquad(2.30)$$

In the most general case of the scattering amplitude $\langle p_1\sigma_1, p_2\sigma_2|a|p_3\sigma_3, p_4\sigma_4\rangle$ with incoming quasi-particles with $p_3\sigma_3$, $p_4\sigma_4$ and outgoing ones $p_1\sigma_1$, $p_2\sigma_2$, the antisymmetry of the fermionic wave function imposes a constraint:

$$\langle p_1\sigma_1, p_2\sigma_2|a|p_3\sigma_3, p_4\sigma_4\rangle = -\langle p_1\sigma_1, p_2\sigma_2|a|p_4\sigma_4, p_3\sigma_3\rangle.\qquad(2.31)$$

In the particular case of the forward scattering, we obtain $a(p\sigma, p\sigma) = 0$, which is equivalent to

$$A + B = 0.\qquad(2.32)$$

This relation is an example of the so-called forward scattering sum rule [16]. Therefore, we are left with a single independent parameter to characterize the quasi-particle interaction in the Anderson model.

Since the low-lying excitations of the system are quasi-particles, they should determine the specific heat and the susceptibility. In the thermally excited case δn_p is an odd function of $p - p_F$. Therefore, the increase of energy is dominated by the term linear in δn_p in eqn (2.26), as in the translationally invariant system. This gives $O(T^2)$ correction, while the quasi-particle interaction gives $O(T^4)$ correction. We can extract the impurity contribution to the specific heat from the change in the density of states. Thus, we introduce $\alpha = O(1/N)$ by

$$\rho^* = \rho_c(1 + \alpha).\qquad(2.33)$$

Then we obtain the impurity specific heat C as

$$C = \tfrac{1}{3}\pi^2 \rho_c \alpha T \equiv \gamma T, \tag{2.34}$$

which is of $O(1)$ since ρ_c is of $O(N)$. On the other hand, the spin susceptibility χ_s^{total} of the whole system is calculated as

$$\chi_s^{\text{total}} = \frac{\rho^*}{4(1+Z)} = \frac{1}{4}\rho_c(1 + \alpha - Z) + O\left(\frac{1}{N}\right). \tag{2.35}$$

Thus, the impurity contribution χ_s of $O(1)$ is given by

$$\chi_s = \tfrac{1}{4}\rho_c(\alpha - Z). \tag{2.36}$$

Similarly, the impurity contribution to the charge susceptibility χ_c is calculated as

$$\chi_c = \rho_c(\alpha - F). \tag{2.37}$$

We can now deduce an important relation between the specific heat and susceptibilities. Because of the constraint $A + B = F + Z + O(1/N^2) = 0$, we obtain

$$4\chi_s + \chi_c = 6\gamma/\pi^2, \tag{2.38}$$

or in units of noninteracting counterparts with suffix 0:

$$\chi_s/\chi_{s0} + \chi_c/\chi_{c0} = 2\gamma/\gamma_0. \tag{2.39}$$

In the case where the f level is singly occupied and U is large, the charge susceptibility is almost zero. In this limit, the ratio R of χ_s/χ_{s0} to γ/γ_0 becomes 2. Otherwise, R takes a value between 1 and 2. The quantity R measures the strength of the electron correlation at the impurity, and is often referred to as the Wilson ratio.

There is also a remarkable dynamical relation as a consequence of the Fermi-liquid ground state. Namely, the impurity contribution to the dynamical susceptibility is proportional to the square of the static susceptibility. This relation is called the Korringa–Shiba relation. The rigorous proof of the relation requires a detailed diagrammatic analysis, for which we refer to the original papers [15–18]. In this book, we give a less rigorous but more intuitive proof in § 2.2.2 leading to eqn (2.71).

2.1.4 *The $1/n$ expansion*

The fixed point of the Anderson model is a spin singlet, and is connected continuously to the trivial case of the vacant f state. It is found that in the limit of large degeneracy n of the f electron level, the singlet state has an energy lower than the singly occupied state with a local moment. Therefore, perturbation theory in terms of $1/n$ should converge. Actually, the small parameter $1/n$ appears as scaling of hybridization. There is thus a chance of having a perturbation theory valid for all temperatures T, since at high T the expansion with respect to hybridization guarantees that the atomic limit is recovered.

Let us assume that the spin index σ in the Anderson model H_A takes n different values and that U is infinite. The resultant model, called the SU(n) Anderson model, is written as

$$H_{\mathrm{SU}(n)} = H_{\mathrm{c}} + H_{\mathrm{f}} + H_{\mathrm{hyb}}. \tag{2.40}$$

We introduce the hybridization intensity $W_0(\epsilon)$ by

$$W_0(\epsilon) = \frac{1}{N} \sum_k |V_k|^2 \delta(\epsilon - \epsilon_k), \tag{2.41}$$

which is taken to be a constant W_0 in the range $-D < \epsilon < D$, and zero otherwise. We take nW_0 as the unit of energy, and use the Brillouin–Wigner perturbation theory to calculate the ground-state energy [19]. In the case of the singlet state, the effective interaction (or rather potential in this case) is of $O(1)$ because of summation over all channels. This is shown in Fig. 2.3. Taking the direct product of the vacant f state and the Fermi sea as the model space $|0\rangle$, the ground-state energy E_0 relative to the Fermi sea is then given by

$$E_0 = \langle 0| H_{\mathrm{hyb}} (E_0 - H_{\mathrm{c}} - H_{\mathrm{f}})^{-1} H_{\mathrm{hyb}} |0\rangle = nW_0 \int_{-D}^{0} \frac{d\epsilon}{E_0 + \epsilon - \epsilon_{\mathrm{f}}}. \tag{2.42}$$

In the extreme case of $nW_0 \ll |\epsilon_f| \ll D$, one can solve this equation analytically. Namely, one obtains

$$E_0 - \epsilon_f \equiv -T_0 = -D \exp\left(\frac{\epsilon_f}{nW_0}\right). \tag{2.43}$$

On the other hand, the lowest energy of the multiplet state has at most $O(1/n)$ corrections because summation over n channels is absent for the intermediate states. Thus, it remains ϵ_f to the leading order and is higher than E_0 by T_0.

It is remarkable that such a simple calculation can identify the characteristic energy scale T_0 in the singlet ground state. The scale T_0 corresponds to the Kondo temperature T_K in the case where the average occupation n_f of the f state is almost 1. In the more general case, n_f is calculated as

$$n_{\mathrm{f}} = \frac{\partial E_0}{\partial \epsilon_{\mathrm{f}}} = nW_0 \int_{-D}^{0} d\epsilon \frac{1 - n_{\mathrm{f}}}{(E_0 + \epsilon - \epsilon_{\mathrm{f}})^2} \sim \frac{nW_0}{T_0} (1 - n_{\mathrm{f}}); \tag{2.44}$$

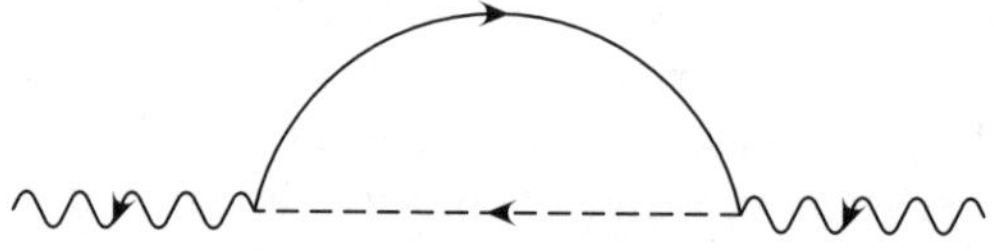

FIG. 2.3. Lowest-order effective potential in the $1/n$ expansion. For the f states, the wavy line shows the vacant state, while the dashed line shows singly occupied states.

thus, we obtain

$$n_{\mathrm{f}} = \left(1 + \frac{T_0}{n W_0}\right)^{-1}. \tag{2.45}$$

We can also obtain the magnetic susceptibility by the second derivative of $E_0(H)$ with respect to the magnetic field H. The ground-state energy E_0 in the presence of H is given by

$$E_0 = W_0 \sum_{J_z} \int_{-D}^{0} \frac{d\epsilon}{E_0 + \epsilon - \epsilon_{\mathrm{f}} - g_J \mu_{\mathrm{B}} J_z H}. \tag{2.46}$$

A straightforward calculation gives

$$\chi = C_J n_{\mathrm{f}} / T_0, \tag{2.47}$$

where $C_J = (g_J \mu_{\mathrm{B}})^2 J(J+1)/3$ is the Curie constant with $J = (n-1)/2$, the angular momentum of the f state. Thus, the susceptibility is also determined by T_0.

One should recognize here that the excitation gap T_0 between the singlet and the multiplet is an artefact of the leading-order theory. In reality, there is no gap since the multiplet wave function can extend to a macroscopic distance. As a result, the lowest multiplet wave function has no difference from the local singlet one near the origin. The presence of the gap is not a problem if one is interested only in static properties, as we have seen above. However, it becomes serious if one discusses the dynamical properties.

2.1.5 *Effects of spin–orbit and CEF splittings*

In order to understand the experimental results, it is very important to take into account the spin–orbit and the crystal electric field (CEF) splittings of the f shell. According to the energy scale of excitations or temperature, the apparent Kondo temperature changes. Thus, the Kondo temperature derived from the specific heat or the susceptibility at low temperatures is generally smaller than the apparent Kondo temperature derived, for example, by photoemission spectroscopy. Such changes of the Kondo scale explain the temperature dependence of the magnetic relaxation rate measured by neutron scattering. In this section, we explain how the real and apparent Kondo temperatures are influenced by the presence of splittings.

Let us assume that each split level E_α in the f^1 state has a degeneracy n_α and hybridization intensity W_α between the band edges $[-D, D]$. In the Brillouin–Wigner perturbation theory, the singlet ground-state energy $E_{\mathrm{s}} \equiv E_0 - T_{\mathrm{K}}$ is given by

$$E_{\mathrm{s}} = \sum_\alpha W_\alpha \int_{-D}^{0} \frac{d\epsilon}{E_{\mathrm{s}} - E_\alpha + \epsilon}. \tag{2.48}$$

When D is much larger compared to the other energies, such as the level splittings and T_{K}, we may replace E_{s} on the left-hand side by E_0 to obtain

$$E_0 \sim n_0 W_0 \ln \frac{T_{\mathrm{K}}}{D} + \sum_{\alpha \neq 0} n_\alpha W_\alpha \ln \frac{\Delta_\alpha}{D}, \tag{2.49}$$

where $\Delta_\alpha = E_\alpha - E_0$. From eqn (2.49), we can solve for T_K as

$$T_K = D \exp\left(\frac{E_0}{n_0 W_0}\right) \prod_{\alpha \neq 0} \left(\frac{D}{\Delta_\alpha}\right)^{n_\alpha W_\alpha/(n_0 W_0)}. \tag{2.50}$$

This result was first obtained by Ref. [20] by a different method.

It is clear that T_K in eqn (2.50) is enhanced by the presence of higher multiplets. However, if one compares with the hypothetical Kondo temperature without any splitting, the actual T_K is much reduced. This observation becomes important when one consistently interprets experimental data taken with different resolutions and temperatures.

2.2 Dynamics of the Kondo impurity

2.2.1 *Mean-field theory*

As an approach from the ground state, the simplest is a mean-field theory which starts from the Fermi-liquid ground state. We introduce boson operators b and $b^\dagger$ to represent the vacant f state in the n-fold degenerate Anderson model. The X-operators introduced in § 1.3.2 are then represented by

$$X_{00} = b^\dagger b, \quad X_{\sigma\sigma} = f_\sigma^\dagger f_\sigma, \quad X_{\sigma 0} = f_\sigma^\dagger b, \tag{2.51}$$

with the constraint $\sum_\sigma f_\sigma^\dagger f_\sigma + b^\dagger b = 1$. Namely, $b^\dagger$ creates the physical vacant state by operating on the fictitious vacuum of f states. If the constraint is regarded as an operator relation, f_σ and $f_\sigma^\dagger$ do not obey the commutation rule for fermions. The same applies to b and $b^\dagger$. Another viewpoint is that one uses the standard commutation rules for fermions and bosons and imposes the constraint as the relevant subspace to be picked out from the full Fock space. The auxiliary particle representing the vacant state is often called the slave boson . This kind of artifice was in fact used already for the Kondo model [9] to represent the spin operators. The constraint was handled by a fictitious chemical potential which goes to infinity at the end of the calculation. The use of the boson in the case of the Anderson model was proposed later [21,22], again with use of the Abrikosov technique. Because of the limiting procedure, however, one cannot use the standard linked cluster theorem which leads to the Feynman diagram expansion [23]. If one carefully selects a part of the Feynman diagrams, the result becomes identical to the perturbation theory using resolvents, as will be discussed later in this section.

In dealing with the Fermi-liquid ground state and low-energy excitations, a mean-field theory was introduced [24] in which the constraint is satisfied only as an average:

$$\sum_\sigma \langle f_\sigma^\dagger f_\sigma \rangle + \langle b^\dagger b \rangle = 1. \tag{2.52}$$

The average $\langle b \rangle = r$ should vanish in the exact theory because of the fluctuation of the phase of b. If one neglects this aspect and determines r variationally, the model maps to another Anderson model without the Coulomb repulsion. Thus, we try to simulate the low-energy physics by the mean-field Hamiltonian:

$$H_{\text{MF}} = \sum_{k\sigma} \left[\epsilon_k c_{k\sigma}^\dagger c_{k\sigma} + \frac{1}{\sqrt{N}} V_k r (c_{k\sigma}^\dagger f_\sigma + f_\sigma^\dagger c_{k\sigma}) \right]$$
$$+ \epsilon_f \sum_\sigma f_\sigma^\dagger f_\sigma + \lambda(n_{\text{f}} + r^2 - 1), \tag{2.53}$$

where λ is the Lagrange multiplier chosen so as to satisfy the constraint on the average, and V_k and r are taken to be real. If the right-hand side of eqn (2.52) were n instead of 1, the number fluctuations of $O(1)$ are negligible as compared with the average of $O(n)$. Then the mean-field theory would become exact in the limit of large n. Because of the actual constraint of $O(1)$, the mean-field theory is different from the $1/n$ expansion [25]. However, the static property agrees with that derived by the lowest-order $1/n$ expansion as explained below.

The mean-field parameters λ and r are determined by a variational principle. We write the statistical average of an operator O in terms of the density operator $\exp(-\beta H_{\text{MF}})$ as $\langle O \rangle$. Then the exact thermodynamic potential Ω of the SU(n) Anderson model $H_{\text{SU}(n)}$ is bounded by the Feynman inequality [26]:

$$\Omega \leq \Omega_{\text{MF}} + \langle H_{\text{SU}(n)} - H_{\text{MF}} \rangle. \tag{2.54}$$

Thus, we can optimize the parameters by minimizing the right-hand side. The second term is simply given by $-\lambda(n_{\text{f}} + r^2 - 1)$. The variation with respect to λ gives the constraint $n_{\text{f}} + r^2 = 1$. The optimization of r yields

$$\lambda r = \sum_{k\sigma} V_k \langle f_\sigma^\dagger c_{k\sigma} \rangle. \tag{2.55}$$

The stationary condition gives only a trivial solution $r = 0$ above a critical temperature T_{B} ($\sim T_{\text{K}}$). Below T_{B}, the nontrivial solution $r \neq 0$ is lower in free energy. The phase transition at $T = T_{\text{B}}$ is a fictitious one, which shows the inadequacy of the mean-field theory. However, at zero temperature the theory gives the correct fixed point of the Anderson model. Therefore, we restrict in the following to the case of $T = 0$.

It is convenient to use the Green function to calculate the average. The f electron Green function $G_{\text{f}}^*(z)$ is easily derived for H_{MF} as follows:

$$G_{\text{f}}^*(z) = [z - \tilde{\epsilon}_{\text{f}} + i\tilde{\Delta}\,\text{sgn}(\text{Im}\,z)]^{-1}, \tag{2.56}$$

with $\tilde{\epsilon}_{\text{f}} = \epsilon_{\text{f}} + \lambda$ and $\tilde{\Delta} = \pi r^2 W_0$. For the mixed-type Green function $G_{\text{cf}}(k, z) = \langle\langle c_{k\sigma}^\dagger, f_\sigma \rangle\rangle(z)$, we obtain

$$G_{\text{cf}}(k, z) = G_{\text{fc}}(k, z) = V_k r (z - \epsilon_k)^{-1} G_{\text{f}}^*(z). \tag{2.57}$$

The resultant spectrum is interpreted as that of quasi-particles, hence the asterisk attached to relevant quantities. For example, the density of f electron states, $\rho_{\text{f}}^*(\epsilon)$, for each spin is given by

$$\rho_{\text{f}}^*(\epsilon) = \frac{\tilde{\Delta}}{\pi} \frac{1}{(\epsilon - \tilde{\epsilon}_{\text{f}})^2 + \tilde{\Delta}^2}. \tag{2.58}$$

The occupation number n_f is given by integration of $\rho_f^*(\epsilon)$ up to the Fermi level. The result is

$$n_f = \frac{n}{\pi} \arctan\left(\frac{\tilde{\Delta}}{\tilde{\epsilon}_f}\right). \tag{2.59}$$

On the other hand, a little algebra using $G_{cf}(k, z)$ gives

$$\sum_{k\sigma} V_k \langle f_\sigma^\dagger c_{k\sigma}\rangle = n W_0 r \ln\left(\frac{\sqrt{\tilde{\epsilon}_f^2 + \tilde{\Delta}^2}}{D}\right), \tag{2.60}$$

where we have assumed that ρ_c and V_k are constant for $D > \epsilon > -D$ and 0 otherwise. Thus, eqn (2.55) is equivalent to

$$\sqrt{\tilde{\epsilon}_f^2 + \tilde{\Delta}^2} = D \exp\left(\frac{\lambda}{nW_0}\right), \tag{2.61}$$

where λ in the exponent can be approximated by $-\epsilon_f$ in view of the relation $|\tilde{\epsilon}_f| \ll |\epsilon_f|$. The left-hand side sets the energy scale of the system. In fact, the energy is the same as T_0 obtained in § 2.1.4.

One can derive the static properties approximately using the mean-field theory. For example, the density of states $\rho_f^*(0)$ for quasi-particles at the Fermi level determines the specific heat at low temperatures. We obtain

$$\rho_f^*(0) = \frac{n\tilde{\Delta}}{\pi(\tilde{\epsilon}_f^2 + \tilde{\Delta}^2)} = \frac{n}{\pi\tilde{\Delta}} \sin^2\left(\frac{\pi n_f}{n}\right). \tag{2.62}$$

The specific heat coefficient γ due to the impurity is given by

$$\gamma = \tfrac{1}{3}\pi^2 \rho_f^*(0), \tag{2.63}$$

while the magnetic susceptibility is computed as

$$\chi = C_J \rho_f^*(0) \to C_J n_f / T_0 \quad (n \to \infty), \tag{2.64}$$

where C_J is the Curie constant. Therefore, in the large-n limit, the mean-field theory gives the correct result. For finite n, however, the mean-field theory cannot derive the correct Wilson ratio. This is simply because all the Landau parameters vanish, by construction, in the mean-field theory.

Let us give the relationship between the Green function $G_f^*(z)$ derived by the mean-field theory and the formally exact one $G_f(z)$. In terms of the self-energy $\Sigma_f(z)$, the latter is written as

$$G_f(z) = [z - \epsilon_f + i\Delta \, \text{sgn}\,(\text{Im}\,z) - \Sigma_f(z)]^{-1}, \tag{2.65}$$

where $\Delta = \pi W_0$. In order to extract the quasi-particle dynamics, one expands $\Sigma_f(z)$ near the Fermi level $z = 0$ as

$$\Sigma_f(z) = \Sigma_f(0) + \frac{\partial \Sigma_f(z)}{\partial z}\bigg|_{z=0} z + O(z^2), \tag{2.66}$$

where the derivative is real in the Fermi-liquid state. Then we obtain, near the Fermi level,

$$G_f(z) = a_f G_f^*(z), \tag{2.67}$$

where $a_f = [1 - \Sigma_f'(0)]^{-1}$ is called the renormalization factor. The quasi-particle Green function $G_f^*(z)$ corresponds to the one obtained in the mean-field theory provided one makes the identification

$$\Sigma_f(0) = \lambda, \qquad a_f = r^2. \tag{2.68}$$

Thus, the mean-field theory can be regarded as the simplest approximation to perform the renormalization explicitly.

2.2.2 Dynamical susceptibility in the local Fermi liquid

As the degeneracy n of the f orbital decreases, the interaction among quasi-particles becomes stronger. Then one can improve the mean-field theory to incorporate the Fermi-liquid effects. We apply the quasi-particle RPA explained in Chapter 1 to the present system. This is in fact exact in the low-energy limit, and constitutes a convenient approximation in the spherically symmetric case because there is only a single parameter to describe the interaction effect. The dynamical susceptibility of f electrons is written as

$$nC_J \chi(\omega)^{-1} = nC_J \chi_1(\omega)^{-1} - U_{\text{eff}}, \tag{2.69}$$

where $-U_{\text{eff}}$ is the effective interaction corresponding to $Z_0/\rho^*(\mu)$ in eqn (1.78), and $\chi_1(\omega)$ is the polarization function of quasi-particles with the f component given by

$$\chi_1(\omega) = nC_J \int d\epsilon_1 \int d\epsilon_2 \rho_f^*(\epsilon_1)\rho_f^*(\epsilon_2)\frac{f(\epsilon_1) - f(\epsilon_2)}{\omega - \epsilon_1 + \epsilon_2 + i\delta}, \tag{2.70}$$

with $\rho_f^*(\epsilon_1)$ being the density of states of f quasi-particles. The important point here is that the effective interaction is real in the limit of small ω. Taking the imaginary parts of both sides of eqn (2.69), we obtain

$$\lim_{\omega \to 0} \text{Im}\frac{\chi(\omega)}{\omega\chi^2} = \frac{\pi}{nC_J}, \tag{2.71}$$

where the right-hand side is obtained by noting that $\chi_1(0) = nC_J \rho_f^*(0)$.

Equation (2.71) is called the Korringa–Shiba relation. The remarkable feature is that its right-hand side is independent of the interaction. The original proof was provided by diagrammatic analysis for the spin 1/2 Anderson model [15]. ·

2.2.3　*Self-consistent theory at finite temperatures*

We now turn to a microscopic approach to the dynamics at finite temperatures. We have seen in the previous section that the $1/n$ expansion is a powerful scheme which can describe correctly the singlet ground state. Once we obtain the approximate wave function of the ground state by the $1/n$ expansion, we can obtain a restricted class of excited states by applying the hybridization operator successively. Within this manifold of states, the matrix elements of physical operators are easily obtained, and then the excitation spectrum such as the density of states can be derived at zero temperature. This direct $1/n$ expansion approach to the dynamics has been made extensively to derive the photoemission spectrum [27]. However, the direct expansion cannot eliminate the spurious energy gap of the order of T_K from the singlet ground state. We refer to a review article for further details of the direct expansion method [28].

In a self-consistent perturbation theory, it is practical to perform partial summation of relevant perturbation terms to infinite order. Since the spin operators (or more generally the X-operators) do not obey the commutation rules of fermions or bosons, use of the standard many-body technique with Feynman diagrams and the linked cluster expansion cannot be applied. As an alternative, we shall present a resolvent method which permits time-ordered diagrams to represent perturbation terms [29]. The resolvent method leads to an effective atomic picture which eliminates the degrees of freedom of the conduction electrons. At zero temperature, the effective atom acquires the same effective Hamiltonian as given by the Brillouin–Wigner perturbation theory.

In order to pursue the effective atomic picture, we try to factorize the partition function Z of the SU(n) Anderson model into the conduction electron part Z_c and the local part Z_f, which includes the effect of interaction with conduction electrons [30]. We divide the Hamiltonian H into the unperturbed part $H_0 = H_c + H_f$ and the perturbation $H_1 = H_{hyb}$. In terms of the integration along a contour C encircling the real axis in the counter-clockwise direction, Z is given by

$$Z = \int_C \frac{dz}{2\pi i} \exp\left(-\beta z\right) \mathrm{Tr}\, \frac{1}{z - H}. \tag{2.72}$$

The trace is over the direct product state $|\gamma, c\rangle$ from an f state $|\gamma\rangle$ and a conduction electron state $|c\rangle$. We introduce a creation operator $a_\gamma^\dagger$ of the f state and represent the direct product state as $a_\gamma^\dagger |c\rangle$. In order to make a compact description, we introduce the Liouville operator L by the relation $LA \equiv [H, A]$ for any operator A. Then we obtain

$$H a_\gamma^\dagger |c\rangle = ([H, a_\gamma^\dagger] + a_\gamma^\dagger H)|c\rangle = (L + E_c)a_\gamma^\dagger |c\rangle, \tag{2.73}$$

where E_c is the energy of $|c\rangle$. Repeated application H^n leads to $(L + E_c)^n$ operating on $a_\gamma^\dagger |c\rangle$. Thus, we rewrite eqn (2.72) as

$$Z = \int_C \frac{dz}{2\pi i} \exp(-\beta z) \sum_{c\gamma} \exp(-\beta E_c)\langle c|a_\gamma \frac{1}{z - L} a_\gamma^\dagger |c\rangle \equiv Z_c Z_f, \tag{2.74}$$

where for each $|c\rangle$ a shift of the integration variable by E_c has been made. Factoring out $Z_c = \sum_c \exp(-\beta E_c)$ for the conduction electron part, the effective partition function

Z_f of f states is given by [18]

$$Z_f = \int_C \frac{dz}{2\pi i} \exp(-\beta z) \sum_\gamma \left\langle a_\gamma \frac{1}{z - L} a_\gamma^\dagger \right\rangle_c, \qquad (2.75)$$

with $\langle \cdots \rangle_c$ being the thermal average over conduction states. The average defines the resolvent $R_\gamma(z) = \langle a_\gamma (z - L)^{-1} a_\gamma^\dagger \rangle_c$.

The density of states of the effective atom is given by the spectral intensity

$$\eta_\gamma(\epsilon) = -\frac{1}{\pi} \operatorname{Im} R_\gamma(\epsilon + i\delta) = -\frac{1}{2\pi i} [R_\gamma(\epsilon + i\delta) - R_\gamma(\epsilon - i\delta)]$$

$$= \langle a_\gamma \delta(\epsilon - L) a_\gamma^\dagger \rangle_c. \qquad (2.76)$$

Then Z_f is also written as

$$Z_f = \sum_\gamma \int_{-\infty}^{\infty} d\epsilon \, \exp(-\beta\epsilon) \eta_\gamma(\epsilon), \qquad (2.77)$$

which fits naturally with the effective atomic picture. In terms of the exact eigenstates $|t\rangle$ of H, R_γ can be represented as

$$R_\gamma(z) = \frac{1}{Z_c} \sum_{c\gamma} \exp(-\beta E_c) \frac{|\langle t | \gamma, c \rangle|^2}{z - E_t + E_c}. \qquad (2.78)$$

A perturbation treatment with respect to hybridization H_1 can be performed by the expansion

$$\frac{1}{z - L} = \frac{1}{z - L_0} + \frac{1}{z - L_0} L_1 \frac{1}{z - L_0} + \cdots, \qquad (2.79)$$

where $L = L_0 + L_1$, in accordance with $H = H_0 + H_1$. Since $H_1 |c\rangle = 0$, it can be shown easily that

$$L_0 L_1 a_\gamma^\dagger |c\rangle = L_0 [H_1, a_\gamma^\dagger] |c\rangle = L_0 H_1 a_\gamma^\dagger |c\rangle$$
$$= [H_0, H_1 a_\gamma^\dagger] |c\rangle = (H_0 - E_c) H_1 a_\gamma^\dagger |c\rangle. \qquad (2.80)$$

Thus, when acting on $a_\gamma^\dagger |c\rangle$, L_1 can be replaced by H_1 and L_0 by $H_0 - E_c$; hence, the denominator $z - L_0$ depends only on excitation energies from a state with E_c. This feature is very convenient in taking the average over conduction states. The perturbation series generates a product of creation and annihilation operators of conduction electrons. The average over the conduction states can be performed by using the factorization property

$$\langle c_\alpha^\dagger c_\beta c_\gamma^\dagger c_\delta \rangle_c = \langle c_\alpha^\dagger c_\beta \rangle_c \langle c_\gamma^\dagger c_\delta \rangle_c + \langle c_\alpha^\dagger c_\delta \rangle_c \langle c_\beta c_\gamma^\dagger \rangle_c, \qquad (2.81)$$

which is an example of Wick–Bloch–de Dominicis theorem. A similar factorization property holds for cases with more operators.

One of the simplest perturbation processes for $R_0(z)$ is shown by the same diagram as Fig. 2.3. This is the only elementary process which survives in the limit of large n. The repetition of the process is accounted for by the self-energy $\Sigma_0(z)$ defined by

$$R_0(z) = [z - \Sigma_0(z)]^{-1}, \tag{2.82}$$

$$\Sigma_0(z) = nW_0 \int_{-D}^{D} \frac{f(\epsilon)\,d\epsilon}{z + \epsilon - \epsilon_{\mathrm{f}}}. \tag{2.83}$$

Note that the ground-state energy E_0 relative to the Fermi sea in eqn (2.83) is given by $E_0 = \Sigma_0(E_0)$ at $T = 0$. This leads to a threshold singularity in $R_0(z)$. At finite T or for higher-order perturbation, the resolvent has a cut singularity along the real z-axis and is analytic otherwise. In the exact theory at $T = 0$, the cut extends from the ground-state energy to infinity, as is clear from eqn (2.78).

The resolvent $R_1(z)$ is common to any of the degenerate f^1 states. In the next leading order, $R_1(z)$ also acquires the self-energy $\Sigma_1(z)$, as shown in Fig. 2.4. The resolvent $R_0(z - \epsilon)$ to be used in the intermediate state should not be a bare one, but should be a renormalized one in order to be consistent with the $1/n$ arrangement of perturbation terms. The use of renormalized resolvents in the intermediate states leads to the following integral equations [29–31]:

$$\Sigma_0(z) = nW_0 \int_{-D}^{D} d\epsilon \, f(\epsilon) R_1(z + \epsilon), \tag{2.84}$$

$$\Sigma_1(z) = W_0 \int_{-D}^{D} d\epsilon \, [1 - f(\epsilon)] R_0(z - \epsilon). \tag{2.85}$$

The solution of this set of equations corresponds to a summation of all perturbation diagrams without crossing the conduction-electron lines, and is called the non-crossing approximation (NCA) [32].

The NCA can derive not only the thermodynamics through the partition function, but dynamical quantities such as the dynamical magnetic susceptibility $\chi(\omega)$ due to the impurity [30] and the density of f states $\rho_{\mathrm{f}}(\omega)$ for single-particle excitations [30,33,34]. It can be shown that the NCA satisfies the conservation laws and sum rules required for response functions [30]. Let us first consider the dynamical magnetic susceptibility $\chi(\omega)$. It is given in the imaginary frequency domain by

$$\chi(i\nu_{\mathrm{m}}) = \frac{nC_J}{Z_{\mathrm{f}}} \int_C \frac{dz}{2\pi i} \exp(-\beta z) R_1(z + i\nu_{\mathrm{m}}) R_1(z), \tag{2.86}$$

where the contour C encircles all singularities of the integrand in a counter-clockwise manner. After analytically continuing the Matsubara frequency to the real axis, we are

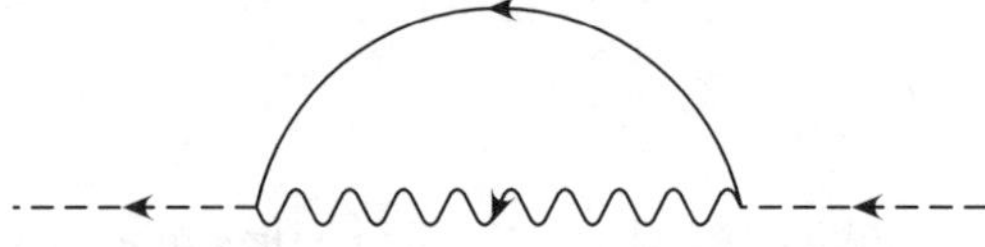

FIG. 2.4. The simplest self-energy for $R_1(z)$.

left with integration along the real axis with the Boltzmann factor $e^{-\beta\epsilon}$, which becomes singular at $T = 0$. Then the following spectral intensity is conveniently introduced as a quantity supplementary to resolvent:

$$\xi_\gamma(\epsilon) = Z_{\mathrm f}^{-1} e^{-\beta\epsilon} \eta_\gamma(\epsilon) = \frac{1}{Z} \sum_{c,t} \exp(-\beta E_t) |\langle t | \gamma, c \rangle|^2 \delta(\epsilon - E_t + E_c)$$

$$= \langle a_\gamma^\dagger \delta(\epsilon + L) a_\gamma \rangle. \tag{2.87}$$

From the last expression, we understand that $\xi_\gamma(\epsilon)$ corresponds to the spectral intensity after annihilation of the impurity state by a_γ [18]. The spectral function $-\mathrm{Im}\,\chi(\omega)/\pi$ (see eqn (B.15) in Appendix B) is given by

$$\mathrm{Im}\,\chi(\omega) = (1 - e^{-\beta\omega}) \frac{nC_J}{Z_{\mathrm f}} \int_{-\infty}^{\infty} d\epsilon\, e^{-\beta\epsilon} \eta_1(\epsilon) \eta_1(\epsilon + \omega) \tag{2.88}$$

$$= nC_J \int_{-\infty}^{\infty} d\epsilon\, [\xi_1(\epsilon)\eta_1(\epsilon + \omega) - \eta_1(\epsilon)\xi_1(\epsilon + \omega)]. \tag{2.89}$$

This expression naturally follows the effective atomic picture, and it is easy to see that the line width of $\mathrm{Im}\,\chi(\omega)$ is given in terms of $\eta_1(\epsilon)$ and $\xi_1(\epsilon)$. Thus, in the absence of hybridization, the width vanishes since we have $\eta_1(\epsilon) = \delta(\epsilon - \epsilon_{\mathrm f})$ and $\xi_1(\epsilon) = (n_{\mathrm f}/n)\delta(\epsilon - \epsilon_{\mathrm f})$. The actual width is controlled by the Kondo temperature, as will be discussed later. It is clear that use of $\xi_\gamma(\epsilon)$ permits us to take the zero-temperature limit without the singular Boltzmann factor.

The static susceptibility χ, on the other hand, is given by

$$\chi = \frac{nC_J}{\pi Z_{\mathrm f}} \int_{-\infty}^{\infty} d\epsilon\, e^{-\beta\epsilon} \mathrm{Im}\,[R_1(\epsilon - i\delta)^2] \tag{2.90}$$

$$= 2nC_J \int_{-\infty}^{\infty} d\epsilon\, \xi_1(\epsilon) \mathrm{Re}\,R_1(\epsilon). \tag{2.91}$$

Of course, the results given by eqns (2.89) and (2.91) are consistent with the Kramers–Kronig relation.

The single-particle Green function $G_{\mathrm f}(z)$ is given by

$$G_{\mathrm f}(z) = -i \int_0^\infty dt\, e^{izt} \langle \{ f_\sigma(t), f_\sigma^\dagger \} \rangle = \int_{-\infty}^\infty d\epsilon\, \frac{\rho_{\mathrm f}(\epsilon)}{z - \epsilon}, \tag{2.92}$$

where the density of states $\rho_{\mathrm f}(\omega)$ has been used. In the NCA, $G_{\mathrm f}(i\epsilon_n)$ in the Matsubara frequency domain is given in terms of the resolvents by

$$G_{\mathrm f}(i\epsilon_n) = \frac{1}{Z_{\mathrm f}} \int_C \frac{dz}{2\pi i} \exp(-\beta z) R_1(z + i\epsilon_n) R_0(z). \tag{2.93}$$

After analytic continuation, we obtain the spectral intensity as

$$\rho_{\mathrm f}(\omega) = n(1 + e^{-\beta\omega}) Z_{\mathrm f}^{-1} \int_{-\infty}^{\infty} d\epsilon\, e^{-\beta\epsilon} \eta_0(\epsilon) \eta_1(\epsilon + \omega) \tag{2.94}$$

$$= n \int_{-\infty}^{\infty} d\epsilon\, [\xi_0(\epsilon)\eta_1(\epsilon + \omega) + \eta_0(\epsilon)\xi_1(\epsilon + \omega)]. \tag{2.95}$$

The second equation indicates explicitly the excitation from the vacant f state to the filled one (electron addition) and vice versa (electron removal).

At high temperatures, the dynamical susceptibility can be calculated analytically since perturbation theory with respect to hybridization is applicable. We introduce the magnetic relaxation rate Γ measured by NMR as follows:

$$\operatorname{Im} \chi(\omega)/\omega = \chi(0)/\Gamma. \qquad (2.96)$$

If the dynamical susceptibility can be approximated by a Lorentzian, i.e.

$$\chi(\omega) = \frac{\chi(0)\Gamma}{-i\omega + \Gamma}, \qquad (2.97)$$

the rate Γ appears also as the half-width in the neutron scattering spectrum.

In order to calculate Γ, we note the identity

$$R_1(z)R_1(z') = \frac{R_1(z) - R_1(z')}{z' - z - \Sigma_1(z') + \Sigma_1(z)}. \qquad (2.98)$$

Then we use eqn (2.76) in making eqn (2.89) suitable for perturbation theory for finite but small ω. We obtain

$$\lim_{\omega \to 0} \operatorname{Im} \frac{\chi(\omega)}{\omega} = nC_J\beta \int_{-\infty}^{\infty} d\epsilon \frac{\xi_1(\epsilon)}{\operatorname{Im}\Sigma(\epsilon)}. \qquad (2.99)$$

At high temperatures where the Curie law $\chi(0) \sim C_J n_{\mathrm{f}}/T$ holds, we can make an approximation to set $\xi_1(\epsilon) \sim (n_{\mathrm{f}}/n)\delta(\epsilon - \epsilon_{\mathrm{f}})$. Then we obtain a simple result

$$\Gamma = -2\operatorname{Im}\Sigma_1(\epsilon_{\mathrm{f}}). \qquad (2.100)$$

We can evaluate eqn (2.100) by perturbation theory. In the lowest order with respect to hybridization, we obtain

$$\Gamma = 2\pi W_0[1 - f(\epsilon_{\mathrm{f}})] = 2n\pi W_0 \left(n_{\mathrm{f}}^{-1} - 1\right). \qquad (2.101)$$

This quantity is independent of temperature T provided that the average occupation n_{f} does not change with T. In the valence fluctuation regime, this is the relevant case. The result was first obtained using the memory function method [35], which is explained in Appendix D. In the Kondo regime $n_{\mathrm{f}} \sim 1$, on the other hand, the lowest-order result is exponentially small. One should therefore go to the fourth order in hybridization. As discussed in §2.1, the evaluation amounts to second order in J. Then the result reproduces eqn (2.21) obtained by the Born approximation. For general T, we can perform numerical calculations to evaluate the integrals for response functions. Many results have been reported [25] which compare favourably with experimental observations by neutron scattering or by photoemission spectroscopy.

Figure 2.5 shows an example of the density of states $\rho_{\mathrm{f}}(\epsilon)$ computed in the NCA [36]. A sharp peak develops near the Fermi level at low temperatures, and is called the Kondo resonance. Note that the density of states $\rho_{\mathrm{f}}^*(\epsilon) = \rho_{\mathrm{f}}(\epsilon)/a_{\mathrm{f}}$ for quasi-particles should be even larger because the renormalization factor a_{f} is much smaller than unity. The NCA can derive the density of states in the presence of spin–orbit and/or the CEF splittings.

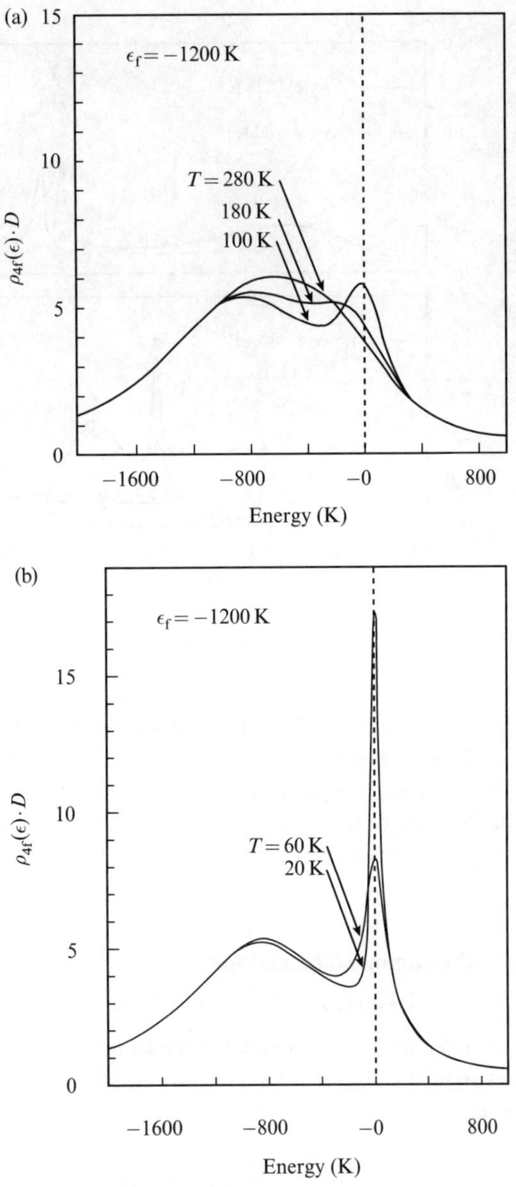

FIG. 2.5. The f electron density of states at various temperatures.

Figure 2.6 shows an example of numerical results [37]. We note that the overall intensity of the density of states is much larger than that expected from the Kondo temperature. In the energy scale larger than the spin–orbit splitting, the effective degeneracy of the $4f^1$ configuration amounts to 14. Thus, according to eqn (2.50), there is a large difference

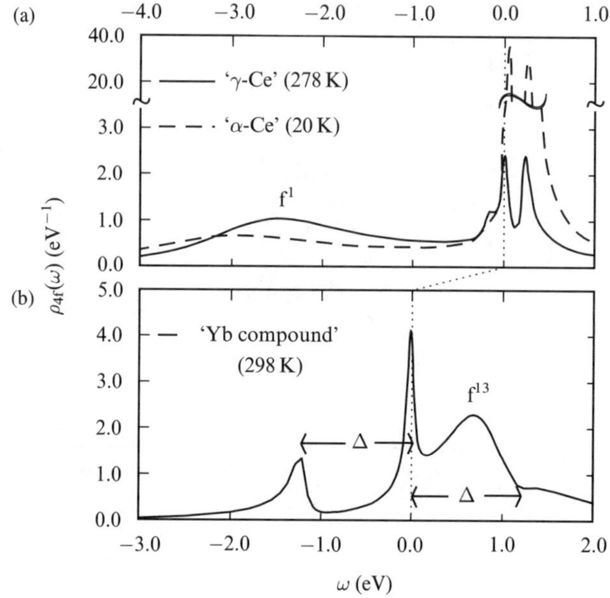

FIG. 2.6. The f electron density of states in the presence of spin–orbit interaction.

between the actual T_K determined by low-temperature properties and the apparent Kondo temperature determined by the spectral shape. This observation is important in order to avoid confusion about the correct model to account for spectroscopic results.

At $T = 0$ and in the limit of large n, one can derive nontrivial results analytically [38]. The lowest order in $1/n$ expansion coincides with the mean-field theory using the slave boson, which was explained in § 2.2.1.

2.3   Deviation from the canonical behaviour

2.3.1   *Non-Fermi-liquid ground state*

The ground state of the Kondo model is not always the Kondo singlet. In order to explore a possible non-singlet state, we consider H_K given by eqn (2.6), but with the number of orbital channels $n > 1$:

$$s_c = \frac{1}{2N} \sum_{kk'} \sum_{m=1}^{n} \sum_{\alpha\beta} c_{km\alpha}^{\dagger} \sigma_{\alpha\beta} c_{k'm\beta}, \tag{2.102}$$

where m denotes each channel. The magnitude S of the localized spin S is now not restricted to 1/2.

There are three cases which give different ground states: (i) $n = 2S$, (ii) $n < 2S$, and (iii) $n > 2S$. In the case (i), as in $n = 2S = 1$, the ground state is the Kondo singlet. Each channel screens spin 1/2 so that the entire local spin S is completely screened by n channels. The fixed point corresponds to $J = \infty$. In the case (ii), even though each conduction channel couples strongly to the impurity spin, the screening is still incomplete because of the large spin S. As a result, there remains a localized spin of $S - n/2$ in the ground state. Although this fixed point corresponds to $J = 0$, the ground state cannot be reached by perturbation theory since the localized spin has the magnitude $S - n/2$.

A new situation arises in the case of (iii). The fixed point corresponds to $J = J_c$ which is neither 0 nor infinite [39]. In order understand why this is so, let us consider a hypothetical case of $J = \infty$. Then the conduction electrons overscreen the impurity spin by the amount $n/2 - S$. This extra amount now acts as a new impurity spin which couples with the remaining conduction electrons. The resultant coupling is antiferromagnetic since the Pauli principle allows only conduction electrons with spin antiparallel to the new spin $n/2 - S$ to hop into the impurity site and to gain the perturbation energy. Hence, $J = \infty$ cannot be a fixed point. It is obvious that $J = 0$ cannot be a fixed point either. Hence, the fixed point is forced to have a nonzero J_c. The scaling is shown schematically in Fig. 2.7.

One can confirm this qualitative reasoning by renormalization to third order. Following the argument of Appendix C, one obtains the scaling equation for the model given by eqn (2.102) as follows:

$$\frac{dg}{d \ln D} = -g^2 + \frac{n}{2}g^3, \tag{2.103}$$

where $g = J\rho_c$ is the effective exchange interaction. The right-hand side becomes zero when $g = g_c = 2/n$, which corresponds to the fixed point of renormalization. The large n means small g_c. Then one can rely on the perturbative renormalization to this order, since higher-order contributions can be shown to be smaller by $1/n$. The linearized scaling equation for $\delta g \equiv g - g_c$ is given by

$$\frac{d\delta g}{d \ln D} = \frac{2}{n}\delta g. \tag{2.104}$$

The plus sign on the right-hand side means that the fixed point is stable.

2.3.2 *Mapping to one-dimensional models*

From the Wilson form of the Hamiltonian, eqn (2.24), it is obvious that the system of a magnetic impurity in a metal can be mapped to a one-dimensional model where the radial distance from the impurity plays the role of the one-dimensional coordinate. Thus, the impurity is on the left end of the system, and the right end grows as the renormalization step proceeds. The s-wave part of conduction band states with momentum k has the incoming and outgoing wave functions, which are interpreted as left-going and right-going waves, respectively. Alternatively, a left-going wave $\exp(-ikx)$ with k and x positive is reinterpreted as a right-going wave with momentum k and coordinate $-x$.

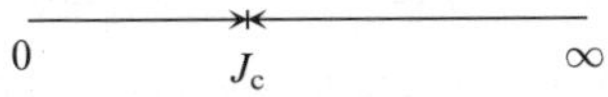

FIG. 2.7. Scaling of the exchange interaction in the overscreened case.

In the latter picture the conduction electrons are all right-going with velocity v_F. Their Hamiltonian is written as

$$H_c = v_F \sum_{k\sigma} k c_{k\sigma}^\dagger c_{k\sigma}, \qquad (2.105)$$

where $k \in [-k_F, k_F]$ is measured relative to the Fermi momentum. In order to understand the nature of the spectrum with an impurity, it is necessary to analyse the spectrum of H_c with various kinds of internal degrees of freedom.

The spectrum of H_c depends on the boundary condition. For the system size L, the periodic boundary condition (PBC) and the antiperiodic boundary condition (A-PBC) requires, respectively,

$$k = \frac{2\pi}{L} \times \begin{cases} n & \text{(PBC)} \\ (n + 1/2) & \text{(A-PBC)} \end{cases} \qquad (2.106)$$

with n integers. We introduce a quantum number Q which measures the change of the electron number from a reference state. In the case of spinless fermions, the change ΔE of the ground-state energy is given for the PBC by

$$\Delta E = \frac{2\pi v_F}{L} \sum_{n=0}^{Q-1} n = \frac{\pi v_F}{L} Q(Q-1), \qquad (2.107)$$

where we take the vacant zero-energy state as the reference state. If one takes the A-PBC, one should replace $Q(Q-1)$ by Q^2. Actual electrons have spin degrees of freedom. In the absence of orbital degeneracy, we put $Q = Q_\uparrow + Q_\downarrow$ and use the identity

$$Q_\uparrow^2 + Q_\downarrow^2 = \frac{1}{2} Q^2 + 2S_z^2, \qquad (2.108)$$

where $2S_z = Q_\uparrow - Q_\downarrow$. Then we obtain the energy for the A-PBC by

$$\Delta E = \frac{2\pi v_F}{L} \left[\frac{1}{4} Q^2 + \frac{1}{3} S^2 \right]. \qquad (2.109)$$

For simplicity, here the particle–hole excitations that appeared in eqn (2.25) have been omitted. For the PBC one has

$$\Delta E = \frac{2\pi v_F}{L} \sum_\sigma \sum_{n=0}^{Q_\sigma - 1} n = \frac{2\pi v_F}{L} \left[\frac{1}{4} Q^2 - \frac{1}{2} Q + \frac{1}{3} S^2 \right]. \qquad (2.110)$$

With the shift $Q \to (Q-1)$ in the above expression, one has the correspondence to the A-PBC result apart from a constant term. The shift corresponds to take the new reference state as having one electron in the zero-energy level. With the PBC, we always take this new reference state in the following discussion. Thus, the boundary conditions for electrons require a particular combination of quantum numbers as

$$(Q, S) = (\text{even, integer}) \oplus (\text{odd, half-integer}) \quad \text{(A-PBC)}, \qquad (2.111)$$

$$(Q, S) = (\text{even, half-integer}) \oplus (\text{odd, integer}) \quad \text{(PBC)}. \qquad (2.112)$$

The NRG calculation in the Kondo model shows that the spectrum of the model turns out to be the same as that of free conduction electrons either with the PBC or the A-PBC, depending on the odd or even number of the NRG steps. The reason for this simple behaviour is that the fixed point corresponds to infinitely strong exchange interaction [40]. Namely, the right-going electrons acquire the phase shift $\delta = \pm\pi/2$ after passing through the impurity because of the exchange interaction. This phase shift is equivalent to multiplying the factor $\exp(2i\delta) = -1$ to the wave function. Then the change from the PBC to the A-PBC, or vice versa, is equivalent to having this phase shift. In other words, the amount 1/2 of conduction-electron spin is swallowed by the impurity by the Kondo screening without changing the total number of conduction electrons. This results in the shift $S \to S \pm 1/2$ in the combination of (Q, S).

Now we turn to the spectrum of the free conduction band with orbital degeneracy. The Hamiltonian is given by

$$H_{\mathrm{c}} = v_{\mathrm{F}} \sum_{k\sigma} \sum_{m=1}^{n} k c_{km\sigma}^{\dagger} c_{km\sigma}, \qquad (2.113)$$

where m denotes the same orbital channel as appears in eqn (2.102). After some algebra, the spectrum is derived as

$$\Delta E = \frac{2\pi v_{\mathrm{F}}}{L} \left[\frac{1}{4n} Q^2 + \frac{1}{2+n} C_l + \frac{1}{2+n} S^2 + n_Q + n_l + n_S \right], \qquad (2.114)$$

where C_l denotes the eigenvalue of the second-order Casimir operator $\sum_{\alpha}(l^{\alpha})^2$ where l^{α}'s are $n^2 - 1$ generators of the SU(n) group. The latter describes symmetry of the orbital channel. In the case of doubly degenerate orbitals with $n = 2$, each generator l^{α} is reduced to the Pauli spin matrix with $\alpha = x, y, z$, and the squared sum of them gives $l(l + 1)$ with l a non-negative integer or a positive half-integer. The quantum numbers n_Q, n_l, n_S, which are non-negative integers, represent particle–hole excitations from the ground state. In the terminology of the conformal field theory, the spectra associated with various values of n_Q, n_l, n_S are called descendants or conformal towers. The ground state of the free electrons has a set $(Q, l, S) = (0, 0, 0)$ with the A-PBC. With the PBC, on the other hand, the ground state is 16-fold degenerate: the degeneracy 2^4 corresponds to filling or not filling the zero-energy level for each species of fermions. One of the ground states has a set $(Q, l, S) = (0, 1, 0)$, for example. Table 2.1(a), (b) shows the spectrum of two-channel free fermions either with PBC or A-PBC.

We have seen in § 2.3.1 that the multichannel Kondo model has a fixed point which has a finite value of the exchange interaction. As a result, one cannot expect that the spectrum of the model is described by a simple change of the boundary condition. The NRG calculation indeed reveals more complex structure of the spectrum. However, a description in terms of the spectrum given by eqn (2.114) is still valid provided one attaches particular combinations of quantum numbers. For example, the NRG calculation shows that the ground state has a set of quantum numbers $(Q, l, S) = (0, 0, 1/2)$. Obviously, this combination of quantum numbers is impossible in the free-electron case, whatever the boundary condition chosen. Physically, the quantum number $S = 1/2$ corresponds to overscreening of the impurity spin. Table 2.1(c) shows the resultant spectrum given by eqn (2.114) with $S = 1/2$ and $n = 2$. It has been confirmed that the low-energy spectrum derived by the NRG can be nicely reproduced by this analytical formula. We refer to review papers [40,41] for more detailed discussions.

Table 2.1 *Spectra of free right-going electrons with the antiperiodic boundary condition (a), and with the periodic one (b). In (b), the origin $Q = 0$ is so chosen that the zero-energy level is filled by two electrons. Also shown is the spectra for the two-channel Kondo model (c).*

Q	S	l	$\Delta E/(2\pi v_{\mathrm{F}}/L)$	Degeneracy
(a) A-PBC				
0	0	0	0	1
0	1	1	1	9
± 1	1/2	1/2	1/2	4
2	0	1	1	3
2	1	0	1	3
(b) PBC				
0	1	0	0	3
0	0	1	0	3
± 1	1/2	1/2	0	8
± 2	0	0	0	2
2	1	1	1	9
(c) Two-channel Kondo model				
0	1/2	0	0	2
0	1/2	1	1/2	6
± 1	0	1/2	1/8	2
± 1	1	1/2	5/8	6
2	1/2	0	1/2	2
2	1	1	1	6

According to the conformal field theory, only a part of the degeneracy in the ground state is associated with the impurity states. This is related to the order of taking the thermodynamic limit and the zero-temperature limit. For example, the ground-state degeneracy of the two-channel impurity is derived to be $2^{-1} \ln 2$ instead of $\ln 2$. This feature is consistent with the exact solution by the Bethe Ansatz theory [42,43].

2.3.3 *Properties of the non-Fermi-liquid state*

The ground state in the overscreened impurity has many anomalous properties. It is, however, difficult to derive these properties by elementary theoretical methods. Therefore, we state only final results referring to the literature [40–43] for details of derivation. For example, it has been shown that the specific heat is given by the power law

$$C = C_0 \left(\frac{T}{T_{\mathrm{K}}} \right)^{\alpha}, \tag{2.115}$$

where $\alpha = 4/(n+2)$. In the case of $n = 2$, the specific heat acquires the logarithmic term $T \ln T$. The temperature dependence of the susceptibility is also given by the power law with exponent $\alpha - 1$ or by a term $\ln T$ with $n = 2$. The impurity magnetization at $T = 0$ is given for small fields by

$$M \sim (H/T_K)^{2/n} \quad \text{for } n > 2 \tag{2.116}$$

and

$$M \sim (H/T_K) \ln(H/T_K) \quad \text{for } n = 2. \tag{2.117}$$

The resistivity behaves as $\rho(T)/\rho_0 \sim 1 - cT^{\alpha/2}$. The zero-temperature limit ρ_0 is given by

$$\rho_0/\rho_{\max} = \frac{1}{2}\left(1 - \frac{\cos[2\pi/(2+n)]}{\cos[\pi/(2+n)]}\right), \tag{2.118}$$

where $\rho_{\max}$ is the maximum value realized with $n = 1$, and is called the unitarity limit. The temperature coefficient c is positive if the bare exchange interaction is smaller than the fixed point value J_c, and is negative otherwise. In the latter case, the resistivity should decrease with decreasing temperature in the low-temperature limit, which contrasts with the Kondo effect.

There are efforts to investigate the non-Fermi liquid state in real systems. A candidate is a crystal field ground state which is a non-Kramers doublet [44]. An example of this is Γ_3 of $U^{4+}(5f^2)$ ion within the cubic symmetry. If one introduces a quasi-spin to represent a pair of wave functions constituting the non-Kramers doublet, the conduction electrons get two screening channels since they also have real spins. The quasi-spin actually represents the dynamics of the quadrupole moment. Hence, this kind of overscreening is called the quadrupolar Kondo effect [44]. It has been shown that the non-Fermi-liquid ground state can be realized even for the Anderson model with a dominant f^2 configuration and with proper crystal field [45]. We discuss the experimental situation in § 2.4.3.

2.3.4 *Pair of local moments with hybridization*

Heavy electrons emerge as a result of interaction between many Kondo centres. As a necessary step toward investigation of how the heavy electrons are formed, we consider the two-impurity Anderson model where the local moments can form a singlet either by mutual coupling or by the Kondo effect. Suppose that two Anderson impurities are separated from each other by a distance R in a metallic matrix. We assume that the occupation of each impurity is almost unity because of large U and $|\epsilon_f|$. In the limit of large R, the ground state is regarded as two Kondo singlets which are almost independent of each other. The characteristic energy is given by the Kondo temperature T_K.

The Kondo exchange interaction H_{ex} induces a new interaction between the local moments. The new interaction is mediated by propagation of conduction electrons, and is called the RKKY interaction [46]. The effective Hamiltonian up to second order in

H_{ex} is given by

$$H_{\text{RKKY}} = P H_{\text{ex}} (E_0 - H_{\text{c}})^{-1} H_{\text{ex}} P, \tag{2.119}$$

$$H_{\text{ex}} = J[S_1 \cdot s_{\text{c}}(R/2) + S_2 \cdot s_{\text{c}}(-R/2)], \tag{2.120}$$

$$s_{\text{c}}(\pm R/2) = \frac{1}{2N} \sum_{kk'} \sum_{\alpha\beta} c_{k\alpha}^\dagger \sigma_{\alpha\beta} c_{k'\beta} \exp[\pm i(k - k') \cdot R/2], \tag{2.121}$$

where S_i with $i = 1, 2$ represents the local spin at site i, and each wavenumber is a vector instead of the radial component. The projection operator P is onto the ground state of H_{c} plus all states of local spins, and E_0 is its ground-state energy.

Calculation shows that H_{RKKY} is written in the form

$$H_{\text{RKKY}} = J_{\text{RKKY}}(R) S_1 \cdot S_2, \tag{2.122}$$

$$J_{\text{RKKY}}(R) = -\frac{J^2 V}{8N^2} \sum_q \chi_0(q, 0) \exp(i q \cdot R), \tag{2.123}$$

where $\chi_0(q, 0)$ is the static susceptibility defined by eqn (1.46). It can be shown that $J_{\text{RKKY}}(R)$ decays as $\cos(2k_{\text{F}} R + \theta)/R^3$ in the free-electron-like conduction band. Thus, at large R, $|J_{\text{RKKY}}(R)|$ becomes much smaller than the Kondo temperature. The ground state in this case is that of two almost independent Kondo impurities, and the wave function is written as $\psi_{2\text{K}}$.

In the opposite limit of small R, the intersite interaction becomes significant. If the intersite interaction J_{RKKY} is ferromagnetic, the pair forms the triplet. Then the problem is reduced to the spin-1 Kondo problem. If the interaction is antiferromagnetic, on the contrary, the two impurities form a pair singlet by J_{RKKY}. Then the Kondo effect does not occur. The wave function of the ground state is written as ψ_{PS}. We now ask whether the two kinds of singlet states, ψ_{PS} and $\psi_{2\text{K}}$, can be connected continuously as the distance R changes.

The pair singlet is analogous to the Heitler–London picture for the ground state of the hydrogen molecule. The zeroth-order wave function for ψ_{PS} is given by

$$\psi_{\text{PS}} = \left(f_{1\uparrow}^\dagger f_{2\downarrow}^\dagger - f_{1\downarrow}^\dagger f_{2\uparrow}^\dagger \right) \phi_{\text{F}}, \tag{2.124}$$

where ϕ_{F} represents the ground state of conduction electrons. In the presence of hybridization, ψ_{PS} mixes with various states as shown in Fig. 2.8. In particular, the effective transfer from site 1 to site 2 becomes possible at the cost of U. In analogy with the hydrogen molecule, one can think of a molecular orbital by a linear combination of the two f orbitals at different sites. The effective transfer causes a splitting of levels for bonding and antibonding molecular orbitals. In the limit of large transfer as compared to U, the picture for the ground state is such that two electrons with opposite spins are accommodated in the bonding orbital. With $U = 0$ the two-impurity Anderson model can be diagonalized easily. The analogy to the molecular orbital for f electrons applies in this case. Actually, the electron correlation mixes the antibonding orbital in the ground state to some extent. As U increases, the ground state has more component of the antibonding orbital, and it changes continuously to the one described by the Heitler–London picture, namely by ψ_{PS}.

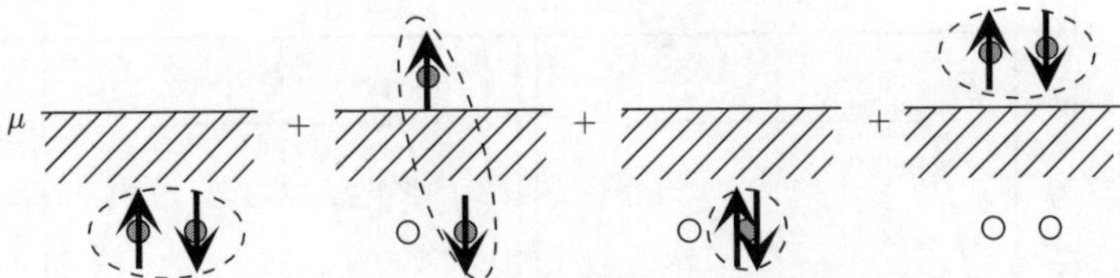

FIG. 2.8. Perturbation processes for ψ_{PS}. The two spins inside the dotted oval represent a singlet pair. Circles show localized f sites which can have double occupation with finite U.

On the other hand, the Kondo state is connected continuously to the $U = 0$ limit. Therefore, ψ_{PS} and ψ_{2K} can be connected by continuous deformation through the $U = 0$ limit, namely the molecular orbital picture. If one starts from the $U = 0$ limit and increases U, keeping the total f electron number 2 by appropriate shift of ϵ_{f}, the charge fluctuation is gradually suppressed. The resultant state is close to ψ_{2K} if J is small, while it tends to ψ_{PS} if J is large enough. This smooth change between ψ_{PS} and ψ_{2K} rests heavily on the presence of charge degrees of freedom. Namely, it is important that the splitting between bonding and antibonding orbitals does not completely vanish even with large U. On the other hand, if one uses the s–d (Kondo) model, the bonding–antibonding splitting does not have a meaning since there is no charge fluctuation. Then in ψ_{2K} the occupation of each orbital is strictly unity. According to the local Fermi-liquid theory, the density of states at the Fermi level is equal to $(\pi\Delta)^{-1}$ for each orbital. However, ψ_{PS} can have a much smaller density of states at the Fermi level. Therefore, the two kinds of states cannot connect continuously. In other words, if one neglects the bonding–antibonding splitting and increases J/T_{K}, a discontinuous change should occur. This is an example of the quantum phase transition at zero temperature.

Explicit calculations have been done using the numerical renormalization group [47,48], and the quantum Monte Carlo [49]. In the calculation by Jones *et al* [47] the s–d model is used and the discontinuity is observed in going from ψ_{2K} to ψ_{PS}. On the contrary, Fye and Hirsch [49] use the Anderson model and find no discontinuity. Sakai and Shimizu [12,48] clarified how these contrasting behaviours are reconciled by changing the strength of the bonding–antibonding splitting, which is also referred to as parity splitting. Figure 2.9 shows the f-electron density of states in the two-impurity symmetric Anderson model [48]. In the absence of the parity splitting, the density of states at the Fermi level should be the same as the one without the two-body interactions. This is a consequence of the Friedel sum rule in the Fermi-liquid side. As the intersite exchange increases, the density of states near the Fermi level becomes narrow and vanishes at the quantum phase transition to the pair singlet state. The width becomes infinitely narrow just before the transition.

If the f–f hopping t is introduced in the two-impurity symmetric Anderson model, the density of states can be nonsymmetric around the Fermi level. Furthermore, there is no constraint like the Friedel sum rule at the Fermi level. Figure 2.10 shows the result of the NRG calculation [48]. The result in Fig. 2.10 shows that a smooth crossover from the local-moment limit and the itinerant Fermi-liquid limit occurs with the f–f transfer.

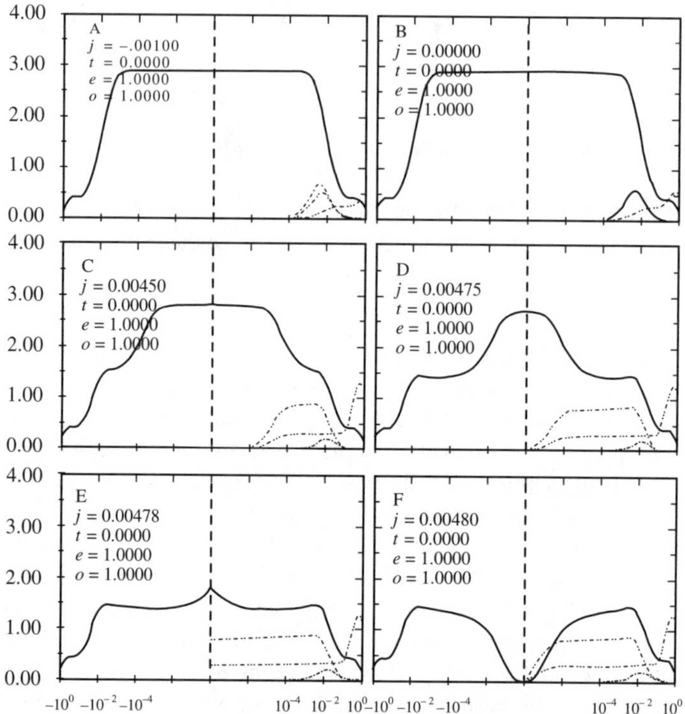

FIG. 2.9. Density of states of f electrons (solid line) for the model without the parity-splitting terms [48]. The parameter J denotes the intersite exchange in units of the half-width of the conduction band. The other parameters are $\epsilon_f = -0.4$ (f-electron level), $W_i = 0.03$ (hybridization intensity for channels $i =$ even, odd), and $U = 0.8$. The direct f–f transfer energy t is set to zero here, and the even-parity occupation number e, and the odd-parity occupation number o are the same without the parity splitting. The dot-dashed line represents the imaginary part of the uniform magnetic susceptibility, the two-dot-dashed line the staggered susceptibility (both scaled to 1/4), the three-dot-dashed line the superconducting response function (scaled to 2).

If this kind of smooth change occurs in heavy-electron systems also, the Fermi surface should also change without discontinuity. This subject will be treated in more detail in the next chapter on heavy electrons.

2.4 Experimental signatures of local spin dynamics in f-electron systems

At temperatures much higher than the characteristic energy of hybridization or intersite exchange interaction, many of the lanthanide (Ce, Yb, etc.) and the actinide (U, Pu, etc.) compounds behave as if they consist of an assembly of local moments. This quasi-independent behaviour continues sometimes down to low temperatures, as shown in Figs 1.1 and 1.2. In this chapter, we deal with this regime of heavy-electron systems

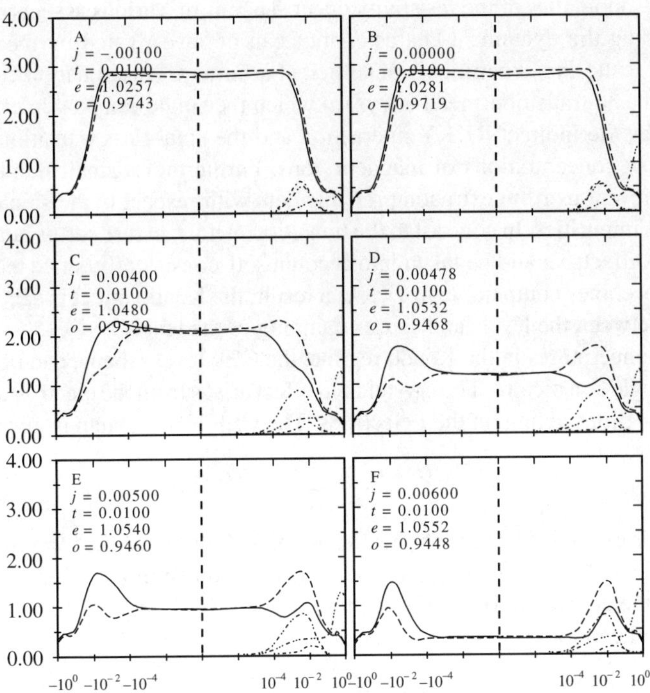

FIG. 2.10. Spectral intensities (solid line) of the model with the parity-splitting terms [48]. The solid line (dashed line) denotes the even (odd) component of the f-electron density of states. The other notations are the same as those used in Fig. 2.9. The transfer is taken to be $t = 0.01$ in all cases.

and valence fluctuation systems. In the case where the hybridization energy is larger than the CEF level splittings, the splittings are smeared out. Then the large degeneracy n associated with the J multiplet plays an important role in determining the characteristic energy scale which is typically $100\,\text{K}$. For example, we have $n = 6$ for Ce^{3+} and $n = 8$ for Yb^{3+}. In many lanthanide compounds, the valence fluctuating state has been probed by means of X-ray photoemission spectroscopy, measurement of lattice constant and Mössbauer isomer shift. In the valence fluctuation regime, the strong hybridization effect leads to both charge and spin fluctuations. As a consequence, the number n_f of f electrons is less than one.

On the contrary, if the hybridization between the f- and the conduction electrons is relatively weak, the charge fluctuation is suppressed since the Coulomb repulsion between f electrons is strong. In the latter case, the hybridization generates the exchange interaction J_{cf} between the f- and the conduction electron spins. This case is referred to as the Kondo regime. The characteristic energy scale, T_K, in the Kondo regime is typically $10\,\text{K}$. In most cases, the CEF splitting becomes larger than T_K.

At high temperatures, an assembly of local moments behaves as independent Kondo scattering centres. For example, the Curie–Weiss behaviour of the susceptibility appears and the log T anomalies of the resistivity occur. Therefore, various aspects of the Kondo effect including the dynamical characteristics can be investigated by means of NMR and neutron scattering experiments of sufficient accuracy. This situation contrasts with the case of the 3d transition-metal alloys, in which the single-ion Kondo effect is easily suppressed by the indirect RKKY interaction and the spin-glass transition takes place even with low concentrations of magnetic ions. Furthermore, small number of dilute impurities prevent us from extracting reliable data with respect to the spin dynamics of the magnetic impurities. In contrast to the transition metals, in rare earths both the Kondo and the CEF effects should be taken into account self-consistently, since the splitting of CEF levels becomes comparable to T_K. As a result, the Kondo effect is accompanied by transitions between the levels and causes damping of the levels.

At low temperatures in the Kondo regime, the CEF level scheme can be analysed in terms of a single-ion picture. The crystal field effect arises from the electrostatic potential of surrounding ions acting on the f electrons. The CEF Hamiltonian is given by

$$H_{\mathrm{CEF}} = e \sum_i V(\mathbf{r}_i). \tag{2.125}$$

Since $V(\mathbf{r}_i)$ is expanded by the spherical harmonics $Y_l^m(\theta, \phi)$, where $l=3$ for f electrons, H_{CEF} is expressed by a polynomial of total angular momentum J_z, J_+, J_-, and J^2. By introducing the equivalent operators such as $O_2^0 = 3J_z^2 - J(J+1)$, $O_2^2 = J_+^2 + J_-^2$, H_{CEF} is calculated [50,51] as

$$H_{\mathrm{CEF}} = \sum_{n,m} B_n^m O_n^m, \tag{2.126}$$

where B_n^m are the crystal field parameters to be determined experimentally, e.g. from inelastic neutron scattering. The CEF Hamiltonian H_{CEF} for a rare-earth ion has the maximum number $n = 6$, which is twice the orbital angular momentum $l = 3$ of the f orbital. For Ce^{3+} ($J = 5/2$), the maximum is $n = 4$ because $2J < 6$. By diagonalizing eqn (2.126), one can obtain the CEF energies E_n as functions of the CEF parameter B_2^0, B_4^0, and B_4^4. It is also possible to derive the matrix elements. For neutron scattering, the relevant matrix elements are $|\langle n| \mathbf{J}_\perp |m\rangle|^2$, where $\mathbf{J}$ is the projection of the total angular momentum onto the plane perpendicular to the scattering vector $\mathbf{Q}$.

The difference in the characteristic energy scales between the valence fluctuation and the Kondo regimes manifests itself in the T dependence of the magnetic relaxation rate. If the dynamical susceptibility is isotropic and can be approximated by a Lorentzian with the relaxation rate Γ, then $1/T_1$ is expressed as

$$\frac{1}{T_1} = 2\gamma_n^2 T |A_{\mathrm{hf}}|^2 \frac{\chi(T)}{\Gamma}, \tag{2.127}$$

where A_{hf} is the average hyperfine field discussed in Chapter 1. For quasi-elastic neutron scattering, the magnetic cross section is derived from the combination of eqns (1.155), (1.156), and (1.85) as

$$\frac{d^2\sigma}{d\Omega\, d\omega} = A^2 \frac{k_1}{k_0} |F(\mathbf{q})|^2 \chi(T) \frac{\omega}{1 - \exp(-\omega/T)} \frac{\Gamma}{\Gamma^2 + \omega^2}. \tag{2.128}$$

Thus, Γ appears as the half-width in the quasi-elastic neutron scattering spectrum.

In the limit of small ω, the imaginary part of $\chi(\omega)$ obeys the Korringa–Shiba relation given by eqn (2.71). Accordingly, with the law $T_1 T = \text{constant}$, at low temperatures we can estimate Γ by the relation

$$\Gamma = 2\gamma_{\rm n}^2 T_1 T \chi(0)|A_{\rm hf}|^2. \tag{2.129}$$

The NMR relaxation rate and the half-width of the quasi-elastic magnetic neutron scattering spectrum to be presented below can be understood in a consistent way.

2.4.1 *Valence fluctuating regime*

NMR In most cases, NMR experiments on the valence fluctuating materials have thus far been restricted to the non-lanthanide nuclei. The first NMR experiment using ^{139}Yb was reported by Shimizu *et al.* [52]. The valencies in YbAl$_2$ and YbAl$_3$ are determined from the experiments as 2.4 and 2.7–3.0, respectively. The magnetic susceptibility of YbAl$_2$ is T-independent below about 200 K and exhibits a broad maximum near 850 K, which is assigned as T_0. The susceptibility of YbAl$_3$ also exhibits a maximum at $T_0 = 125$ K above which the Curie-Weiss behaviour is observed with the effective moment nearly equal to that of a free Yb^{3+} ion.

From the observed spin-echo spectra of ^{139}Yb, the Knight shifts (K) are obtained as $K = 7.7\%$ for YbAl$_2$ and 100% for YbAl$_3$. Both are T-independent in the measured temperature range of 4.2–80 K and 1.4–4.2 K, respectively. The Knight shift as well as the susceptibility consists of both contributions from the 4f electrons ($K_{\rm 4f}$ and $\chi_{\rm 4f}$) and the conduction electrons ($K_{\rm ce}$ and $\chi_{\rm ce}$). Since both $K_{\rm ce}$ and $\chi_{\rm ce}$ are negligible compared to $K_{\rm 4f}$ and $\chi_{\rm 4f}$, the hyperfine field $H_{\rm hf} = K_{\rm 4f}\mu_{\rm B}/\chi_{\rm 4f}$ due to 4f electrons is dominant. Then the hyperfine interaction is obtained as $A_s = 1.1$ and 1.2 MOe/$\mu_{\rm B}$ for YbAl$_2$ and YbAl$_3$, respectively. On the other hand, $A_s(\text{Yb}^{3+})$ of a free Yb^{3+} ion is the sum of the orbital and dipole fields. Using the Hartree–Fock value of $\langle 1/r^3 \rangle_{\rm 4f}$, they can be calculated as $A_s = 1.15$ and 2.07 MOe/$\mu_{\rm B}$ for $J = 7/2$ and 5/2, respectively. It is notable that the experimentally deduced A_s values for both compounds are consistent with the calculated values for the $J = 7/2$ ground state of a free Yb^{3+} ion. Furthermore, an experimental value obtained from the hyperfine splitting of Yb^{3+} ESR in CaF$_2$ is also consistent with these results ($A_s(\text{ESR}) = 1.04$ MOe/$\mu_{\rm B}$). It should be noted that the Yb^{2+} state has no contribution to $K_{\rm 4f}$ and $\chi_{\rm 4f}$. Accordingly, it is deduced that the character of the 4f ground state wave function specified by the total angular momentum J is still conserved in these valence fluctuation compounds in spite of the strong hybridization effect.

Adequate knowledge of the hyperfine interaction allows us to extract the magnetic relaxation rate Γ for 4f spins in valence fluctuation compounds from the measurement of $1/T_1$. The T dependence of $1/T_1$ in YbAl$_2$ and YbAl$_3$ is shown in Fig. 2.11. For both compounds, $1/T_1$ is proportional to the temperature, having $T_1 T = (3.5 \pm 0.1) \times 10^{-2}$ s K for YbAl$_2$ and $(1.6 \pm 0.1) \times 10^{-4}$ s K for YbAl$_3$. From eqn (2.129), Γ's are evaluated as 1.5×10^4 K for YbAl$_2$ and 9.7×10^2 K for YbAl$_3$, respectively, both being temperature-independent. By combining eqn (2.71) and the Knight shift, the Korringa

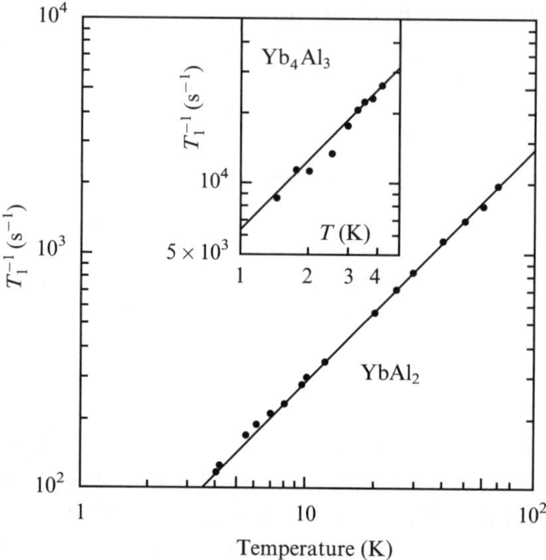

FIG. 2.11. Temperature dependence of $^{139}(1/T_1)$ in YbAl$_2$ and YbAl$_3$ (inset) [52].

relaxation for nuclei at impurity or f ion sites is obtained at $T \ll T_\mathrm{K}$ as

$$T_1 T K^2 = \frac{C_J (2J + 1)}{2\pi \gamma_\mathrm{n}^2}. \tag{2.130}$$

The Korringa constant on the right-hand side in eqn (2.130) is calculated to be 2.29×10^{-4} s K for the $J = 7/2$ ground state of Yb^{3+} ions. Experimentally, the Korringa constant on the left-hand side in eqn (2.130) is obtained as $T_1 T K^2 = 2.0 \times 10^{-4}$ s K for YbAl$_2$, and 1.6×10^{-4} s K for YbAl$_3$. The good agreement between theory and experiment for YbAl$_2$ implies that the spin fluctuations possess a local character and are described well in the framework of a single-impurity model, even though the Yb ions form a periodic lattice. For YbAl$_3$, however, the theoretical value is somewhat larger than the experimental value. This may be because the dynamical response function is different between dilute and periodic 4f ion systems and/or the degeneracy of $J = 7/2$ ground states is lifted partially, so that the application of eqn (2.130) becomes unjustified.

Neutron scattering The Ce-based compounds in the valence fluctuation regime were studied extensively by magnetic neutron scattering experiments. The quasi-elastic relaxation rate Γ is very large, up to several tens of meV. In the valence fluctuation regime, the intersite magnetic correlation driven by the RKKY interaction is masked by the local hybridization with conduction electrons. No CEF transitions are resolved in these compounds. By integrating the magnetic scattering cross section given by eqn (2.128) over ω for fixed $\boldsymbol{Q}$, one can obtain $g_J^2 J(J + 1) F(\boldsymbol{Q})^2$.

Equation (2.128) holds well for all rare-earth metals with a stable valence at high temperatures. However, the energy spectrum becomes more complex upon cooling, as

$\Gamma(T)$ is decreased to the order of the CEF splittings or magnetic ordering energies. On the other hand, the scattering profiles in the Ce-based compounds are quite different. As an example, we take $CePd_3$, which is compared with the isostructural reference compounds, diamagnetic YPd_3, which has no 4f electrons, and $TbPd_3$ with a stable $4f^8$ configuration [53]. Figure 2.12 indicates the energy dependence of the inelastic scattering cross section for those compounds. All three compounds exhibit incoherent elastic scattering at $\omega = 0$ due to nuclear isotopic and/or nuclear spin disorder. This contribution is shown by the shaded part in Fig. 2.12. For YPd_3, there is a phonon peak between $\hbar\omega = -20\,meV$ and $-10\,meV$, while $CePd_3$ has a considerable intensity in this energy window in addition to phonon scattering. In contrast to the coherent phonon scattering, the integrated intensity in this energy window decreases with increasing scattering angle and has the angular dependence expected from the magnetic 4f form factor.

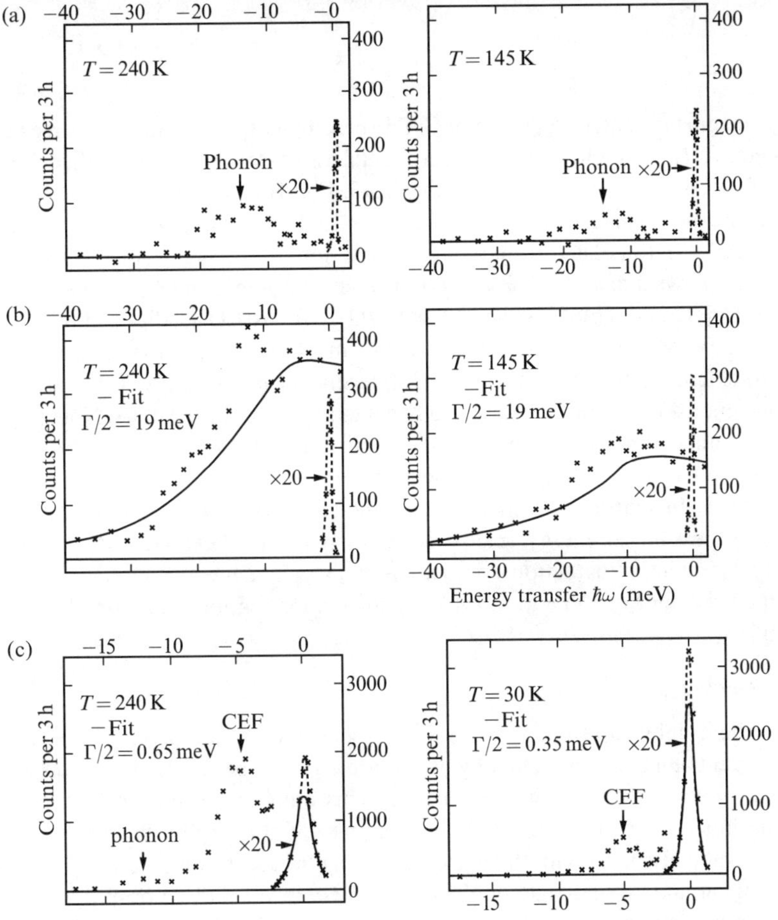

FIG. 2.12. Scattered neutron intensity as function of energy for the isostructural compounds, (a) YPd_3, (b) $CePd_3$, and (c) $TbPd_3$ [53].

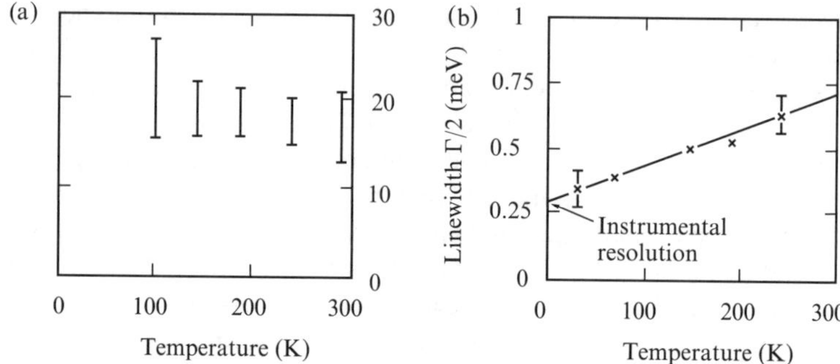

FIG. 2.13. Temperature dependence of the quasi-elastic magnetic half-width $\Gamma/2$ for (a) the valence fluctuation system $CePd_3$ and (b) the local-moment system $TbPd_3$ [53].

The magnetic scattering profile in $CePd_3$ is actually fitted by the solid lines according to eqn (2.128) employing the experimental values of the susceptibility. From this fitting, $\Gamma(T)$ is deduced as shown in Fig. 2.13. For $TbPd_3$, the quasi-elastic spectrum becomes sharper upon cooling, as seen in Fig. 2.13. Its linewidth decreases linearly with temperature, if extrapolated from the resolution limit of the spectrometer. We also note that $TbPd_3$ shows a distinct CEF transition. Thus, difference in the magnetic scattering between $CePd_3$ and $TbPd_3$ is clear. For $CePd_3$, a quasi-elastic linewidth amounts to about 20 meV, which is nearly three orders of magnitude larger than that in $TbPd_3$. It is remarkable that this width is nearly T-independent in the range 200–300 K. This is again in contrast to the normal Korringa behaviour observed for $TbPd_3$. Note that below 200 K, the linewidth actually exceeds $k_B T$. Therefore, the decay of local 4f moments is not driven thermally but quantum mechanically. The hybridization effect between f and conduction electrons is responsible for the magnetic ordering or the development of CEF spectra on an energy scale smaller than Γ. All these features are a general property in valence fluctuation compounds with nonmagnetic ground states. Apparently, the behaviour of $\Gamma(T)$ in the valence fluctuation regime is described well by eqn (2.101) [54].

2.4.2 *Kondo regime*

NMR   The Knight shift for non-transition elements such as Al, Si, Sn, As, etc. in heavy-electron compounds is dominated by an isotropic hyperfine interaction, even when the magnetic properties are highly anisotropic. This empirical result enables us to estimate the low-energy scale in the Kondo regime by comparing the NMR spectra of non-lanthanide elements with the quasi-elastic neutron scattering intensity. In the low-energy region, the ground-state doublet in the CEF levels plays a primary role. In this case, from the measurement of T_1, the magnetic relaxation rate Γ of the f electron can be extracted by eqn (2.127). In order to compare the results of the NMR and neutron scattering experiments, we present the ^{29}Si NMR investigation in $CeRu_2Si_2$ [55].

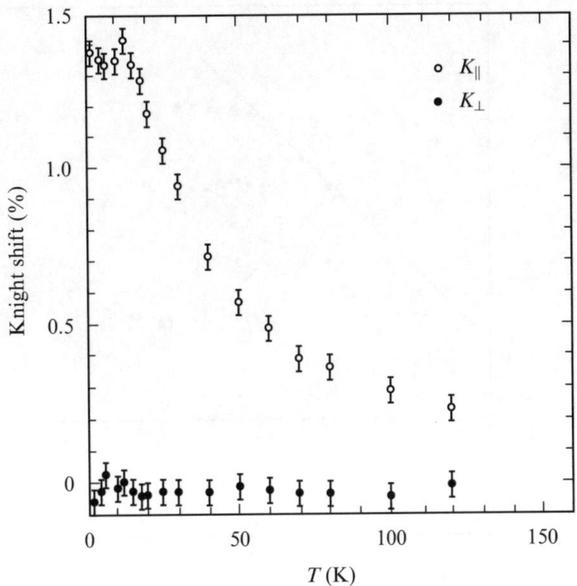

FIG. 2.14. Temperature dependence of the Knight shift $K_\parallel(T)$ parallel to the tetragonal c-axis, and $K_\perp(T)$ perpendicular to the c-axis in CeRu$_2$Si$_2$ [55].

From a characteristic powder pattern of the NMR spectrum, the T dependence of the Knight shift, parallel ($K_\parallel$) and perpendicular ($K_\perp$) to the tetragonal c-axis in CeRu$_2$Si$_2$ is obtained as displayed in Fig. 2.14. There appears a huge uniaxial anisotropy in the Knight shift. In fact, the anisotropy of susceptibility measured for the single crystal is as large as $\chi_\parallel/\chi_\perp \simeq 15$. From plotting an isotropic Knight shift defined by $K_{\mathrm{iso}} = (K_\parallel + 2K_\perp)/3$ against the susceptibility measured for the powder, we estimate a parallel component of the hyperfine field $A_\parallel$ and the anisotropy of the susceptibility at 4.2 K as $A_\parallel = 0.91 \pm 0.06\,\mathrm{kOe}/\mu_\mathrm{B}$ and $\chi_\parallel/\chi_\perp \simeq 18$, respectively. Thus, we conclude that the anisotropy of the Knight shift does not originate from the hyperfine interaction, but from the susceptibility.

We now have an adequate knowledge of the transferred hyperfine interaction of ^{29}Si, which is isotropic and dominated by the hybridization effect with five nearest-neighbour Ce-4f electrons. Next, we examine the relaxation result shown in Fig. 2.15. Part (a) of the figure indicates the T dependence of $(1/T_1)_\parallel$ and $(1/T_1)_\perp$. Both $(1/T_1)_\parallel$ and $(1/T_1)_\perp$ exhibit a weak T dependence above the temperature at which both the susceptibility and the Knight shift have a peak. However, below 8 K, both follow the relation $T_1 T$=constant. We emphasize that the value of $(T_{1\perp}T)^{-1} = 1.0 \pm 0.05\,\mathrm{s}^{-1}\,\mathrm{K}^{-1}$ is two orders of magnitude larger than $(T_1 T)^{-1} = 0.014 \pm 0.005\,\mathrm{s}^{-1}\,K^{-1}$ for LaRu$_2$Si$_2$. This verifies that the relaxation process in the former is governed by spin fluctuations of the f electrons. If we take into account the anisotropic magnetic property in eqn (2.127), $(1/T_1)_\parallel$ and $(1/T_1)_\perp$ are related to the dynamical susceptibility components $\chi_\parallel(\omega)$ and

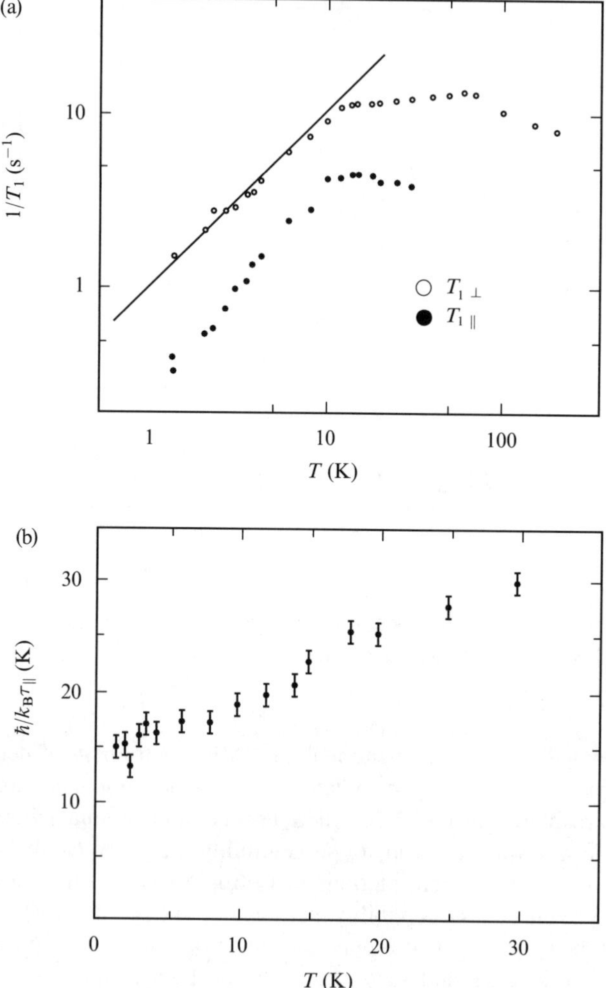

FIG. 2.15. (a) Temperature dependence of $(1/T_1)_\perp$ and $(1/T_1)_\parallel$. (b) Temperature dependence of Ce spin fluctuation rate, $\hbar/(\tau_\parallel k_{\mathrm{B}}) = \Gamma_\parallel(T)$ parallel to the c-axis [55].

$\chi_\perp(\omega)$ of the 4f electrons as

$$\frac{1}{T_{1\parallel}} = 2\gamma_{\mathrm{n}}^2 T \sum_i [A_{\perp,i}]^2 \frac{\mathrm{Im}\,\chi_\perp(\omega_{\mathrm{n}})}{\omega_{\mathrm{n}}} \tag{2.131}$$

and

$$\frac{1}{T_{1\perp}} = 2\gamma_{\mathrm{n}}^2 T \sum_i \left([A_{\parallel,i}]^2 \frac{\mathrm{Im}\,\chi_\parallel(\omega_{\mathrm{n}})}{\omega_{\mathrm{n}}} + [A_{\perp,i}]^2 (1/2) \frac{\mathrm{Im}\,\chi_\perp(\omega_{\mathrm{n}})}{\omega_{\mathrm{n}}} \right), \tag{2.132}$$

respectively. By combining the above two formulae, we obtain

$$\frac{1}{T_{1\perp}} - \frac{1}{2T_{1\parallel}} = 2\gamma_n^2 T \sum_i [A_{\parallel,i}]^2 \frac{\mathrm{Im}\,\chi_\parallel(\omega_n)}{\omega_n}, \qquad (2.133)$$

where $A_{\parallel,i}$ is the transferred hyperfine field parallel to the c-axis due to five nearest-neighbour 4f spins. In the present case, one can safely approximate $\sum_i [A_{\parallel,i}(R_i)]^2$ by $z(A_\parallel/z_n)^2 = A^2/z_n$, where $A_\parallel = 0.91 \pm 0.06\,\mathrm{kOe}/\mu_B$ is determined from the plot of K_{iso} vs χ_{iso}, with z_n the number of nearest neighbour 4f ions ($z_n = 5$ for $^{29}\mathrm{Si}$ of $\mathrm{CeRu_2Si_2}$).

By combining eqns (2.127) and (2.133), $(1/T_1)_\parallel$ and $(1/T_1)_\perp$ can be related to the magnetic relaxation rate $\Gamma_\parallel$ as

$$\Gamma(T)_\parallel = 2\gamma_n^2 T \left[\frac{1}{T_{1\perp}} - \frac{1}{2T_{1\parallel}} \right]^{-1} \frac{K_\parallel A_\parallel}{\mu_B z_n}, \qquad (2.134)$$

where the experimental relation $K_\parallel(T) = A_\parallel \chi_\parallel(T)/N\mu_B$ is utilized with $A_\parallel = 0.91 \pm 0.06\,\mathrm{kOe}/\mu_B$. Part (b) of Fig. 2.15 shows the T dependence of $\Gamma_\parallel$ thus estimated. It is difficult to evaluate $\Gamma_\perp$ since $K_\perp$ is almost zero. As seen in the figure, $\Gamma_\parallel$ is nearly T-independent with $\Gamma_\parallel = 16.0 \pm 1.0\,\mathrm{K}$ up to 8 K and increases gradually to 30 K.

This value of $\Gamma_\parallel$ extracted from NMR is very close to that obtained from neutron scattering: $\Gamma_\parallel^{\mathrm{neutron}} = 20 \pm 3\,\mathrm{K}$ as deduced from the quasi-elastic spectrum at 1.4 K [56]. This may suggest that the spin relaxation is governed by fluctuations of the f electrons, which are almost independent of the wave vector. However, as described in more detail in Chapter 3, the intersite magnetic correlations become significant in $\mathrm{CeRu_2Si_2}$ at low temperatures. Thus, it is likely that, below 8 K, the law $T_1 T = \mathrm{constant}$ does not originate from the single-site Kondo effect but from heavy itinerant electrons.

Neutron scattering   Experimentally, the magnetic neutron scattering spectra at low T consist of the quasi-elastic and several inelastic spectra. From the former, the magnetic relaxation rate Γ is extracted. On the other hand, the inelastic part originates from transitions between the ground-state multiplet and several excited CEF multiplets of the 4f shells. Thus, one can extract the CEF level scheme by analysing the spectra.

As a typical example, we present a systematic study of $\mathrm{CeM_2X_2}$ (M = Cu, Ag, Au, Ru, Ni; X = Si, Ge) by means of magnetic neutron scattering experiments [56–59]. These Ce-based ternary compounds with $\mathrm{ThCr_2Si_2}$-type structure involve systems with different ground states: $\mathrm{CeNi_2Ge_2}$ is paramagnetic, $\mathrm{CeCu_2Si_2}$ is the first heavy-electron superconductor and $\mathrm{CeCu_2Ge_2}$ is a heavy-electron antiferromagnetic compound, which, interestingly, undergoes a superconducting transition under an application of high pressure. $\mathrm{CeRu_2Si_2}$ and $\mathrm{CeNi_2Ge_2}$ are characterized by heavy-electron effective masses, which reveal neither a magnetic nor a superconducting phase transition. In $\mathrm{CeRu_2Si_2}$, at low temperatures there occurs a pseudo-metamagnetic transition at a critical magnetic field $H_c = 80\,\mathrm{kOe}$.

We first show the result in $\mathrm{CeCu_2Si_2}$ [58]. The neutron scattering experiment has revealed that the inelastic magnetic part of the spectrum consists of a well-pronounced peak at $\hbar\omega = 31.5\,\mathrm{meV}$ and, in addition, a less pronounced peak at $\hbar\omega = 12\,\mathrm{meV}$. Both peaks are fitted reasonably by two Lorentzians centred at $E_1 = \hbar\omega = k_B(135 \pm 15\,\mathrm{K})$ and $E_2 = \hbar\omega = k_B(360 \pm 20\,\mathrm{K})$, respectively. These two transitions are assigned to two CEF transitions from the ground-state doublet, $|0\rangle$, to the excited doublets, $|1\rangle$ and $|2\rangle$, expected for Ce^{3+} ions occupying the tetragonal sites. Thus, the CEF parameters such as B_2^0, B_4^0, and B_4^4 are obtained by simultaneously fitting the splittings E_1 and E_2 and the ratio of the intensities. As shown in Fig. 2.16, we can find the CEF level scheme.

On the other hand, the width of the quasi-elastic spectrum Γ is a measure of the strength of the hybridization or the exchange interaction between the f and the conduction electrons. In rare-earth compounds with a magnetically stable 4f configuration, one expects a Korringa behaviour for the quasi-elastic linewidth, as demonstrated for TbPd$_3$ (see Fig. 2.13), namely $\Gamma = \alpha T$, where α is typically 10^{-3}, while the valence fluctuation compounds, such as CePd$_3$ and CeSn$_3$, show an almost T-independent Γ, typically 20–30 meV, as represented by CePd$_3$ in Fig. 2.13.

In contrast to these, the relaxation rate in the Kondo regime exhibits a characteristic T dependence and probes the presence of a very low-energy scale of 10–30 K. Figure 2.17 shows the summary of the magnetic relaxation rate $\Gamma(T)$ vs temperature for $T \geq T_M$, where T_M is the magnetic ordering temperature for a series of CeM$_2$X$_2$ compounds with the tetragonal structure. The upper part of Fig. 2.17 shows $\Gamma(T)$ of the nonmagnetic CeRu$_2$Si$_2$ and CeNi$_2$Ge$_2$, and superconducting CeCu$_2$Si$_2$. Then the presence of such a low-energy scale is related to the large linear term of the specific heat, which amounts to about 1 J/mol K^2. Namely, at low temperatures, the system can be described by renormalized heavy quasi-particles.

The lower part of Fig. 2.17 indicates the results for magnetically ordered compounds with significantly lower values of Γ. Most remarkably, except for CeRu$_2$Ge$_2$ and CeAu$_2$Ge$_2$ in which $\Gamma(T)$ follows the Korringa behaviour as observed for TbPd$_3$, all other systems show a distinct deviation from a linear dependence on T. Instead, $\Gamma(T)$ seems to follow roughly a square-root dependence, $\Gamma = A\sqrt{T}$, as shown by the solid lines in Fig. 2.17. These systems are regarded as heavy-electron antiferromagnets with heavy quasi-particles where quasi-elastic Lorentzian intensities still survive even below T_M.

In cubic crystals such as Ce$_{1-x}$La$_x$Al$_2$ and CeB$_6$, the CEF levels are split into a doublet and a quartet. The ground state is the doublet Γ_7 in Ce$_{1-x}$La$_x$Al$_2$ and the quartet Γ_8 in CeB$_6$, respectively. On the other hand, in the tetragonal crystal CeCu$_2$Si$_2$,

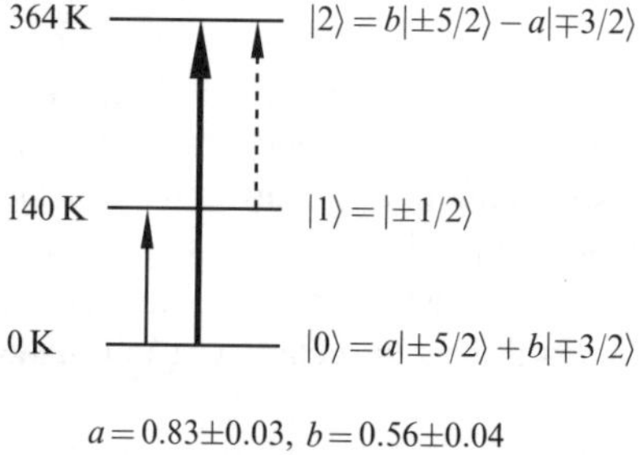

FIG. 2.16. CEF level scheme of Ce^{3+} in tetragonal CeCu$_2$Si$_2$ determined by the neutron scattering experiment [59].

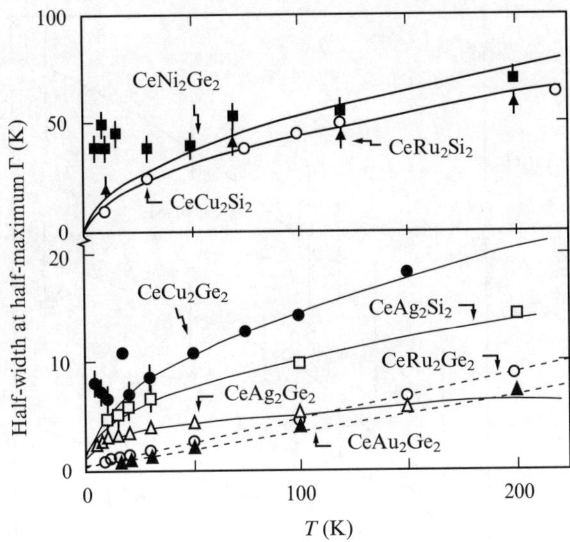

FIG. 2.17. Temperature dependence of the quasi-elastic Lorentzian linewidth, $\Gamma/2$ in isostructural compounds CeM$_2$X$_2$: (a) nonmagnetic CeNi$_2$Ge$_2$, CeRu$_2$Si$_2$, and superconducting CeCu$_2$Si$_2$; (b) antiferromagnetic CeCu$_2$Ge$_2$, CeAg$_2$Si$_2$, CeAg$_2$Ge$_2$, CeAu$_2$Ge$_2$ and ferromagnetic CeRu$_2$Ge$_2$. Solid lines represent fits using a square-root dependence of $\Gamma(T)$ [58].

the CEF levels consist of three doublets, as presented above. Hence, these three crystals have CEF states different from one another. By approximating the inelastic neutron scattering spectra to be of the Lorentzian form, the linewidths of the quasi-elastic spectra in CeCu$_2$Si$_2$, Ce$_{1-x}$La$_x$Al$_2$, and CeB$_6$ are displayed in Fig. 2.18. As seen in the figure, an NCA-type calculation using the Coqblin–Schrieffer model reproduces the experiment fairly well, the results of which are indicated by the solid lines [60]. In CeB$_6$ with the larger CEF splitting, on the other hand, the relaxation rate is closer to the linear behaviour.

The $T^{1/2}$-like behaviour of Γ seen in CeCu$_2$Si$_2$ and Ce$_{0.7}$La$_{0.3}$Al$_2$ is due to a combination of the CEF and the Kondo effects. As we have shown in § 2.1.5, the Kondo temperature is influenced by the presence of CEF splittings. If the temperature T is higher than the overall splitting, the effective Kondo temperature is increased to the value without the splitting. Thus, the relaxation rate depends on T as if T_K varies when T is comparable to the CEF splittings. In the case of CeB$_6$, the large CEF splitting ($\sim$500 K) does not produce this behaviour.

2.4.3 *Non-Fermi-liquid behaviour*

In contrast to the 4f electron systems, the spatial extent of the 5f electron states in the uranium-based compounds is larger than that of the 4f electrons and the hybridization with the valence conduction electrons is stronger. Then the valency of a U ion in a crystal can be either U^{3+}(5f^3), U^{4+}(5f^2), or U^{5+}(5f^1). It was predicted that the interaction

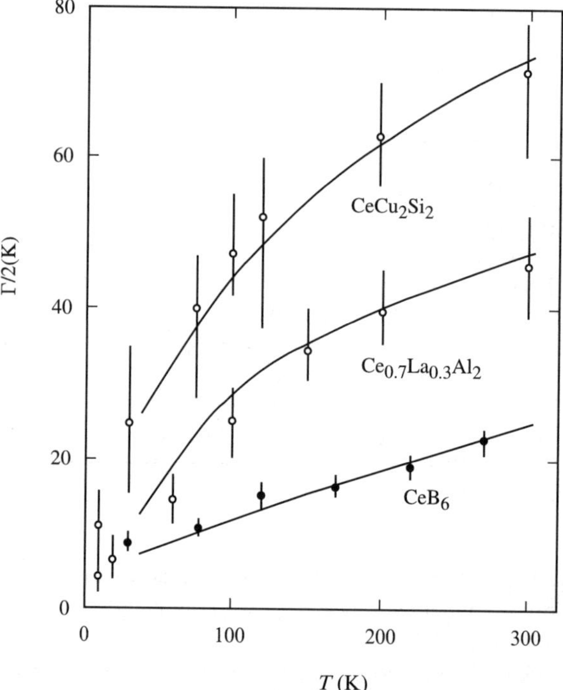

FIG. 2.18. Comparison between theoretical calculation (solid lines) and experimental data of $\Gamma(T)/2$ in tetragonal CeCu$_2$Si$_2$, cubic Ce$_{0.7}$La$_{0.3}$Al$_2$, and CeB$_6$ [60].

between the non-Kramers U^{4+} ion and the conduction electrons leads to non-Fermi-liquid behaviour, which is characterized by logarithmic or power-law divergence of the physical quantities at low temperatures. The two-channel Kondo effect ($n = 2, S = 1/2$) discussed in § 2.3 is invoked with screening channels induced by virtual transitions between the lowest CEF doublet of the 5f^2 configuration and an excited CEF doublet of the 5f^1 configuration [41,44]. A possible experimental signature for the non-Fermi-liquid behaviour was observed in dilute U systems U$_x$Th$_{1-x}$Ru$_2$Si$_2$ [61] and U$_{1-x}$Y$_x$Pd$_3$ [62]. As shown in Fig. 2.19, the magnetic susceptibility $\chi_{\text{imp}}(T)$ due to U impurity follows the log T dependence for over two orders of T below 10 K. The reason for this behaviour is not yet understood. We describe some models proposed so far, although none of them are completely successful.

In the two-channel screening model, the T^{-1} divergence of the paramagnetic susceptibility for the non-Kramers doublet is marginally suppressed. The numerical solution predicted a simple logarithmic function of the susceptibility with $\chi(T) = a/T_K \ln(T/bT_K)$ for $T \ll T_K$ where the constants a and b are given as $\sim$0.05 and 2.2 [63], respectively. The best fit to the experiment was obtained with T_K=11.1 K. This impurity model is convenient for explaining the fact that $\chi(T)$ is almost proportional to the U concentration. However, the residual entropy, which should be released by the application of magnetic field, has not been observed.

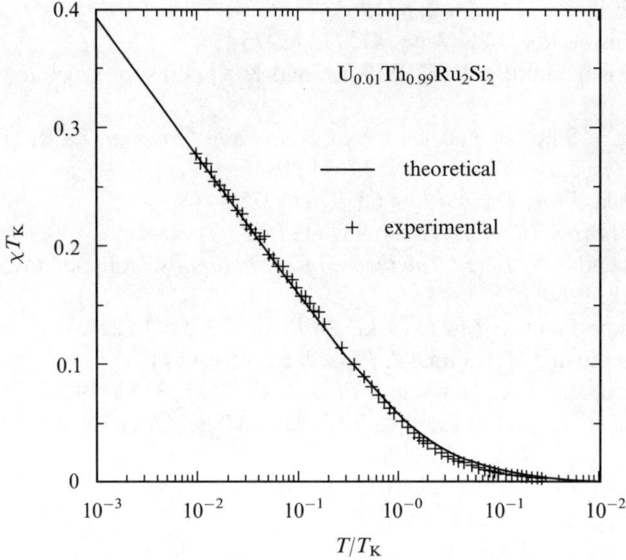

FIG. 2.19. Comparison between the U contribution to χ_{imp} in $U_{0.01}Th_{0.99}Ru_2Si_2$ and the numerical calculation based on the $S = 1/2$ two-channel Kondo model (solid line). Parameters used for fitting are $\mu = 1.7\mu_{\mathrm{B}}$ and $T_{\mathrm{K}} = 11.1$ K [61,63].

Alternative explanations for the non-Fermi-liquid behaviour are provided based on the distribution of the Kondo temperatures [64], and a spin-glass transition very near the zero temperature [65].

To conclude Chapter 2, we note that the spin dynamics probed by NMR and neutron scattering techniques can be semiquantitatively interpreted within impurity models such as the degenerate Anderson and the Coqblin–Schrieffer models in so far as the high-temperature behaviour is concerned. In the following chapters, we shall see how the strong-interaction effects among the electrons lead to richer and more fascinating behaviour at low temperatures.

Bibliography

[1] P. W. Anderson, *Phys. Rev.* **124**, 41 (1961).

[2] J. Kondo, *Solid state physics* Vol. 23 (Academic Press, New York, 1969), p. 183.

[3] P. Fulde, J. Keller, and G. Zwicknagl, *Solid state physics* Vol. 41 (Academic Press, New York, 1988), p. 2.

[4] A. C. Hewson, *The Kondo problem to heavy fermions* (Cambridge University Press, Cambridge, 1993).

[5] N. Andrei, K. Furuya, and J. H. Lowenstein, *Rev. Mod. Phys.* **55**, 331 (1983).

[6] A. M. Tsvelick and P. B. Wiegmann, *Adv. Phys.* **32**, 453 (1983).

[7] B. Coqblin and J. R. Schrieffer, *Phys. Rev.* **185**, 847 (1969).

[8] P. W. Anderson, *J. Phys. C* **3**, 2439 (1970).

[9] A. A. Abrikosov, *Physics* **2**, 5 (1965).

[10] K. G. Wilson, *Rev. Mod. Phys.* **47**, 773 (1975).

[11] H. R. Krishnamurthy, J. W. Wilkins, and K. G. Wilson, *Phys. Rev. B* **21**, 1003 (1980).

[12] O. Sakai, Y. Shimizu, and T. Kasuya, *Solid State Commun.* **75**, 81 (1990).

[13] P. Nozieres, *J. Low Temp. Phys.* **17**, 31 (1975).

[14] K. Yamada, *Prog. Theor. Phys.* **53**, 970 (1975).

[15] H. Shiba, *Prog. Theor. Phys.* **54**, 967 (1975).

[16] D. Pines and P. Nozieres, *The theory of Fermi liquids:* Volume I (W.A. Benjamin, New York, 1966).

[17] A. Yoshimori and A. Zawadowski, *J. Phys. C* **15**, 5241 (1975).

[18] Y. Kuramoto and H. Kojima, *Z. Phys. B* **57**, 95 (1984).

[19] J. W. Rasul and A. C. Hewson, *J. Phys. C* **17**, 2555, 3332 (1984).

[20] K. Hanzawa, K. Yamada, and K. Yoshida, *J. Magn. Magn. Mater.* **47**, 357 (1985).

[21] S. E. Barnes, *J. Phys. F* **6**, 1375 (1976).

[22] P. Coleman, *Phys. Rev. B* **29**, 3035 (1984).

[23] H. Keiter, *Z. Phys. B* **214**, 466 (1968).

[24] N. Read and D. M. Newns, *J. Phys. C* **16**, 3273 (1983).

[25] N. E. Bickers, *Rev. Mod. Phys.* **59**, 845 (1987).

[26] R. P. Feynman, *Statistical mechanics* (W.A. Benjamin, Reading, U.S.A., 1972), p. 67.

[27] O. Gunnarsson and K. Schönhammer, *Phys. Rev. B* **28**, 4315 (1983).

[28] O. Gunnarsson and K. Schönhammer, *Handbook on the physics and chemistry of rare earths* (Eds, K.A. Gschneider *et al.*, North-Holland, Amsterdam, 1987) Vol. 10, p. 103.

[29] H. Keiter and J. C. Kimball, *Int. J. Magn.* **1**, 233 (1971).

[30] Y. Kuramoto, *Z. Phys. B* **53**, 37 (1983).

[31] S. Inagaki, *Prog. Theor. Phys.* **62**, 1441 (1979).

[32] Y. Kuramoto, *J. Magn. Magn. Mater.* **31–34**, 463 (1983).

[33] F. C. Zhang and T. K. Lee, *Phys. Rev. B* **30**, 1556 (1983).

[34] N. Grewe, *Z. Phys. B* **53**, 271 (1983).

[35] Y. Kuramoto, *Z. Phys. B* **37**, 299 (1980).

[36] H. Kojima, Y. Kuramoto, and H. Tachiki, *Z. Phys. B* **54**, 293 (1984).

[37] N. E. Bickers, D. L. Cox and J. W. Wilkins, *Phys. Rev. Lett.* **54**, 230 (1987).

[38] Y. Kuramoto and E. Müeller-Hartmann, *J. Magn. Magn. Mater.* **52**, 122 (1985).

[39] P. Nozieres and A. Blandin, *J. Physique* **41**, 193 (1980).

[40] I. Affleck, *Acta Physica Polonica* **26**, 1826 (1995).

[41] D. L. Cox and A. Zawadowski, *Adv. Phys.* **47**, 599 (1998).

[42] A. M. Tsvelick and P. B. Wiegmann, *Z. Phys. B* **54**, 201 (1984); *J. Stat. Phys.* **38**, 125 (1985).

[43] N. Andrei and C. Destri, *Phys. Rev. Lett.* **52**, 364 (1984).

[44] D.L. Cox, *Phys. Rev. Lett.* **59**, 1240 (1987).

[45] O. Sakai, S. Suzuki, and Y. Shimizu, *Solid State Commun.* **101**, 791 (1997).

[46] T. Kasuya, *Magnetism* Vol. IIB (Eds, G.T. Rado and H. Suhl, Academic Press, New York, 1965).

[47] B. A. Jones, C. M. Varma, and J. W. Wilkins, *Phys. Rev. B* **61**, 125 (1988).

[48] O. Sakai and Y. Shimizu, *J. Phys. Soc. Jpn.* **61**, 2333, 2348 (1992).

[49] R. M. Fye and J. E. Hirsch, *Phys. Rev. B* **40**, 4780 (1989).

[50] K. W. H. Stevens, *Proc. Phys. Soc. A* **65**, 209 (1952).

[51] N. T. Hutchings, *Solid State Phys.* **16**, 227 (1964).

[52] T. Shimizu, M. Takigawa, H. Yasuoka, and J. H. Wernick, *J. Magn. Magn. Mater.* **52**, 187 (1985).

[53] E. Holland-Moritz, M. Loewenhaupt, and W. Schmatz, *Phys. Rev. Lett.* **38**, 983 (1977).

[54] E. Holland-Moritz and G. H. Lander, *Handbook on the physics and chemistry of rare earths*, Vol. 19 (Eds, K.A. Gschneider, L. Eyring, G.H. Lander, and G.R. Choppin, North-Holland, Amsterdam, 1993), Ch. 130.

[55] Y. Kitaoka, H. Arimoto, Y. Kohori, and K. Asayama, *J. Phys. Soc. Jpn.* **54**, 3236 (1985).

[56] J. Rossat-Mignod, L. P. Regnault, J. L. Jacoud, C. Vettier, P. Lejay, J. Flouquet, E. Walker, D. Jaccard, and A. Amato, *J. Magn. Magn. Mater.* **76 & 77**, 376 (1988).

[57] J. L. Jacoud, L. P. Regnault, J. Rossat-Mignod, C. Vettier, P. Lejay, and J. Flouquet, *Physica B* **156 & 157**, 818 (1989).

[58] A. Loidl, G. Knopp, H. Spille, F. Steglich, and A. P. Murani, *Physica B* **156 & 157**, 794 (1989).

[59] S. Horn, E. Holland-Moritz, M. Loewenhaupt, F. Steglich, H. Scheuer, A. Benoit, and J. Flouquet, *Phys. Rev. B* **23**, 3171 (1981).

[60] S. Maekawa, S. Kashiba, S. Takahashi, and M. Tachiki, *J. Magn. Magn. Mater.* **52**, 149 (1985) .

[61] H. Amitsuka and T. Sakakibara, *J. Phys. Soc. Jpn.* **63**, 736 (1994).

[62] M. B. Maple *et al.*, *J. Phys. Condens. Matter* **8**, 9773 (1996).

[63] P. D. Sacramento and P. Schlottmann, *Phys. Rev. B* **43**, 13 294 (1991).

[64] E. Miranda, V. Dobrosavljevic, and G. Kotliar, *J. Phys.: Condens. Matter* **8**, 9871 (1996); *Phys. Rev. Lett.* **78**, 290 (1997).

[65] A. M. Sengupta and A. Georges, *Phys. Rev. B* **52**, 10 295 (1995).

3

METALLIC AND INSULATING PHASES OF HEAVY ELECTRONS

In the previous chapters, we have seen that the magnetic and thermal properties of heavy electrons are determined basically by strong local correlations. A signature of the local nature is that the susceptibility and specific heat are approximately proportional to the concentration of Ce in systems such as $Ce_xLa_{1-x}Cu_6$ or $Ce_xLa_{1-x}B_6$. The RKKY interaction is operative in bringing about magnetic ordering. Its energy scale is given by $J_{RKKY} \sim J^2\rho_c$, where J is the exchange interaction between f and conduction electrons and ρ_c is the density of conduction band states per site. The RKKY interaction competes with the Kondo effect which points to the nonmagnetic singlet ground state. There exists a rather detailed balance between these two tendencies. It is expected that the magnetic transition temperature T_M, if it exists, is lowered due to the Kondo effect. In the opposite case, $T_K \gg J^2\rho_c$, no magnetic order should occur.

Even in the singlet ground state, the intersite interaction should influence the residual interactions among the renormalized heavy quasi-particles. It is this residual interaction that leads to a rich variety of phenomena in the ground states, such as heavy-electron superconductivity, heavy-electron band magnetism, etc. Short-range magnetic correlation might survive even if the dominating Kondo effect prevents the development of long-range order. We have already seen in Fig. 1.3 that the electrical resistivity reveals growing importance of coherence upon increasing the concentration of rare-earth ions at low temperatures.

Direct evidence for heavy quasi-particles comes from the de Haas–van Alphen (dHvA) effect, which is the oscillation of the magnetic susceptibility arising from the Landau quantization of electron orbits [1]. The geometry of the Fermi surface can be obtained from the period of oscillation in the differential magnetic susceptibility as a function of $1/H$, H being the magnetic field. The effective mass can be deduced from the temperature dependence of the oscillation amplitude. In order to observe these oscillations, the mean free path l of a heavy quasi-particle must be larger than the cyclotron radius: $l \gg v_F/\omega_H$ where v_F is the Fermi velocity, and $\omega_H = eH/m^*c$ is the effective cyclotron frequency of a heavy quasi-particle with effective mass m^*. Furthermore, thermal smearing at the Fermi surface must be sufficiently small: $T \ll \omega_H$. With large effective masses and short mean free paths, these conditions require that the experiments should be performed at low temperatures and in high field. Measurements of dHvA oscillations have been made on UPt_3 [2,3], $CeRu_2Si_2$ [3,4], and $CeCu_6$ [5]. In all these cases, heavy effective masses have been observed.

For UPt_3, the Fermi surface obtained by the dHvA effect is in good agreement with that obtained from the energy-band theory [6]. The effective mass on some sheets of the

Fermi surface, however, is about 10–30 times larger than that deduced from the band-structure calculations. This shows that there are significant many-body effects which are not described by the standard band theory.

In the present chapter, we shall review the paramagnetic heavy-electron states which have both metallic and insulating phases. The ordered states such as the magnetic and superconducting phases will be discussed in Chapters 4 and 5, respectively. In the following, we focus on the growth of intersite coherence, or the itinerant character of f electrons, mainly in dynamic magnetic properties. As specific examples for subtle role of the intersite correlation in the heavy-electron state, we take typical systems such as $CeCu_6$ and $CeRu_2Si_2$ where the Fermi liquid remains stable at low temperatures even with significant magnetic correlations. We then discuss UBe_{13} and UPt_3 mainly in their paramagnetic phase to contrast with the Ce systems.

3.1 Formation of heavy-electron metals with magnetic correlation

3.1.1 *Metamagnetic behaviour*

The electronic specific heat coefficient of $CeRu_2Si_2$ is $\gamma = 0.35$ J/(mol K^2), which is much smaller than 1.6 J/(mol K^2) for $CeCu_6$, and a little smaller than 0.4 J/(mol K^2) for UPt_3. A remarkable feature of this compound is the pseudo-metamagnetic transition. This transition has been observed for a magnetic field $H_c = 8$ T applied along the tetragonal c-axis in $CeRu_2Si_2$, as shown in Fig. 3.1 [7]. A similar pseudo-metamagnetic transition has also been reported for UPt_3 at much higher fields of 20 T [8]. A metamagnetic behaviour has also been observed in $CeCu_6$ at 2 T. In this case the metamagnetism is faint and is seen as a peak only in $\partial^2 M/\partial H^2$.

The decay of the heavy-electron state through H_c was found by the dHvA effect [9]. The effective masses up to $120m_0$, which are observed below H_c, decrease significantly around H_c and continue to decrease with increasing field. It has been proposed that the system above H_c may be described in terms of the localized f electron model [9]. A basis of the proposal is that the Fermi surfaces measured above H_c are similar to those with localized f electrons calculated by the band theory [10]. On the other hand, in dHvA experiment for UPt_3, the behaviour above the pseudo-metamagnetic transition does not suggest a localization of f electrons [3]. We note that if two 5f electrons per U^{4+} ion are localized, the size of the Fermi surface can remain the same as that of the itinerant 5f electrons. This aspect will be discussed in § 3.3 .

We mention that the metamagnetic behaviour accompanies a large elastic anomaly. The elastic constants show pronounced softening near the critical magnetic field H_c [11] and the magnetostriction is also anomalous [12].

3.1.2 *NMR on $CeCu_6$ and $CeRu_2Si_2$*

Figure 3.2 shows the temperature dependence of the nuclear spin–lattice relaxation rate $1/T_1$ for $CeCu_6$ and $CeRu_2Si_2$ [13]. In the latter case, $1/T_{1\perp}$ of ^{29}Si is measured under the condition that the c-axis is aligned along the magnetic field. For $CeCu_6$, $1/T_1$ of ^{63}Cu was measured in zero field by Cu NQR. As seen in the figure, $1/T_1$ is almost independent of the temperature above 6 K for $CeCu_6$ and 12 K for $CeRu_2Si_2$. Then it begins to decrease gradually above 50 K for $CeRu_2Si_2$. This relaxation behaviour shares

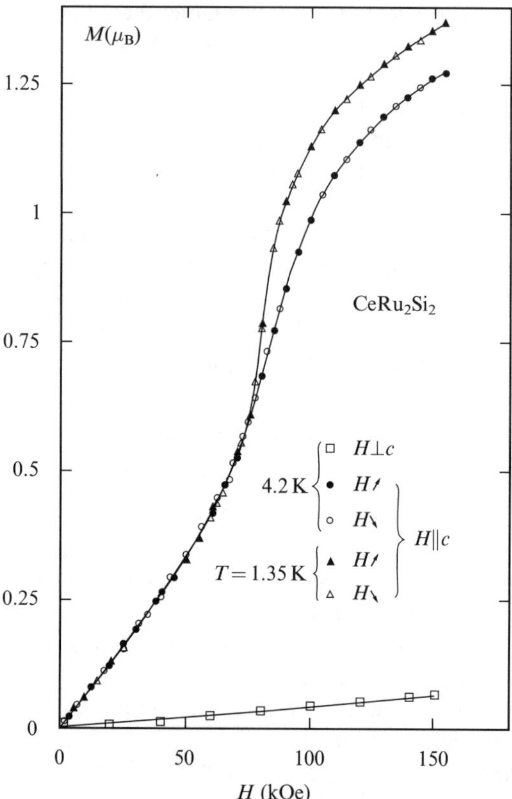

FIG. 3.1. Magnetization curves for CeRu$_2$Si$_2$ up to 150 kOe at 4.2 and 1.35 K [7].

a common feature with other heavy-electron systems at high temperatures. As discussed in the previous chapter, the spin correlation between different sites can be ignored at high temperatures, and $1/T_1$ is described by the local spin susceptibility $\chi(T)$ and the magnetic relaxation rate Γ.

The temperature below which $1/T_1$ begins to depend on T is close to the Kondo temperature T_K, which is extracted from the analysis of the resistivity and the magnetic specific heat. With further decrease of temperature, $1/T_1$ follows the behaviour $T_1 T =$ constant, which is called the Korringa law. To see this clearly, Fig. 3.3 shows $(T_1 T)^{-1}$ vs T. It turns out that the Korringa law is valid below 0.2 K for CeCu$_6$ and 8 K for CeRu$_2$Si$_2$. The characteristic temperature in each case is sometimes called the coherence temperature T^*. As discussed in Chapter 1, the Fermi-liquid ground state leads to the Korringa law at low T. This is independent of whether the system is homogeneous or a local Fermi liquid. In the temperature range where the local Fermi liquid description is valid, the Korringa relation given by eqn (1.140) should hold.

In contrast to the expectation based on the single-ion model of spin fluctuations, the ratio of T^*/T_K is quite different in CeCu$_6$ from that in CeRu$_2$Si$_2$;

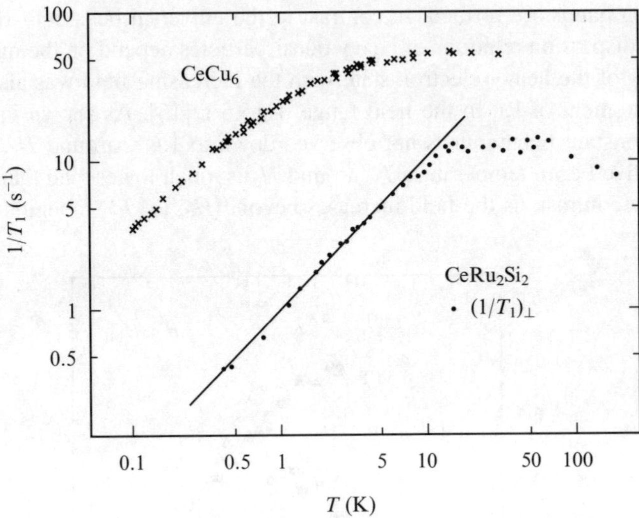

FIG. 3.2. Temperature dependence of $(1/T_1)$ of ^{63}Cu in $CeCu_6(\times)$ and ^{29}Si in $CeRu_2Si_2(\bullet)$ [13].

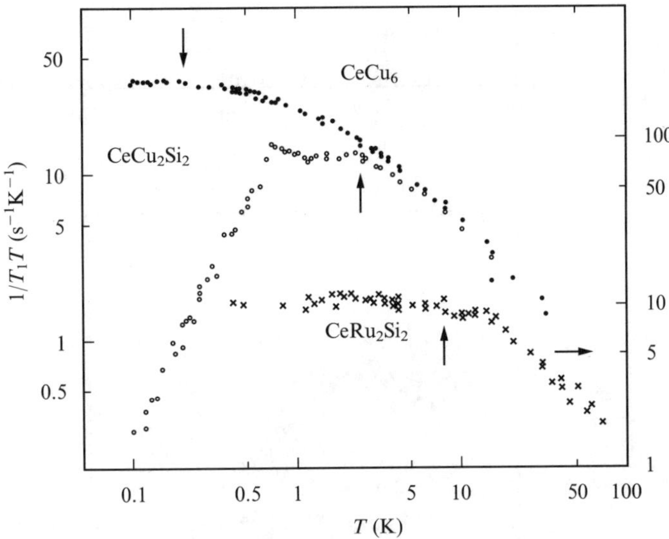

FIG. 3.3. Temperature dependence of $1/(T_1T)$ of ^{63}Cu in $CeCu_6(\bullet)$ and $CeCu_2Si_2(\circ)$ and ^{29}Si in $CeRu_2Si_2(\times)$ [13].

$T^*/T_K = 0.2/6 = 0.03$ for $CeCu_6$ and $8/12 = 0.67$ for $CeRu_2Si_2$. Furthermore, T^* decreases to 0.08 K in $Ce_{0.75}La_{0.25}Cu_6$ [14]. These results show that the energy scale T_K is not sufficient to describe the heavy-electron systems. It should also be specified by the coherence temperature, or the effective Fermi temperature T^* below which the heavy-electron bands are formed. In contrast to the universal behaviour of the Kondo impurity, the dispersion relations of heavy quasi-particles depend on the material.

The decay of the heavy-electron state with the increasing field was also probed by the T_1 measurement of Ru in the field range 0–15.5 T [15]. As shown in Fig. 3.4(a), the $T_1 T = $ constant behaviour is not observed down to 1.6 K around H_c. This means that the effective Fermi temperature T^* around H_c is much lower than that at $H < H_c$; $T^* \sim 8$ K. By contrast, as the field increases beyond H_c, $(T_1 T)^{-1}$ begins to decrease,

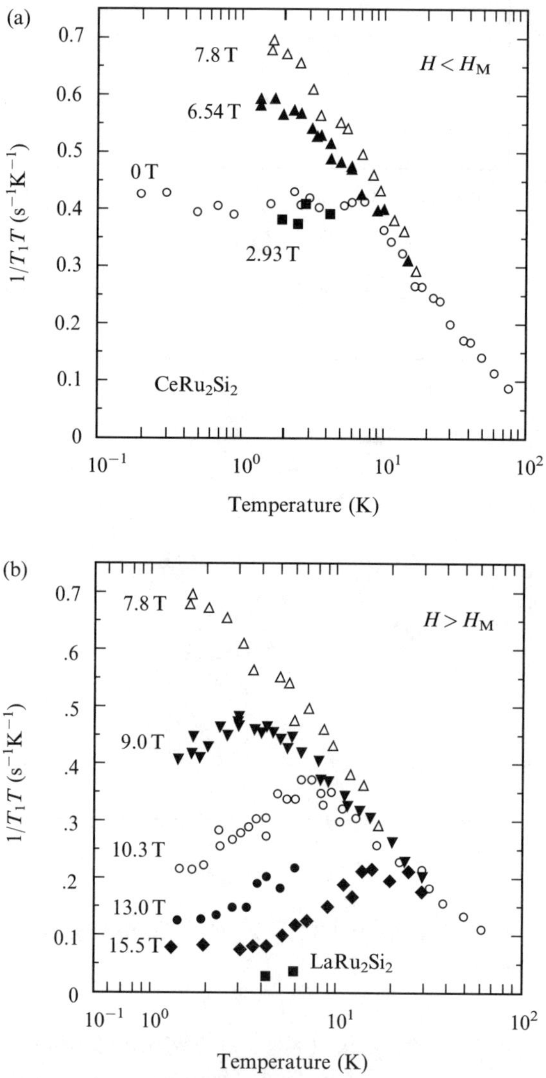

FIG. 3.4. Temperature dependence of $(T_1 T)^{-1}$ of Ru in CeRu$_2$Si$_2$ at various fields H: (a) $H \leq H_{\mathrm{M}} = 7.8$ T; (b) $H \geq H_{\mathrm{M}}$ [15].

as indicated in Fig. 3.4(b). Then, the $T_1 T = $ constant behaviour is observed only below
$\sim$4 K at 15.5 T.

3.1.3   *Neutron scattering on CeCu₆ and CeRu₂Si₂*

Neutron scattering can extract information on the single-site fluctuations and intersite
magnetic correlations separately by scanning the wavenumber and energy. Inelastic neutron scattering experiments have been performed on single crystals of $CeRu_2Si_2$ and
$CeCu_6$. At high temperatures, the magnetic scattering can be described by a single
quasi-elastic Lorentzian peak corresponding to eqn (2.128). With decreasing temperatures, antiferromagnetic and incommensurate magnetic correlations develop below 70 K
in $CeRu_2Si_2$ and 10 K in $CeCu_6$.

An inelastic scattering study with applied magnetic field has shown that the pseudo-
metamagnetic transition corresponds to the collapse of the incommensurate magnetic
correlation. As shown in Fig. 3.5 [16], magnetic scattering at incommensurate wave
vector $\boldsymbol{Q} = (0.7, 0.7, 0)$ and energy transfer $\hbar\omega = 1.6\,\mathrm{meV}$ decreases slightly for H
up to 70 kOe, but more rapidly at higher field. The inflection point for a decrease of
the neutron intensity corresponds to the threshold field $H_c = 83\,\mathrm{kOe}$ at 1.4 K, in good
agreement with the magnetization experiment (see Fig. 3.1). At $H = 98\,\mathrm{kOe}$, it is found

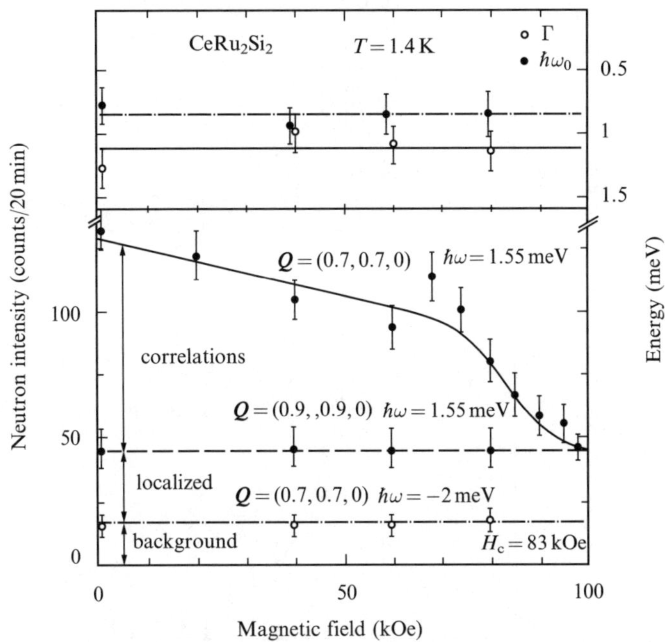

FIG. 3.5. Scattered neutron intensities at 1.4 K as a function of the magnetic field
applied along the [001] direction for $CeRu_2Si_2$; they correspond to the peak
maximum at $\boldsymbol{Q} = (0.7, 0.7, 0)$ with $\hbar\omega = 1.6\,\mathrm{meV}$, intensity in between the
peaks at $\boldsymbol{Q} = (0.9, 0.9, 0)$ with $\hbar\omega = 1.6\,\mathrm{meV}$, and the background intensity at
$\boldsymbol{Q} = (0.7, 0.7, 0)$ with $\hbar\omega = -2\,\mathrm{meV}$ [16].

that the magnetic scattering does not exhibit any q dependence, and the spectrum is quite similar to that observed at $Q = (0.9, 0.9, 0)$ which is within the intensity maximum. The latter spectrum represents the single-site contribution in zero field (see Fig. 3.7). Namely, a single-site contribution of the quasi-elastic type is not affected by a magnetic field of 98 kOe, which exceeds H_c of the pseudo-metamagnetic transition. On the other hand, the intersite contribution of inelastic type is completely suppressed, as seen clearly in the energy scans in Fig. 3.6.

A similar feature was also observed in CeCu$_6$ at $H_c = 25$ kOe, although only a small anomaly appears in dM/dH [17]. This is probably because the weight of the intersite contribution relative to the single-site contribution is much smaller in CeCu$_6$ than in CeRu$_2$Si$_2$. In both cases, the Zeeman energy $\mu_B H_c$ corresponding to the critical field H_c is of the order of $\hbar\omega_0$, which is the peak energy of the inelastic neutron scattering spectrum. The finite energy $\hbar\omega_0$ suggests the presence of a pseudo-gap in the magnetic excitation spectrum. In this case, the correlation length saturates naturally when the temperature or the Zeeman energy is of the order of the pseudo-gap.

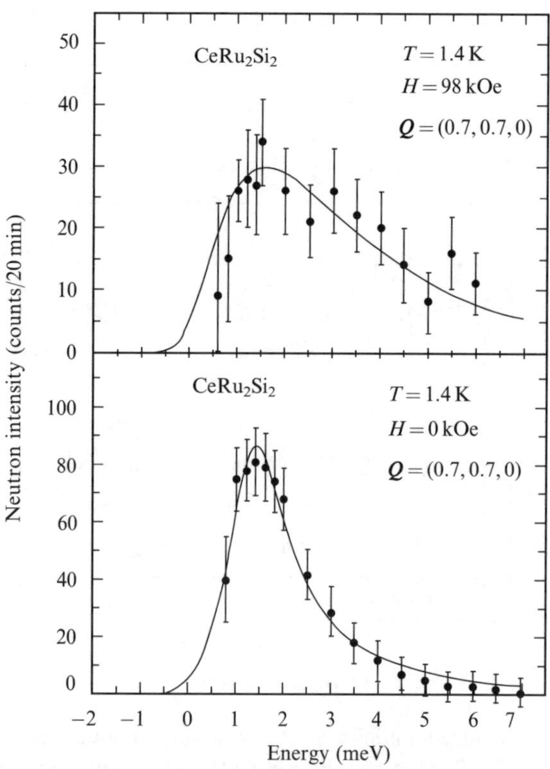

Fig. 3.6. Typical fits of energy scans at $Q = (0.7, 0.7, 0)$ and $T = 1.4$ K showing the quasi-elastic contribution ($H = 98$ kOe) and the intersite inelastic contribution ($H = 0$) for CeRu$_2$Si$_2$ [16].

Figure 3.7 shows the typical q-scan spectra for CeRu$_2$Si$_2$ obtained at 4.2 K with an energy transfer $\hbar\omega = 1.6$ meV [16]. In this subsection, we take the unit of the wavenumber as π/l, with l the lattice constant along each principal direction. The peaks of intensity at incommensurate wave vectors $\mathbf{k}_1 = (0.3, 0, 0)$ and $\mathbf{k}_2 = (0.3, 0.3, 0)$ indicate the presence of competing in-plane couplings between first and second nearest neighbours. In addition, there is a significant q-independent contribution. In these two compounds, both the single-site and the intersite magnetic correlations seem to coexist at low temperatures. The former represents about 60% and 90% of the weight of the signal integrated over the whole q-space for CeRu$_2$Si$_2$ and CeCu$_6$, respectively. As shown in Fig. 3.8 for CeCu$_6$, the magnetic scattering is independent of T below T^*. This indicates a saturation of the in-plane correlation length. From the energy scans performed for various scattering vectors $\mathbf{Q}$, the single- and intersite contributions are analysed separately. The single-site contribution can be described by a quasi-elastic

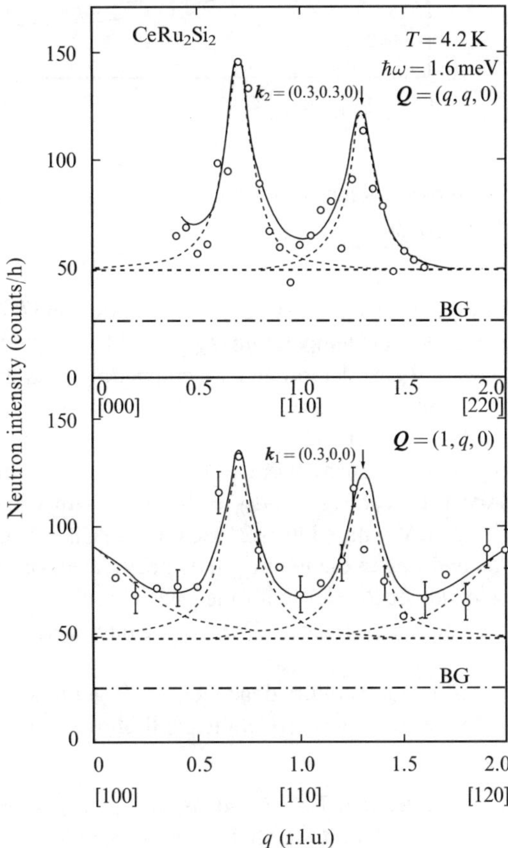

FIG. 3.7. q-scans at a finite energy transfer $\hbar\omega = 1.6$ meV along the directions [1̄10] and [010] at $T = 4.2$ K for CeRu$_2$Si$_2$, showing the two incommensurate wave vectors $\mathbf{k}_1 = (0.3, 0, 0)$ and $\mathbf{k}_2 = (0.3, 0.3, 0)$ [16].

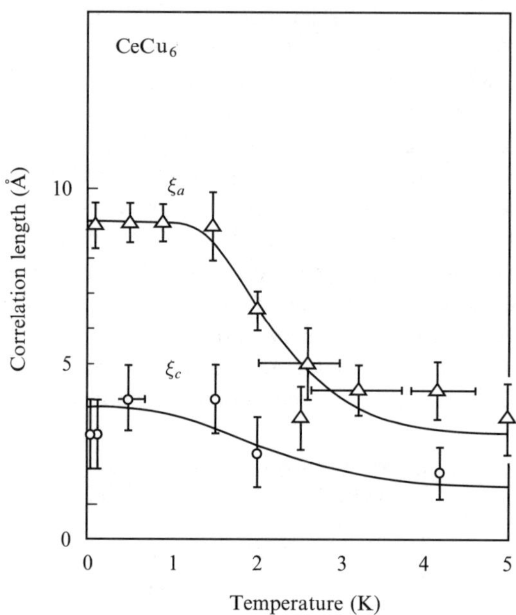

FIG. 3.8. Magnetic correlation lengths along the a- and c-directions as a function of temperature for $CeCu_6$ [16].

Lorentzian, the width $\Gamma_{s\text{-}s}$ of which is 2.0 meV for $CeRu_2Si_2$ and 0.42 meV for $CeCu_6$. These values lead to the Kondo temperature $T_K = 23$ K for $CeRu_2Si_2$ and 5 K for $CeCu_6$. The magnitudes of T_K so determined are compatible with those determined by the coefficient γ of the linear specific heat. For γT_K, both compounds possess comparable values, $\gamma T_K \sim 8000$ mJ/(mol K). This suggests that the single-site fluctuations give the main contribution to γ. On the other hand, the intersite magnetic fluctuations give rise to a broad inelastic spectrum centred at $\hbar\omega_0 = 1.2$ meV with width $\Gamma_{i\text{-}s} = 0.9$ meV for $CeRu_2Si_2$, and at 0.2 meV with width 0.21 meV for $CeCu_6$. Apparently, there exist two energy scales $\Gamma_{s\text{-}s}$ and $\Gamma_{i\text{-}s}$ in the nonmagnetic heavy-electron state. Of these, $\Gamma_{s\text{-}s}$ is related to the single-site fluctuation with the energy scale comparable to T_K, and $\hbar\omega_0 \sim \Gamma_{i\text{-}s}$ reflects the intersite interaction. Without the intersite interaction, $\hbar\omega_0$ should tend to zero.

Figure 3.9 represents the temperature dependence of the linewidth $\Gamma_{s\text{-}s}$ and $\Gamma_{i\text{-}s}$ for $CeCu_6$. Together with the result of the correlation length shown in Fig. 3.8, three regimes are identified:

1. The low-temperature regime, $T < T_1$, where T_1 is 1–1.5 K for $CeCu_6$ and 5–6 K for $CeRu_2Si_2$. The correlation length and the linewidths are independent of T, with the former being equal to about the distance of second nearest neighbours. The magnitude of T_1 in $CeCu_6$ is much larger than the coherent temperature $T^* \sim 0.2$ K, which is deduced from the temperature dependence of $1/T_1$. On the other hand, T_1 in $CeRu_2Si_2$ is comparable to $T^* \sim 6$ K determined by the NMR.

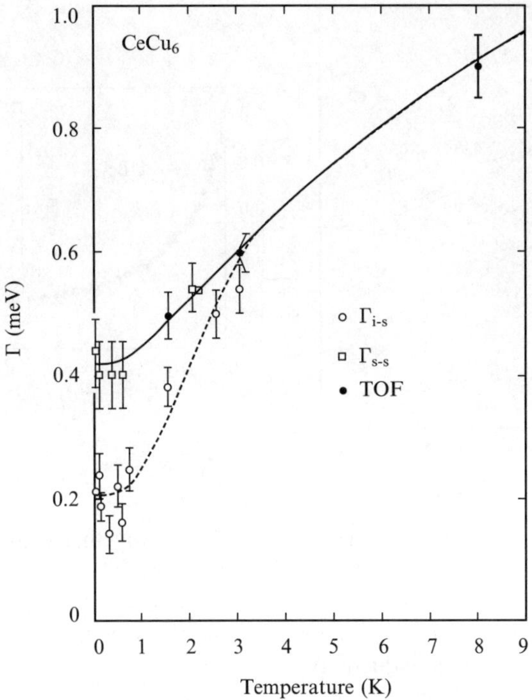

Fig. 3.9. Temperature dependence of the linewidths, deduced from energy scans for the single-site ($\Gamma_{s\text{-}s}$) and the intersite ($\Gamma_{i\text{-}s}$) contributions for CeCu$_6$ [16].

2. The intermediate regime with $T_l < T < T_m$ where $T_m = 3\,\text{K}$ for CeCu$_6$ and 10–15 K for CeRu$_2$Si$_2$. The correlation lengths decrease with T and the linewidth $\Gamma_{i\text{-}s}$ of the intersite contribution increases rapidly with T. As a result, we have $\Gamma_{i\text{-}s} \simeq \Gamma_{s\text{-}s}$ above T_m. The Fermi-liquid description breaks down in this temperature range.

3. The high-temperature regime, $T > T_m$, where the intersite magnetic correlations collapse and the single-site fluctuations prevail. The system in this range appears to be very similar to the dilute Kondo system.

3.1.4 *UBe$_{13}$ and UPt$_3$*

UBe$_{13}$, which undergoes a superconducting transition at $T_c = 0.9\,\text{K}$, has a cubic crystal structure with eight formula units per unit cell. The U–U distance equals 5.13 Å, which is much too large for a direct overlap of 5f wave functions from the adjacent U atoms. The macroscopic physical properties of UBe$_{13}$ are quite unusual. Figure 3.10 shows [18] the electrical resistivity together with that of other systems, scaled to the same value at room temperature, and specific heat. The resistivity increases with temperature decreasing from the room temperature down to 2.4 K where a well-defined peak appears. The resistivity just above T_c is larger than $100\,\mu\Omega$ cm. The magnetic susceptibility χ at high

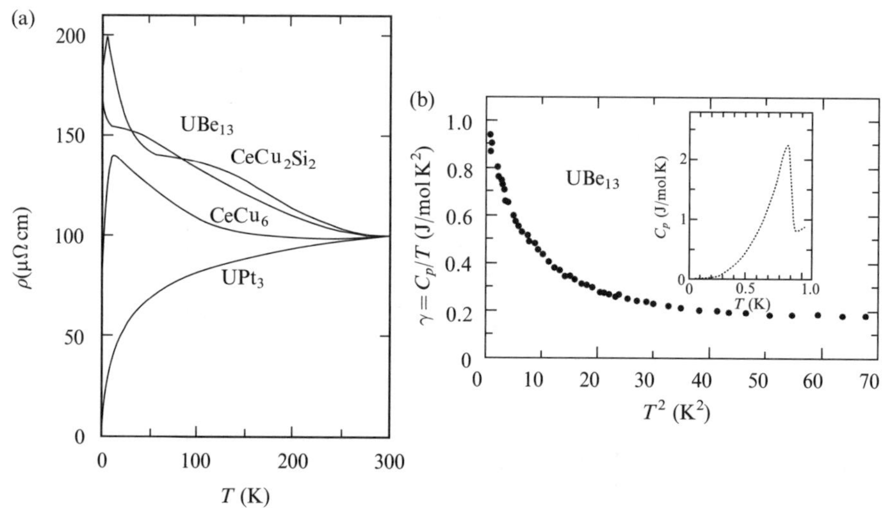

FIG. 3.10. (a) The resistivity vs temperature for UBe_{13} together with that of other compounds. Note that resistivities at room temperature have been normalized to the same value as that of UBe_{13} [18]. (b) The specific heat C_p of UBe_{13} divided by temperature in the normal state [19]. The inset shows the specific heat near the transition temperature T_c of superconductivity.

temperatures is large and obeys the Curie–Weiss law with $\mu_{eff} = 3.1\mu_B$. With decreasing T, χ tends to a large value. The specific heat coefficient γ increases upon cooling below 10 K, as shown in Fig. 3.10(b). It is only below ~ 1.5 K that $T_1 T$ becomes almost constant [20]. The large anomaly of specific heat at T_c indicates that the Cooper pairs consist of heavy electrons. We discuss the anomalous superconducting properties of UBe_{13} in Chapter 5.

The properties of UPt_3 are in contrast with those of UBe_{13} in a number of ways. The crystal has a hexagonal structure with the U–U distance 4.1 Å. The electrical resistivity is large ($\sim 100\,\mu\Omega$) at room temperature, but decreases upon cooling, as in ordinary metals. At high temperatures, the Curie–Weiss law with $\mu_{eff} = 2.6\mu_B$ is observed with large anisotropy in χ. Namely, with H in the basal plane, at low temperatures χ is twice that with H along the c-axis. Although there is a peak in χ at 14 K, no specific heat anomaly is observed.

At temperatures lower than 14 K, the neutron scattering probed some unusual magnetic correlations. There are three different modes of the magnetic correlations with the propagation vectors $Q = (0.5, 0, 1)$, $(0, 0, 0)$, and $(0, 0, 1)$ in units of the primitive reciprocal lattice vectors. The $(0, 0, 1)$ mode begins to develop below ~ 14 K [21], and is observed at high-energy transfers, more than 5 meV. The mode has an out-of-plane antiferromagnetic polarization. At low energies there is another $(0, 0, 0)$ mode which is interpreted in terms of paramagnons, or particle–hole excitations of heavy quasi-particles [22]. This 'slow' quasi-particle component of the spin relaxation process is similar to what has been identified in other systems such as liquid ^{3}He [23], paramagnetic 3d transition metals with strongly enhanced susceptibilities and weak ferromagnetic 3d metals with small spin polarizations. These behaviours in the normal state are consistent with the itinerant model of f electrons. In UPt$_3$, however, there is also a 'fast' contribution with a weakly q-dependent characteristic frequency. The 'slow' component in Im $\chi(q, \omega)$ accounts for about 20% of the total static susceptibility [22], according to the Kramers–Kronig relation. This contrasts with the result in 3d transition metals, where a well-defined quasi-particle contribution to Im $\chi(q, \omega)$ accounts for the total static susceptibility in the limit $q \to 0$. This difference between UPt$_3$ and 3d systems may be ascribed in part to the larger strength of the spin–orbit interaction as well as the stronger tendency to localization in 5f systems. The paramagnon mode may mediate the triplet pairing in UPt$_3$, as discussed in Chapter 5.

We now turn to the unusual behaviour of antiferromagnetism at $Q = (0.5, 0, 1)$. It was reported, based on the neutron scattering results [24], that UPt$_3$ has antiferromagnetic order below 5 K. However, the spin correlation length remains finite, 85–500 Å down to 100 mK. No trace of the magnetic transition was reported in the specific heat [25], the static susceptibility [26], and the Knight shift [27] down to 100 mK. With further decreasing temperature, however, the neutron scattering [28] and specific heat [29] measurements suggest the onset of the long-range order below 20 mK. If this is true, the 'quasi-ordering' below 5 K is characterized by a time scale much larger than that of neutrons (typically 10^{-12} s), but much shorter than that detected by NMR (typically 10^{-6}–10^{-8} s). What is unusual is the spin correlation length of about 300 Å, which is much longer than that in typical magnetic systems above the magnetic transition. The origin of these unconventional magnetic correlations is not understood yet. The relationship of the quasi-ordering with superconductivity is discussed in Chapter 5.

3.2 Semiconducting and semimetallic phases of heavy electrons

3.2.1 SmB_6

There is a class of materials called Kondo semiconductors which are characterized by a small energy gap in the electronic excitation spectrum at low temperatures. In SmB$_6$ and YbB$_{12}$, the energy gap amounts to 50–100 K, as identified by the exponential increase of the resistivity and T_1, or by optical measurements. In contrast with the usual semiconductors, the magnetic response of the Kondo semiconductors shows an unexpected complexity. A typical example is SmB$_6$, a mixed-valence or valence fluctuating compound with an average valency of Sm$^{+2.56}$. In this system, it is established that the two configurations, $|4f^6\rangle$ and $|4f^5 5d\rangle$, of Sm are mixed in the ground state as a result of hybridization. Figure 3.11 shows a set of magnetic neutron scattering spectra obtained at 1.8 K. The Q-vectors point to different directions with respect to the crystal axes, but have the same magnitude $|Q| = 1.5\,\text{Å}^{-1}$ [30].

A sharp peak appears in the spectrum at $\hbar\omega = 14$ meV and $Q = (0.5, 0.55, 0.5)$ in units of the primary reciprocal lattice vectors. At the bottom of Fig. 3.12, the scattering intensities are plotted for different Q values (open circles in the inset). There is a substantial anisotropy that the form factor $F(Q)$ is large along the $Q = (1, 1, 1)$ direction, whereas it is markedly reduced along $Q = (0, 1, 1)$ and vanishes along $Q = (1, 0, 0)$. Furthermore, the intensity is unexpectedly reduced with increasing $|Q|$

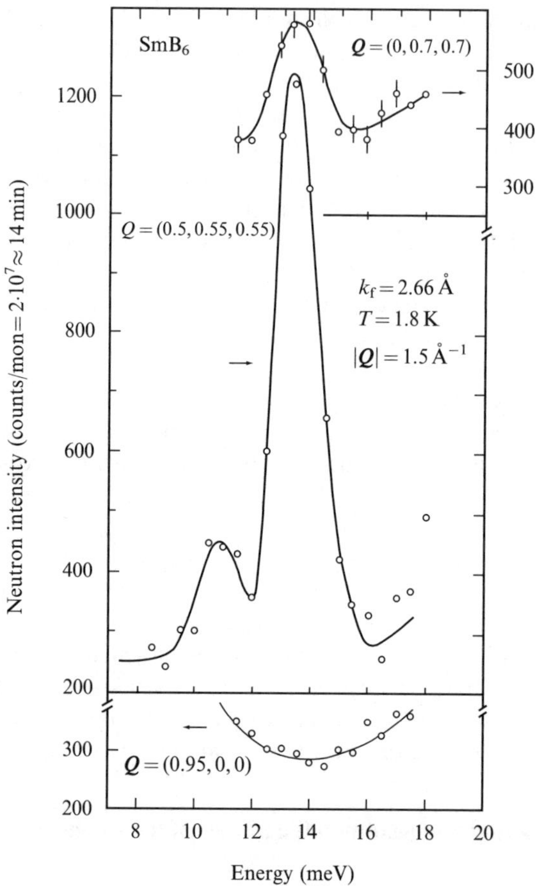

FIG. 3.11. Low-temperature ($T = 1.8$ K) energy spectra of SmB_6 for three different Q values having identical $|Q| = 1.5\,\text{Å}^{-1}$ [30]. The ordinate means the neutron counts in about 14 minutes, and the number $2 \cdot 10^7$ means a technical detail that monitor counting is 20 million per point.

along the direction $Q = (0.5, q, q)$, which is incompatible with any plausible Sm^{2+} form factor as indicated by dash–dot line in the upper part of Fig. 3.12. These features of the magnetic excitation at $\hbar\omega = 14$ meV were ascribed to the $4f^5 5d$ component of the mixed-valence state.

An excitonic bound state has been invoked to explain the spectrum [30]. According to this model, electrons in the 4f orbital are strongly coupled with holes in the six 5d orbitals of the neighbouring Sm sites. Then the ground state has bound states analogous to excitons. The inelastic feature observed at low T corresponds to a breakup of excitons. As shown in Fig. 3.13, the inelastic intensity rapidly falls off in a temperature range of around 20–40 K, which is comparable with the electronic gap Δ. This decrease may

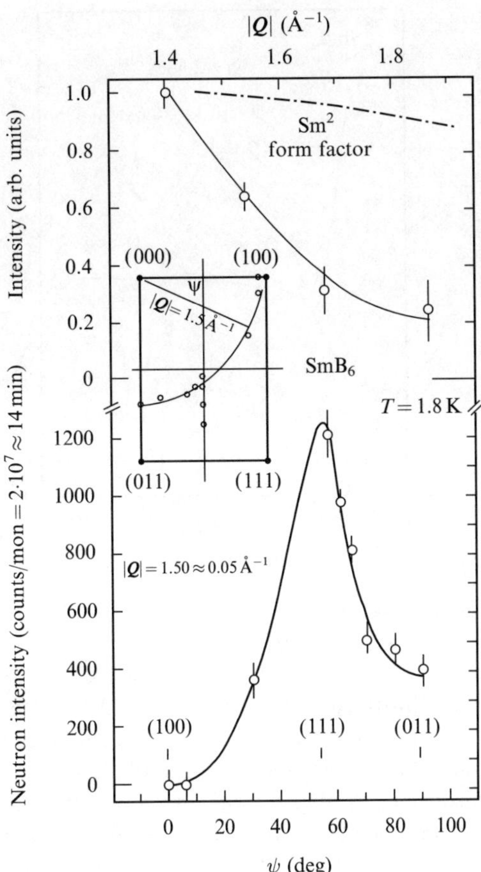

FIG. 3.12. Radial (upper part) and angular (lower part) variations of magnetic scattering intensity associated with the peak at $\hbar\omega_2 \simeq 14\,\mathrm{meV}$ in SmB_6; inset shows trajectories within (011) scattering plane corresponding to these plots [30].

indicate that the excitonic bound state becomes unstable by the thermal population of the itinerant d-band states.

3.2.2 *CeNiSn and CeRhSb*

CeNiSn and CeRhSb crystallize in the orthorhombic ϵ-TiNiSi type structure, and have a very small pseudo-energy-gap of a few K. The magnetic susceptibility, the resistivity, and the thermoelectric power have peaks around $T_{coh} \sim 12$–$20\,\mathrm{K}$. These peaks are ascribed to the formation of heavy-electron states and simultaneous development of antiferro-magnetic correlations. The transport properties are not like those of a semiconductor, but are metallic along all crystal directions [31]. From the temperature dependence of $1/T_1$ of ^{118}Sn and ^{123}Sb in CeNiSn and CeRhSb, the presence of a pseudo-energy-gap is suggested in the spin excitation spectrum. However, the density of states at the Fermi level is finite, as shown in Fig. 3.14. The T dependence of $1/T_1$ is reproduced well by the effective density of states $N_{eff}(E)$, modelled with a V-shaped structure [32] for the

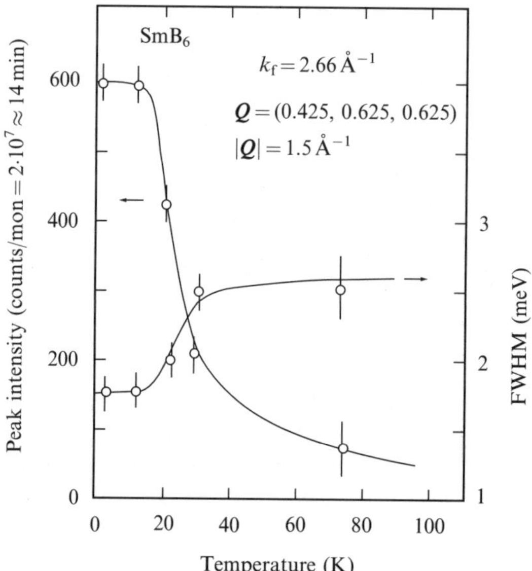

FIG. 3.13. Temperature dependence of neutron intensity and linewidth of the magnetic peak at $\hbar\omega_2 \simeq 14\,\text{meV}$ in SmB$_6$ [30].

quasi-particle bands, as illustrated in the inset of Fig. 3.14. The presence of the tiny density of states at the Fermi level is consistent with a semimetal with very low carrier density. In comparison with the isotropic gap in SmB$_6$, the pseudo-energy-gap with a few K seems to be anisotropic, as judged from the temperature dependence of the resistivity. These correlated semimetals exhibit interesting behaviour that the pseudo-gap of the charge response is larger than that of the spin response. We now turn to theoretical considerations for understanding the metallic and insulating characteristics of heavy electrons.

3.3   Momentum distribution and the Fermi surface

However strong the correlation effect, there is a one-to-one correspondence between a quasi-particle state and a noninteracting counterpart provided that the Fermi liquid is realized. If a single-particle state in the free system is occupied, the corresponding quasi-particle state is also occupied. From this, it follows that the Fermi surface of quasi-particles encloses the same volume (Fermi volume) in the Brillouin zone as that of the free electrons. This property is often referred to as the Luttinger sum rule. We sketch a mathematical proof of the sum rule in Appendix G. In this section, we show that the momentum distribution of bare electrons in strongly correlated metals is very different from that of the quasi-particles. The global distribution is rather similar to the case of the localized electrons. This feature of heavy electrons is important in understanding the consistency between the various experiments with different energy resolutions.

Let us first consider a trivial case where the system consists of monovalent atoms with a single atom in the unit cell of the lattice. The Fermi volume is half the volume of

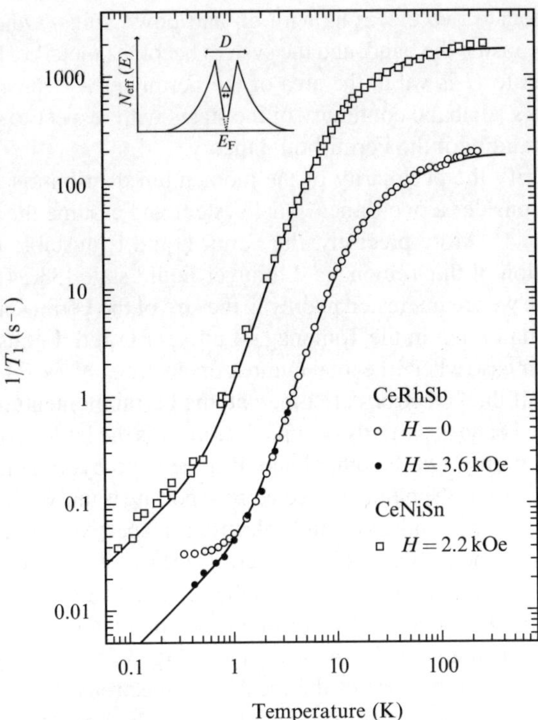

FIG. 3.14. Temperature dependence of $1/T_1$ for ^{119}Sn in CeNiSn and ^{123}Sb in CeRhSb. Inset illustrates the effective density of states $N_{\mathrm{eff}}(E)$ modelled with a V-shaped structure with a finite value near the Fermi level [32].

the Brillouin zone in an ordinary metal. We would like to understand what happens in the case of heavy electrons. If the f electrons participate in forming the quasi-particles, the Fermi volume is determined by the total number of electrons including the f electrons. If the f electrons are localized, on the contrary, only the conduction electrons contribute to the Fermi volume. Thus, information on the size of the Fermi surface gives an important hint on the nature of heavy electrons.

The standard model for investigating heavy electrons is the Anderson lattice. The model is characterized by a periodic arrangement of f-electron sites and is given by

$$H_{\mathrm{AL}} = \sum_{k\sigma}\left[\epsilon_k c_{k\sigma}^{\dagger} c_{k\sigma} + V_k(c_{k\sigma}^{\dagger} f_{k\sigma} + f_{k\sigma}^{\dagger} c_{k\sigma})\right]$$
$$+ \sum_i \left[\epsilon_{\mathrm{f}} \sum_{\sigma} f_{i\sigma}^{\dagger} f_{i\sigma} + U f_{i\uparrow}^{\dagger} f_{i\downarrow}^{\dagger} f_{i\downarrow} f_{i\uparrow}\right], \tag{3.1}$$

where $f_{k\sigma}$ is the annihilation operator of an f electron with momentum k and spin σ, and $f_{i\sigma}$ is the one localized at site i. In the trivial case with $U = 0$, the model can be solved easily and gives two hybridized bands. If the total number of electrons N_{e} is twice the number of lattice sites N_{L}, the system becomes an insulator since each k

state can accommodate two electrons with up and down spins. Otherwise, the Fermi level is somewhere inside the band, and the system becomes metallic. If the perturbation theory with respect to U is valid, the area of the Fermi surface must remain the same as U increases. This adiabatic continuity of the states with respect to U is precisely the condition for the validity of the Fermi-liquid theory.

In order to clarify the peculiarity of the momentum distribution in heavy-electron systems, we first consider a one-dimensional system and assume the presence of large Coulomb repulsion U. More precisely, the Fermi liquid is unstable in one dimension against the formation of the Tomonaga–Luttinger liquid state [33,34]. We neglect this aspect here because we are interested mainly in the size of the Fermi surface, the location of which is common to that in the Tomonaga–Luttinger liquid. Let us assume for definiteness a particular case where the total number of electrons N_e is 7/4 times the number of lattice sites N_L. If the f electrons are itinerant, the Fermi momentum is $7\pi/(8a)$, and the lower hybridized band is partially occupied. Here, a is the lattice constant. The upper hybridized band is empty. On the other hand, if there is no hybridization nor magnetic ordering in the f electrons, the number of f electrons per site is unity. Then the momentum distributions n_k^c, n_k^f for the conduction and f electrons, respectively, become like the ones shown by the dashed line in Fig. 3.15. The Fermi surface of conduction electrons is at $k = 3\pi/(8a)$ without hybridization. The number of conduction electrons N_c is equal to $3N_L/4$. With large U and small hybridization, the momentum distribution becomes like the one shown schematically by the solid line in Fig. 3.15. Globally, the distribution is not very different from the case of the localized f electrons. But the location of the Fermi momentum is the same as in the case $U = 0$. We note that the k dependence

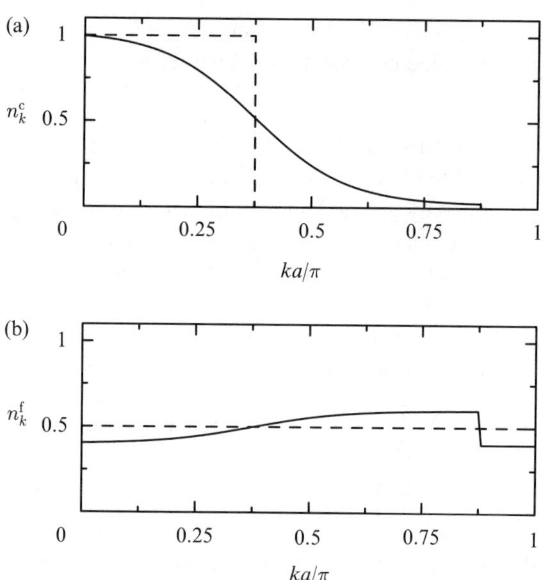

FIG. 3.15. Schematic view of the momentum distributions of (a) c electrons, and (b) f electrons.

of the f-electron distribution means the presence of charge fluctuations. In order to see this, we note that in the case of a pair of Kondo impurities the momentum dependence corresponds to the even–odd splitting of molecular orbitals.

We give a simple example of a variational wave function for the itinerant case. Consider Ψ_B given by [35]

$$\Psi_B = \mathcal{P} \prod_{k\sigma}(1 + \alpha_k f_{k\sigma}^\dagger c_{k\sigma})\phi_{N_e} = \mathcal{P} \exp\left(\sum_{k\sigma}\alpha_k f_{k\sigma}^\dagger c_{k\sigma}\right)\phi_{N_e}, \tag{3.2}$$

where ϕ_{N_e} represents the wave function of N_e conduction electrons and the vacant f orbitals. The operator $\mathcal{P}$ is a projection operator to exclude double occupation of f electrons at any site, and is often called the Gutzwiller projection. The variational parameter α_k should be optimized so as to minimize the energy. In the case of large U, numerical integration using the Monte Carlo sampling [36] yields a momentum distribution similar to the solid line in Fig. 3.15. If α_k is large compared to unity, each site has almost one f electron. Then the discontinuity in n_k^f at the Fermi surface becomes small, but the location remains the same. If $N_e/N_L = 7/4$, the location of the Fermi surface is at $k = 7\pi/(8a)$, which is called 'the large Fermi surface'. Here N_e determines the location. We note that a significant but smooth change of n_c^k appears at the position of 'the small Fermi surface', which is determined only by N_c, as shown in Fig. 3.15(a). The smooth change in n_k^c takes advantage of hybridization. A very simple description of quasi-particles with a large Fermi surface is provided by a mean-field theory. This is sometimes called the renormalized band picture [37] which contains a parameter to describe reduction of hybridization and another to describe the shift in the f-electron level. The parameters are optimized by the same procedure as in the case of the impurity Anderson model.

In the particular case where $N_e = 2N_L$, Ψ_B corresponds to a state where the electrons fill completely the lower hybridized band. In this case, the Fermi surface is absent and the system becomes an insulator at absolute zero. This state goes continuously to the usual band insulator as U decreases. In the opposite case of large U, however, the magnetic and dielectric properties of the system reflect the effect of strong correlation. This should be evident if one considers the high-temperature regime where the localized picture becomes better than the band picture. The system in the large-U limit is called the Kondo insulator.

Let us come back to the metallic case with U and $|\epsilon_f|$ large as compared to $V^2\rho_c$. Then the ground state has almost unit occupation of f electrons at each site. The effective model excluding the empty and doubly occupied f sites is given by

$$H_{KL} = \sum_{k\sigma}\epsilon_k c_{k\sigma}^\dagger c_{k\sigma} + J\sum_i \boldsymbol{S}_i \cdot \boldsymbol{s}_i, \tag{3.3}$$

where $\boldsymbol{S}_i$ is the spin operator of f electrons, and $\boldsymbol{s}_i$ of conduction electrons at site i. The latter is given by

$$\boldsymbol{s}_i = \frac{1}{2N}\sum_{kp}\sum_{\alpha\beta} c_{k\alpha}^\dagger \boldsymbol{\sigma}_{\alpha\beta} c_{p\beta} \exp\left[-i(\boldsymbol{k} - \boldsymbol{p})\cdot \boldsymbol{R}_i\right]. \tag{3.4}$$

The model (3.3) is referred to as the Kondo lattice. Various numerical results show that the location of the Fermi surface in the Kondo lattice is the same as the Anderson lattice with the same number of total electrons [38]. Thus, the electronic state is continuously connected even though the charge fluctuations of f electrons become infinitesimal. This situation is analogous to the single-impurity case where the Kondo model without the charge fluctuation has the same fixed point as the Anderson model with $U = 0$ [39].

One might check the possibility of a small Fermi surface for the Kondo lattice. With $N_c/N_L = 3/4$, for example, can the discontinuity in n_c^k occur at $k = 3\pi/(8a)$ in a paramagnetic ground state? Such a ground state should be realized if the f electrons were completely decoupled from the conduction electrons, and form a spin chain in the presence of additional intersite exchange interaction. Provided that the hybridization or the Kondo exchange between the f and conduction electrons acts as an irrelevant perturbation to this decoupled state, the perturbed ground state would still have the small Fermi surface. Such a ground state is often called the doped spin liquid and emphasizes the difference from the Fermi liquid that has a large Fermi surface. An exact analysis [40] suggests that the doped spin liquid is unlikely to be realized in one dimension; the density fluctuation has zero excitation energy at the wavenumber $q = 2k_F$, where k_F is the large Fermi surface including the localized f electrons. This result is naturally interpreted in terms of the particle–hole excitation across the large Fermi surface, but is hard to be reconciled with the small Fermi surface. Realizing the doped spin liquid in three dimensions is even more unlikely; in higher dimensions with a large number of neighbours, the localized f electrons favour a magnetically ordered ground state rather than a singlet liquid.

So far, we have assumed that there is at most one f electron per site. In the case of two f electrons per site as in U^{4+}, we encounter a different situation concerning the size of the Fermi surface. Namely, the size can be the same independent of whether the f electrons are itinerant or localized since the f electrons in total can fill an energy band completely. This raises an interesting question as to how one can distinguish between the itinerant and localized states for a system with an f^2 configuration [41]. It cannot be excluded that a band singlet state changes continuously into a localized state with the CEF singlet at each site as parameters of the system are varied [42]. The difference in the dHvA behaviour, as discussed in § 3.1.1, between UPt_3 and $CeRu_2Si_2$ after the pseudo-metamagnetic transition, might be related to different numbers of f electrons per site.

At finite T, the discontinuity in the momentum distribution is absent. Therefore, one needs a criterion other than the momentum distribution to distinguish the itinerant and localized states. There are a few experimental methods to measure the Fermi surface. The most popular is the dHvA effect. A characteristic property in some heavy-electron systems is that the Fermi surface changes rapidly at a particular value of the magnetic field, as discussed in § 3.1.1. This change, though rapid, seems to be smooth even at the lowest accessible temperature [43]. On the other hand, particular care is required if one tries to measure the Fermi surface with limited resolution, say in photoemission. The large (but smooth) change of n_k^c at $k = 3\pi/(8a)$ in Fig. 3.15 might easily be mistaken for the Fermi surface.

In order to treat the correlation effects more quantitatively than presented above, an effective medium theory turns out to be useful, and will be explained below.

3.4 Dynamic effective field theory

As we have seen in Chapter 2, perturbation theory from the limit of infinite degeneracy is very successful in dealing with the Kondo effect. In order to deal with the lattice of Kondo centres, which makes up the heavy-electron state, we also pursue a simplifying limit which makes the problem tractable. The proper limit in this case is the infinite spatial dimension d. The limit makes a kind of effective medium theory exact. This effective medium is not static as in the conventional mean-field theory, but dynamic. The dynamics is to be determined self-consistently [44]. Before considering the dynamic effective field theory for heavy electrons, we take the Ising model to clarify the basic idea of the effective field.

Since the following discussion involves a little advanced mathematics, those who are interested mostly in explicit results for heavy electrons are advised to go straight to § 3.6 .

3.4.1 *Effective field in the Ising model*

In the system of Ising spins, it has long been known that one can regard the reciprocal of the number of neighbouring spins as the expansion parameter [45]. In the case of the nearest-neighbour interaction, the number of neighbouring spins increases as the dimensions of the system increase. Then the mean-field theory with use of the Weiss molecular field corresponds to the lowest-order theory, which becomes exact in infinite dimensions. In the next order, one partly includes the fluctuation of the mean field. This level of the theory is called the spherical model [45].

The ferromagnetic Ising model under a magnetic field h in the z-direction can be written as

$$H = -J \sum_{\langle ij \rangle} \sigma_i \sigma_j - h \sum_i \sigma_i, \qquad (3.5)$$

where $\langle ij \rangle$ are the nearest-neighbour sites, $J > 0$, and the variable σ_i takes values ± 1. The unit of the magnetic field is taken so that h has the dimension of energy. The molecular field h_{W} at a site i is defined by

$$h_{\mathrm{W}} = J \sum_{j \in n(i)} \langle \sigma_j \rangle = J z_{\mathrm{n}} m, \qquad (3.6)$$

where $n(i)$ denotes the set of nearest-neighbour sites, z_{n} in number, and $\langle \sigma_j \rangle = m$ is the average of σ_j to be determined later. In the d-dimensional hypercubic lattice, one has $z_{\mathrm{n}} = 2d$. The Hamiltonian is rewritten as

$$H = -(h_{\mathrm{W}} + h) \sum_i \sigma_i - J \sum_{\langle ij \rangle} (\sigma_i - m)(\sigma_j - m) + \frac{N}{2} J z_{\mathrm{n}} m^2$$

$$\equiv H_0 + H_1 + \frac{N}{2} J z_{\mathrm{n}} m^2, \qquad (3.7)$$

where N is the total number of lattice sites. The term H_0 gives the mean-field Hamiltonian, and H_1 accounts for the contribution of fluctuations. If we neglect H_1, the spins under the effective field $h_W + h$ become independent of each other. Then we can easily calculate the magnetization as

$$m = \tanh\left[\beta(h_W + h)\right], \tag{3.8}$$

with $\beta = 1/T$. Together with eqn (3.6), h_W is determined self-consistently. The susceptibility $\chi = \partial m/\partial h$ in the mean-field approximation (MFA) is calculated as

$$\chi_{\mathrm{MFA}} = \frac{\chi_0}{1 - J z_n \chi_0}, \tag{3.9}$$

in the limit $h_W + h \to 0$. Here $\chi_0 = \beta$ is the susceptibility of an isolated spin. The χ_{MFA} is divergent at the Curie temperature $T_c = J z_n$.

Let us now evaluate the importance of H_1 as the perturbation Hamiltonian. The average of H_1 vanishes by the definition of m. Hence the lowest-order contribution to the free energy becomes

$$-\frac{\beta}{2}\langle H_1^2 \rangle = -\frac{1}{4 z_n} N \beta (J z_n)^2 \langle (\sigma_j - m)^2 \rangle^2. \tag{3.10}$$

From eqn (3.8) we obtain

$$\langle (\sigma_j - m)^2 \rangle^2 = \cosh^{-4}[\beta(h_W + h)]. \tag{3.11}$$

As z_n gets larger with T_c kept constant, the fluctuation contribution has the extra small factor $1/z_n$. Even in the case of $h = 0$ at $T = T_c$, where the fluctuation reaches a maximum, the contribution is $O(N T_c/z_n)$. This is smaller than the mean-field contribution by $1/z_n$. Higher order contributions have higher powers of $1/z_n$. Hence, in the limit of large dimensions, one can neglect the deviation from the mean-field theory for all temperatures.

In the realistic case of finite d, however, the mean-field theory is only approximate. The most serious drawback is the excess counting of the exchange field. Namely, a part of the mean field felt by a spin at certain site originates from this spin itself. A correction of it in dielectric media has led to the Clausius–Mossotti formula for the susceptibility. The correction includes the concept of the reaction field of Onsager [45]. We now derive the $O(1/z_n)$ correction to the susceptibility. For simplicity, we consider only the paramagnetic phase with $m = h = 0$. In order to identify the $O(1/z_n)$ correction, it is convenient to introduce spinless fermion operators $f_i^\dagger$ and f_i at site i by

$$\sigma_i = 2 f_i^\dagger f_i - 1. \tag{3.12}$$

The spin susceptibility in the mean-field theory corresponds to the RPA for the charge susceptibility of fermions. Some lower-order corrections to the elementary bubble are shown in Fig. 3.16. The set of diagrams corresponds to the irreducible susceptibility χ_1, which cannot be separated into two pieces by cutting a line representing the exchange interaction. The exact susceptibility is given by the same form as eqn (3.9) provided the irreducible susceptibility χ_1 is used in place of χ_0. It is important to notice that the vertex part in the lowest order is diagonal in site indices. Thus, to $O(1/z_n)$, we can regard χ_1 as site-diagonal. The fermion representation is in fact used only to show this property.

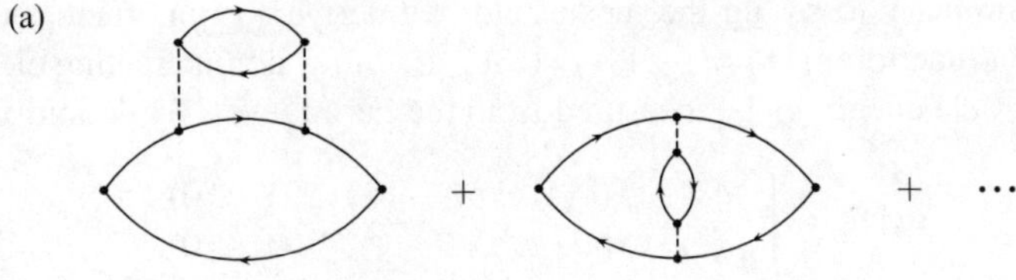
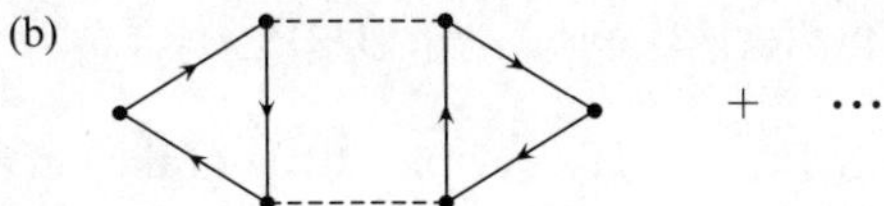

FIG. 3.16. Examples of corrections to the bubble: (a) $O(1/z_n)$ corrections, (b) $O(1/z_n^2)$ corrections.

We parametrize χ_1 as

$$\chi_1^{-1} = \chi_0^{-1} + \lambda, \tag{3.13}$$

where λ represents the local-field correction. The k-dependent susceptibility $\chi(k)$ is given by

$$\chi(k)^{-1} = \chi_1^{-1} - J(k), \tag{3.14}$$

where $J(k) = J \sum_{\delta} \cos(k \cdot \delta)$ is the Fourier transform of the exchange interaction, with δ pointing to each nearest neighbour. Let us now consider the local susceptibility which is given by the k-average of $\chi(k)$. The very special feature of the Ising model is that each spin is a conserved quantity independent of the exchange interaction. Then the local susceptibility is simply given by $\beta\langle\sigma_i^2\rangle = \beta$, which is the same as χ_0. Then we obtain

$$1 = \mathrm{Av}_k \frac{1}{1 - (J_k - \lambda)\beta}, \tag{3.15}$$

which determines λ at a given temperature. Equation (3.15) is known as the spherical model condition [45]. With this condition, the right-hand side of eqn (3.15) without Av_k gives $\chi(k)/\beta$. We note the similarity of the susceptibility given by eqn (3.14) with eqn (1.107) of the mode-coupling theory of itinerant magnetism. However, the different definitions of λ in these two cases should be noted. The reaction field included in the $1/z_n$ correction can also be interpreted as a mode-coupling effect.

3.4.2 Dynamic effective field for fermions

One can naturally ask whether it is possible to set up a mean field for itinerant electrons so that the theory becomes exact at $d = \infty$. Provided such a mean field is found, the problem is reduced to solving an effective single-site problem under the mean field, together with determining the mean field self-consistently. Obviously, the mean field introduced in the Hartree–Fock approximation is inappropriate in this case because the number operator for each spin experiences large fluctuations.

As a preliminary to set up the mean field, we derive an important bound for the thermal Green function $g_k[\tau] = -\langle T_\tau c_{k\sigma}(\tau)c_{k\sigma}^\dagger(0)\rangle_0$ for noninteracting electrons. Let the single-particle energy ϵ_k be measured from the Fermi level. By definition, we have

$$g_k[\tau] = \begin{cases} -[1 - f(\epsilon_k)]\exp(-\epsilon_k\tau) & (\tau > 0), \\ f(\epsilon_k)\exp(-\epsilon_k\tau) & (\tau < 0). \end{cases} \tag{3.16}$$

Combining both signs of τ we obtain the bound [46]

$$|g_k[\tau]| \le \tfrac{1}{2}\exp(\beta|\epsilon_k|/2)\cosh^{-1}(\beta|\epsilon_k|/2) \le 1. \tag{3.17}$$

Therefore, the site representation $g_{ij}[\tau] = -\langle T_\tau c_{i\sigma}(\tau)c_{j\sigma}^\dagger(0)\rangle_0$ of the Green function satisfies the inequality

$$\sum_j |g_{ij}[\tau]|^2 = \frac{1}{N}\sum_k |g_k[\tau]|^2 \le 1. \tag{3.18}$$

We note that the number of lattice sites N is equal to the number of wavenumbers k inside the Brillouin zone. If the number of equivalent sites from a site i is z_n, the inequality leads to the upper bound for the off-diagonal element as $g_{ij}[\tau] = O(1/\sqrt{z_n})$. On the other hand, the diagonal element $g_{ii}[\tau]$ is of order unity. One can prove the same inequality for the exact Green function $G_{ij}[\tau]$ by using the spectral representation. Hence, at $d = \infty$, one can neglect the off-diagonal element $G_{ij}[\tau] = -\langle T_\tau c_{i\sigma}(\tau)c_{j\sigma}^\dagger(0)\rangle$ in comparison with the diagonal one. One should notice that the inequality is valid for all temperatures, and is independent of the choice of the energy scale.

An arbitrary Feynman diagram for the self-energy $\Sigma[\tau]$ has its skeleton consisting of U and $G_{ij}[\tau']$. We can now see that the dominant contribution comes from $G_{ii}[\tau']$ at $d = \infty$. Thus, $\Sigma[\tau]$ does not depend on k [47]. This local property of the self-energy is in common with the result of the coherent potential approximation (CPA) for disordered systems without mutual interactions [48,49]. Indeed, the CPA becomes exact in infinite dimensions, and the Fourier transform of $\Sigma[\tau]$ to the frequency space is called the coherent potential, which is a kind of mean field. In interacting many-body systems it is convenient to introduce the self-energy, or a related quantity $\lambda(z)$, as the dynamic mean field as we shall now explain.

Falicov–Kimball model We explain the concept of the dynamic effective field first by taking the Falicov–Kimball model [50]. Because of the simplicity of the model, the self-consistent solution can be derived analytically. The model is given by

$$H_{\mathrm{FK}} = -\sum_{ij} t_{ij}c_i^\dagger c_j + \sum_i \epsilon_f f_i^\dagger f_i + U\sum_i f_i^\dagger f_i c_i^\dagger c_i. \tag{3.19}$$

It is equivalent to the special case of the Hubbard model where the down-spin electrons have zero transfer and are regarded as spinless f electrons. We first derive intuitively the equation to determine the effective field. Consider the site $i = 0$ and remove for the moment the Coulomb repulsion U only at the site. The on-site Green function in this fictitious system $\tilde{H}$ is written as $\tilde{D}[\tau] = -\langle T_\tau c_{0\sigma}(\tau)c_{0\sigma}^\dagger(0)\rangle_0$. The Fourier transform

$\tilde{D}(z)$ is given by $\tilde{D}(z)^{-1} = z - \lambda(z)$, where $\lambda(z)$ accounts for the hopping motion to and from the other sites. The actual Green function $D(z)$ of H_{FK} should incorporate the effect of U at the site in the form of the self-energy $\Sigma(z)$. Then we obtain the relation

$$D(z)^{-1} = z - \lambda(z) - \Sigma(z). \tag{3.20}$$

On the other hand, $D(z)$ is also regarded as the site-diagonal element of the Green function $G_{ij}(z)$ with full inclusion of U at each site. In other words, $D(z)$ is given by the momentum average of the Green function $G(\boldsymbol{k}, z) = [z - \epsilon_{\boldsymbol{k}} - \Sigma(z)]^{-1}$; namely,

$$D(z) = \mathrm{Av}_{\boldsymbol{k}} \frac{1}{z - \epsilon_{\boldsymbol{k}} - \Sigma(z)} = D_0(z - \Sigma(z)), \tag{3.21}$$

where $D_0(z)$ is the Green function with $U = 0$ at all sites. Note that the appearance of $D_0(z)$ is due to the momentum independence of the self-energy. This relation to $D_0(z)$ simplifies the calculation enormously. Figure 3.17 shows schematically the concept of the effective impurity.

In this way, we have obtained two independent relations eqns (3.20) and (3.21) for each z among the unknowns $D(z)$, $\lambda(z)$, and $\Sigma(z)$. These relations match the effective field and the intersite interaction. If one can solve the effective single-impurity problem and establish another relation between $D(z)$, $\lambda(z)$, and $\Sigma(z)$, then the Green function $G(\boldsymbol{k}, z)$ can be derived explicitly.

The Falicov–Kimball model has the special feature that the f electron occupation number n_{f} is a constant at each site. Because of this, one can solve the effective impurity problem exactly:

$$D(z) = \frac{1 - n_{\mathrm{f}}}{z - \lambda(z)} + \frac{n_{\mathrm{f}}}{z - \lambda(z) - U}. \tag{3.22}$$

The Green function for the Falicov–Kimball model is thus derived using eqns (3.20–3.22).

Examples of density of states often used are the Gaussian $\rho_{\mathrm{G}}(\epsilon)$, the semielliptic $\rho_{\mathrm{E}}(\epsilon)$, and Lorentzian $\rho_{\mathrm{L}}(\epsilon)$ ones. The actual forms are given by

$$\rho_{\mathrm{G}}(\epsilon) = \frac{1}{\sqrt{2\pi}\,\Delta} \exp\left[-\frac{\epsilon^2}{2\Delta^2}\right], \qquad \rho_{\mathrm{E}}(\epsilon) = \frac{1}{\pi}\sqrt{\Delta^2 - \epsilon^2}, \qquad \rho_{\mathrm{L}}(\epsilon) = \frac{1}{\pi}\frac{\Delta}{\epsilon^2 + \Delta^2}. \tag{3.23}$$

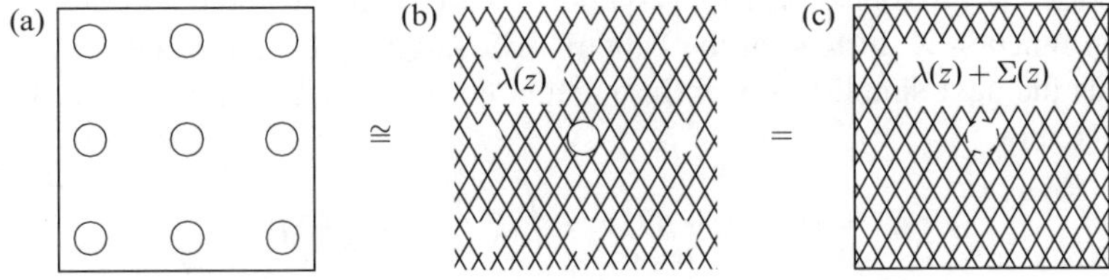

FIG. 3.17. Illustration of an effective impurity and the dynamic effective field. In (b), the interaction with other sites is represented by $\lambda(z)$, and in (c) the Coulomb interaction at the impurity site is further represented by $\Sigma(z)$.

The Gaussian $\rho_G(\epsilon)$ is obtained by extrapolating the simple cubic lattice to $d = \infty$. The nearest-neighbour hopping t is scaled so as to give $\Delta^2 = z_n t^2$ [51]. On the other hand, $\rho_E(\epsilon)$ corresponds to the Bethe lattice of the Cayley tree with infinite number of nearest neighbours [52]. It is known that one can construct the single-particle spectrum from an arbitrary density of states [53].

As the simplest case to derive the analytic solution, we take the Lorentzian density of states $\rho_L(\epsilon)$ for the conduction band. We have $D_0(z) = (z + i\Delta)^{-1}$ with $\text{Im } z > 0$, and by comparing eqns (3.21) and (3.22) we obtain $\lambda(z) = -i\Delta$. Here, Δ has the meaning of the half-width of the band. The Green function is given by

$$D(z) = \frac{1 - n_f}{z + i\Delta} + \frac{n_f}{z + i\Delta - U} \tag{3.24}$$

and the self-energy by

$$\Sigma(z) = U n_f + \frac{U^2 n_f(1 - n_f)}{z + i\Delta - U(1 - n_f)}. \tag{3.25}$$

In eqn (3.25) we see how the higher-order correlation modifies the MFA result given by the term $U n_f$.

It turns out that the self-consistency equations for $D(z)$ are the same as in the CPA [48,49]. In order to see the origin of the equivalence, we may regard ϵ_f as a random variable so that the probability that the conduction electrons have a potential U is given by n_f. For the Hubbard model, the CPA corresponds to an approximation called the alloy analogy [54]. Of course, the alloy analogy is not exact in the Hubbard model, in contrast to the Falicov–Kimball model. This is simply because the occupation number of either spin is not a constant of motion in the Hubbard model.

We comment on the stability of the solution for the Falicov–Kimball model. If the average occupation n_f is between 0 and 1, the homogeneous state corresponding to the above Green function becomes unstable at low temperatures. This instability shows up as divergence of the charge susceptibility with a certain wavenumber, which corresponds to a charge density wave pattern [50]. The instability inevitably arises because the entropy associated with the charge disorder is not removed in the homogeneous state even at $T = 0$.

Dynamic variational principle   In the intuitive derivation above, one may wonder whether the self-energy $\Sigma(z)$ of the effective impurity is indeed the same as that of the whole system. In order to examine the problem on a more solid basis, it is convenient to use the variational principle combined with a path integral representation of the partition function Z. Instead of the Falicov–Kimball model, we now take the Hubbard model as the next simplest model. In a symbolic notation, the partition function Z is represented by

$$Z = \text{Tr} \exp(-\beta H) = \int \mathcal{D}c^\dagger \mathcal{D}c \exp(-\beta\mathcal{L}),$$

$$\mathcal{L} = -\text{Tr}(c^\dagger g^{-1} c) + H_U,$$

where g^{-1} is the inverse matrix of the Green function without the interaction part H_U, and the two-component Grassmann numbers c (spin index omitted) correspond to the annihilation operators of up- and down-spin electrons. The symbol Tr represents the trace over sites, spins, and Matsubara frequencies. The path integral representation is explained in detail in Appendices F and G. With use of the site-diagonal self-energy matrix Σ, we define the renormalized Green function matrix G by $G^{-1} = g^{-1} - \Sigma$. The unknowns Σ and G are to be optimized variationally.

According to the many-body perturbation theory [55,56] as explained in Appendix G, the thermodynamic potential Ω of the system is given by

$$\beta\Omega = \Phi\{G\} - \mathrm{Tr}\,(\Sigma G) - \mathrm{Tr}\ln(\beta G^{-1}), \tag{3.26}$$

where $\Phi\{G\}$ satisfies the relation $\delta\Phi/\delta G = \Sigma$. Since Σ is diagonal in site indices, Φ depends actually only on D, which is the site-diagonal element of G. Hence, the second term on the right-hand side also depends only on D. In order to exploit the situation, we decompose the Lagrangian $\mathcal{L}$ as $\mathcal{L} = \mathcal{L}_G + \mathcal{L}_{\mathrm{int}}$ [57]. Here,

$$\mathcal{L}_G = -\mathrm{Tr}\,(c^\dagger G^{-1} c), \qquad \mathcal{L}_{\mathrm{int}} = H_U - \mathrm{Tr}(c^\dagger \Sigma c). \tag{3.27}$$

We regard $\mathcal{L}_G$ as the unperturbed Lagrangian. If one keeps this part only, the path integral leads to the third term of eqn (3.26). The sum of the first and the second terms represents the correction due to $\mathcal{L}_{\mathrm{int}}$, which depends only on D at $d = \infty$. Hence, in evaluating the interaction parts with the statistical distribution specified by $\mathcal{L}_G$, one may replace every Green function G by D. Namely, we can approximate the partition function as $Z = Z_G(Z_1/Z_D)^N$, where

$$Z_G = \det(\beta G^{-1}), \qquad Z_D = \det(\beta D^{-1}), \tag{3.28}$$

$$Z_1 = \int \mathcal{D}c_0^\dagger \mathcal{D}c_0 \exp\left[\beta \sum_{n\sigma} c_{0\sigma}^\dagger (D^{-1} + \Sigma)c_{0\sigma} - \beta H_{U1}\right]. \tag{3.29}$$

Here, Z_1 and H_{U1} stand for the single-site contribution, and the summation in eqn (3.29) is over spin states and the Matsubara frequency $i\epsilon_n$.

The evaluation of Z_1 is tantamount to solving the effective single-site problem in the presence of a dynamic external field. If one can solve the problem for arbitrary Σ, one can optimize Σ such that the thermodynamic potential $\Omega = -T \ln Z$ is minimized. Then one can get the exact solution at $d = \infty$. In reality, the stationary condition, instead of the minimum condition, is used to obtain the self-consistency equation for the Green function. By differentiating each term constituting $-\beta\Omega = \ln Z$, we get the equation

$$\frac{1}{N}\frac{\delta \ln Z}{\delta\Sigma(i\epsilon_n)} = -D(i\epsilon_n) - D(i\epsilon_n)\frac{\delta D(i\epsilon_n)^{-1}}{\delta\Sigma(i\epsilon_n)} + G_1(i\epsilon_n)\left[1 + \frac{\delta D(i\epsilon_n)^{-1}}{\delta\Sigma(i\epsilon_n)}\right], \tag{3.30}$$

where the third term on the right-hand side comes from Z_1, and $G_1(i\epsilon_n)$ represents the Green function of the effective single impurity. We notice that the Green function without H_{U1} is given by $[D(i\epsilon_n)^{-1} + \Sigma(i\epsilon_n)]^{-1}$. Introducing the self-energy correction due to H_{U1} as $\Sigma_I(i\epsilon_n)$, we obtain

$$G_1(i\epsilon_n)^{-1} = D(i\epsilon_n)^{-1} + \Sigma(i\epsilon_n) - \Sigma_I(i\epsilon_n). \tag{3.31}$$

Then the stationary condition $\delta\Omega/\delta\Sigma = 0$ leads to $G_1(i\epsilon_n) = D(i\epsilon_n)$, i.e. $\Sigma(i\epsilon_n) = \Sigma_I(i\epsilon_n)$. Thus, the relationship between $D(z)$ and $\Sigma(z)$ reproduces eqn (3.21). Note that the stationary condition is independent of the way we solve the effective single-site problem. Equation (3.21), which has been derived for the Falicov–Kimball model on intuitive grounds, is in fact valid also for the Hubbard model [58,59]. It is also valid for the f-electron Green function in the Anderson lattice model [57,60,61], which we now discuss.

The partition function for the Anderson lattice is given by

$$Z = \text{Tr} \exp\left(-\beta H_{AL}\right) = \int \mathcal{D}f^\dagger \mathcal{D}f \mathcal{D}c^\dagger \mathcal{D}c \exp\left(-\int_0^\beta d\tau \mathcal{L}(\tau)\right),$$

$$\mathcal{L} = \sum_{k\sigma}\left(c_{k\sigma}^\dagger \frac{\partial}{\partial\tau} c_{k\sigma} + f_{k\sigma}^\dagger \frac{\partial}{\partial\tau} f_{k\sigma}\right) + H_{AL}.$$

Since the Lagrangian is bilinear in $c^\dagger$ and c, the integration over them can be done explicitly. The result is

$$Z = Z_c \int \mathcal{D}f^\dagger \mathcal{D}f \exp\left(-\int_0^\beta d\tau(\mathcal{L}_f + \mathcal{L}_{\text{hyb}})\right),$$

$$\mathcal{L}_f = \sum_{i\sigma} f_{i\sigma}^\dagger \frac{\partial}{\partial\tau} f_{i\sigma} + H_f,$$

$$\mathcal{L}_{\text{hyb}}(\tau) = V^2 \sum_{ij\sigma} \int_0^\beta d\tau' \, f_{i\sigma}^\dagger(\tau) g_{ij}(\tau - \tau') f_{j\sigma}(\tau'),$$

where Z_c is the partition function of the conduction electrons without hybridization and, for simplicity, the hybridization is assumed to be local.

The f-electron Lagrangian obtained above is reduced to that in the Hubbard model if $\mathcal{L}_{\text{hyb}}$ is replaced by the transfer term without retardation. The presence of retardation in $\mathcal{L}_{\text{hyb}}$ makes no difference, however, in deriving the self-consistent medium. Writing the on-site Green function of f electrons as $D_f(z)$, we again introduce $\lambda(z)$ and $\Sigma_f(z)$, the former representing the effective hopping through hybridization and the latter accounting for the Coulomb interaction at the site. The variational procedure gives the self-consistency equation

$$D_f(z) = [z - \epsilon_f - \lambda(z) - \Sigma_f(z)]^{-1}$$

$$= \frac{1}{N} \sum_k \left(z - \epsilon_f - \frac{V^2}{z - \epsilon_k} - \Sigma_f(z)\right)^{-1}. \tag{3.32}$$

One has to solve the effective single-impurity problem to derive $\Sigma_f(z)$ for given values of $\lambda(z)$ and U.

Reduction to the Anderson impurity model In order to solve the effective single-impurity problem in the Anderson lattice or in the Hubbard model, it is convenient to map to the single-impurity Anderson model [58]. Namely, we use the spectral representation of the effective field $\lambda(z)$ as

$$\lambda(z) = \Delta\epsilon_f + \frac{1}{N} \sum_k \frac{|V_k|^2}{z - \epsilon_c(k)}, \tag{3.33}$$

where the constant part $\Delta\epsilon_f$ represents a shift in the local electron level, V_k is the effective hybridization, and $\epsilon_c(k)$ represents the spectrum of fictitious conduction electrons in the Anderson model. For a given $\lambda(z)$, V_k and $\epsilon_c(k)$ are not determined uniquely. Thus, one has a freedom to choose a combination that is most convenient for the calculation.

In the case of the paramagnetic ground state, the mapping to the Anderson lattice leads to two remarkable identities related to the local response functions. The first is that the Wilson ratio tends to 2 according to eqn (2.39) as the charge fluctuation is suppressed. The second is the Korringa–Shiba relation given by eqn (2.71) is valid. Note, however, that the local static susceptibility can be very different from the homogeneous susceptibility in the presence of intersite interactions. The local dynamic susceptibility, on the other hand, is probed directly by the NMR.

As the special case where the self-consistency equation is solved analytically, we take the Hubbard model at $d = \infty$ with the Lorentzian density of states for the conduction band. The centre of the band is taken to be ϵ_f, namely

$$\rho_L(\epsilon) = (\Delta/\pi)[(\epsilon - \epsilon_f)^2 + \Delta^2]^{-1}.$$

To see why the analytic solution is possible in this case, we observe that eqn (3.21) leads, for $\mathrm{Im}\, z > 0$, to

$$D(z) = [z - \epsilon_f - \Sigma(z) + i\Delta]^{-1}, \tag{3.34}$$

which takes the same form as the f-electron Green function in the single-impurity Anderson model with the constant density of states ρ_c for the conduction band and hybridization V. Thus, the correspondence is $\Delta = \pi V^2 \rho_c$. The resultant Anderson model in this case can be solved exactly by the Bethe Ansatz method [54]. Then one can derive thermodynamic quantities such as the susceptibility and the specific heat of the Hubbard model. Since the ground state of the Anderson model is always the Fermi liquid, the Hubbard model with the Lorentzian density of states also has a paramagnetic Fermi-liquid ground state. As U is increased, the effective bandwidth of the Hubbard model decreases just like the Kondo temperature T_K of the Anderson model. From this solvable case, one can imagine that the Hubbard model in general can have a minute structure in the density of states near the Fermi level as in the Anderson model. This structure represents the formation of a narrow band with a large effective mass.

Except for this particular density of states, one must solve the effective single-impurity problem either approximately, or by numerical methods. One has to determine the effective medium self-consistently, which is usually done by iteration. Thus, the numerical calculation is much more laborious than in the single-impurity case. We should notice that the self-consistent solution does not necessarily correspond to the local minimum of energy, but to a stationary solution. One has to examine fluctuations around the stationary state in order to study the stability of the solution.

Special situation in the half-filled case   Although the effective impurity model to be solved is the same for both Hubbard and Anderson lattice models, the most significant difference between the two models appears if the ground state is insulating. Namely, the Hubbard model with one electron per site, which is conventionally referred to as the half-filled case, has the Mott insulating state explained in Chapter 1. Thus, as the Coulomb interaction U increases, the ground state of the model can change from a paramagnetic metal to an antiferromagnetic insulator with a doubling of the unit cell. We note that the Mott insulating state does not always have a magnetic order. The clearest counter-example is the one-dimensional Hubbard model where the ground state in the half-filled case is a paramagnetic insulator. In the infinite-dimensional theory, however, the insulating paramagnetic state has a residual entropy since the effective impurity spin has no relaxation in the presence of an excitation gap. In order to improve the situation, one has to include spin-fluctuation effects which enter only in finite dimensions.

In the exactly soluble case of the infinite-dimensional Hubbard model with $\rho_L(\epsilon)$, there is no metal–insulator transition even in the half-filled case with large U. The absence of the transition is connected to the divergent energy of the Lorentzian density of states:

$$\int_{-\infty}^{0} \epsilon \rho_L(\epsilon) \, d\epsilon = -\infty. \tag{3.35}$$

In this case, the localization of electrons is energetically unfavourable. The metallic ground state corresponds to the fact that the Anderson model, which is mapped from the Hubbard model, remains a Fermi liquid even in the limit of large U.

On the other hand, the insulating ground state in the Anderson lattice model connects continuously to the band insulator which does not have a residual entropy even without the magnetic order. We note that the Kondo lattice model with half-filled conduction band is also connected continuously to the band insulator.

3.5 Methods for solving the effective impurity model

In this section, we turn to practical ways to solve the model and the corresponding results of the dynamical effective field theory. A large number of investigations have been performed particularly for the Hubbard model using the dynamic effective field theory. Many of them are motivated by the presence of the metal–insulator transition and deal with the half-filled case. For general filling, one has to adjust the chemical potential so as to reproduce the desired number of electrons, and the calculation becomes much more laborious. The same remark applies to the Anderson lattice model. We shall explain below various methods to solve the effective impurity model.

3.5.1 *Perturbation theory with respect to the Coulomb interaction*

Let us consider the half-filled case in the Hubbard model. The Hartree term shown in Fig. 3.18(a) pushes up the localized level from ϵ_f to 0. A particular feature in the half-filled case is that the second-order theory for the Green function fortuitously reproduces the atomic limit $\Delta/U \to 0$. In order to see this, we use the Hartree result as the Green function in Fig. 3.18(b): $G^{(1)}(i\epsilon_n) = (i\epsilon_n)^{-1}$. In the imaginary time domain, this becomes $G^{(1)}[\tau] = -\text{sign}(\tau)/2$. Hence, we obtain the self-energy $\Sigma^{(2)}[\tau] = -U^2\text{sign}(\tau)/8$ whose Fourier transform is $\Sigma(i\epsilon_n) = U^2/(4i\epsilon_n)$. Thus, the Green function $G^{(2)}(i\epsilon_n)$ in the second approximation is given by

$$G^{(2)}(i\epsilon_n) = \frac{1}{i\epsilon_n - U^2/(4i\epsilon_n)} = \frac{1}{2}\left(\frac{1}{i\epsilon_n + U/2} + \frac{1}{i\epsilon_n - U/2}\right), \tag{3.36}$$

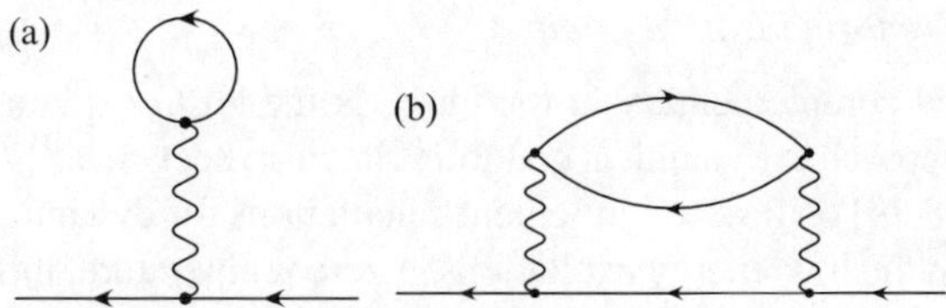

FIG. 3.18. The Feynman diagrams for the self-energy: (a) the first-order and (b) the second-order approximations.

which actually gives the correct atomic limit [58]. If U is small, the low-order perturbation should obviously be reasonable. Thus, if one uses the renormalized Green function which includes both the Hartree field and the dynamic effective field in determining the self-energy, one obtains an interpolation which reproduces both the atomic limit and the band limit. It is also known in the Anderson model [62] that the second-order self-energy constitutes a good approximation when $\epsilon_f + U/2 = 0$.

We emphasize that this fortuitous situation is strictly limited to the half-filled case. In the non-half-filled case, the second-order theory becomes insufficient for large U. Moreover, in order to satisfy the conservation laws in response functions, it is necessary to use the Green function which includes the Coulomb interaction self-consistently [63,64]. Unfortunately, the atomic limit is then no longer reproduced even in the half-filled case.

3.5.2 *Perturbation theory from the atomic limit*

As we have seen in Chapter 2, the NCA gives reasonable dynamical and thermodynamical results in the case of large degeneracy for the local electron level. Thus, it is also applicable for solving the effective impurity problem. The effective hybridization is determined by eqn (3.21). This scheme was first proposed as the extended NCA (XNCA) for the Anderson lattice [57,65,66]. The same theory is applicable to the Hubbard model with the generalization of the NCA to the case of finite U [67,68]. As will be shown in the next section, the numerical result for the density of states is in very good agreement with the quantum Monte Carlo simulation. Since the NCA cannot describe the Fermi-liquid ground state [69], however, very low energy excitations from the metallic ground state of the Hubbard and Anderson models cannot be discussed by the XNCA.

3.5.3 *Quantum Monte Carlo method*

The quantum Monte Carlo (QMC) method has been applied to the single-impurity Anderson model with an infinite number of conduction electrons, and is explained in detail in Ref. [70]. The basic idea of the method is explained very briefly in Appendix G. Dynamical results such as the density of states have also been derived by combination with the maximum-entropy method, which effectively maps a quantity in the imaginary-time domain to real frequencies [71]. Applications of the QMC for solving the effective impurity model have been made [59,72–76]. The method is powerful in giving overall features at finite temperatures. There is, however, a serious technical difficulty in approaching the low-energy range at low temperatures.

3.5.4 *Numerical renormalization group*

A powerful method complementary to the QMC is the NRG explained in § 2.1.2. An extension of the approach to dynamical quantities has also been made [77,78]. We refer to original articles [78,79] on how to implement calculations for dynamics. Since the NRG is most powerful in the low-energy excitations at zero temperature, its application to the infinite-dimensional model shares the same advantage. In the case of the Hubbard model at half-filling, for example, a minute energy scale analogous to the Kondo temperature has been identified near the metal–insulator transition [80].

3.6   Explicit results by dynamic effective field theory

In this section, we show some explicit numerical results for the Hubbard model and the Anderson lattice in infinite dimensions.

3.6.1 *Hubbard model*

The Hubbard model can teach us how the heavy electron energy band is formed within the single band which has both itinerant and localized characters. Figure 3.19 shows the density of states for single-particle excitations [81]. The half-filled case is shown in (I) and a doped case in (II). The Gaussian density of states is used as the noninteracting one to start with. It is remarkable that the agreement of the Monte Carlo result with the XNCA one is excellent. This is because the numerical results are for temperatures well above the critical one where the inadequacy of the NCA appears.

Figure 3.20 shows the result of the NRG [80]. The most significant feature is the very sharp density of states. This means that the energy scale near the metal–insulator transition becomes extremely small. Such a small energy scale is hardly accessible by other methods, including the quantum Monte Carlo method.

Figure 3.21 shows the excitation spectra of spin and charge calculated by the NRG [80]. These spectra correspond to the momentum average of the imaginary part of response functions. The local spin excitation spectrum is relevant to the NMR. It is evident that the characteristic energy of charge excitation is much larger than that of the spin.

3.6.2 *Anderson lattice*

The f-electron density of states in the Anderson lattice with infinite U was computed by the use of the XNCA [66]. Figure 3.22 shows some examples of results where the f-electron level is six-fold degenerate. The parameters in the model are chosen so that the bare f-electron level is $\epsilon_f = -1500$ K, the hybridization intensity $W_0(\epsilon)$ defined in eqn (2.41) is 135 K for $|\epsilon| < 10^4$ K and is zero otherwise. With these parameters, we obtain the Kondo temperature $T_K \sim 16$ K. Here T_K is calculated from the formula

$$T_K = D \left(\frac{n W_0}{D} \right)^{1/n} \exp \left(\frac{\epsilon_f}{n W_0} \right), \tag{3.37}$$

which corresponds to the next-leading-order theory in the $1/n$ expansion discussed in Chapter 2. The f-electron occupation number n_f is computed as $n_f \sim 0.91$ for $T \sim T_K$. The actual calculation was performed for the grand canonical ensemble. It is seen in

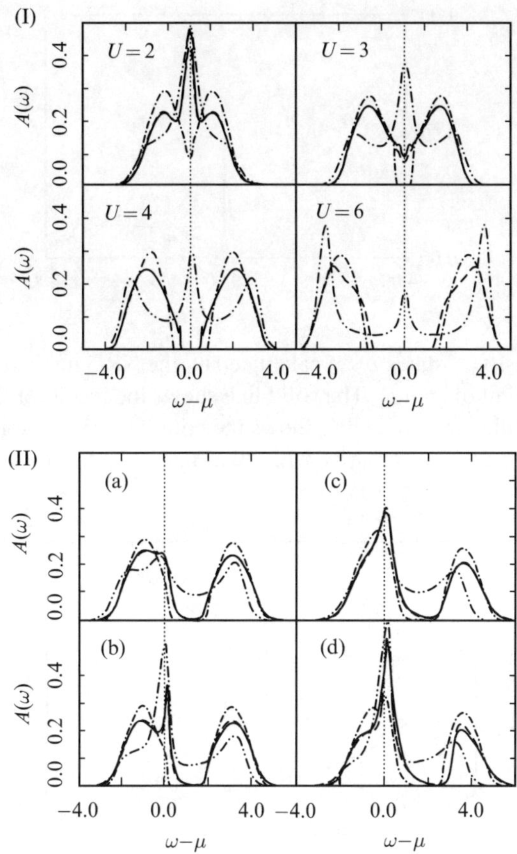

FIG. 3.19. Numerical results for the density of states $A(\omega)$. The half-width Δ is taken as the unit of energy. The temperature in (I) corresponds to $\beta = 7.2$. Each line represents the following: solid line—QMC; dashed line almost overlapping with the solid line—XNCA; double-dot-dashed line—second-order perturbation in U; dot-dashed line—an approximation called LNCA [82]. The temperatures in (II) are $\beta = 3.6$ in (a) and (c), while $\beta = 14.4$ in (b) and (d). The electron number per site is $n_e \sim 0.94$ in (a) and (b), while $n_e \sim 0.8$ in (c) and (d).

Fig. 3.22 that a double-peaked structure develops in the density of states as temperature decreases below about T_K. This dip is interpreted as a remnant of the hybridization gap which would result in the case of electrons without the Coulomb interaction. In this interpretation, the effective f-electron level $\tilde{\epsilon}_f$ lies about 17 K above the Fermi level, although the bare level ϵ_f lies deep below the Fermi level. Because of the finite lifetime of correlated electrons away from the Fermi level, the hybridization gap may not open completely even at $T = 0$. It should be remarked that the characteristic energy in the Anderson lattice is slightly enhanced over that of the impurity Anderson model with the same $W_0(\epsilon)$ and ϵ_f. This is judged [66] from single-particle properties such as $\tilde{\epsilon}_f$ and the renormalization factor a_f defined by eqn (2.67).

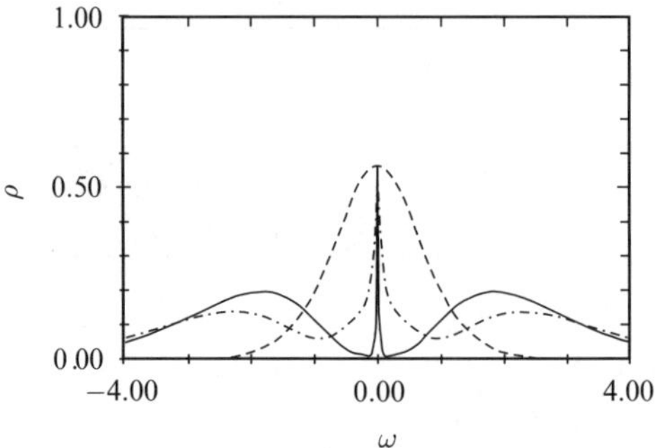

FIG. 3.20. The density of states $\rho(\omega)$ calculated by the NRG method. The half-width Δ is taken as the unit of energy. The solid line shows the result for the Hubbard model with $U = 4$, while the dashed line shows the noninteracting one ρ_0. The dot-dashed line shows the Anderson impurity with hybridization intensity corresponding to ρ_0 with $U = 4$.

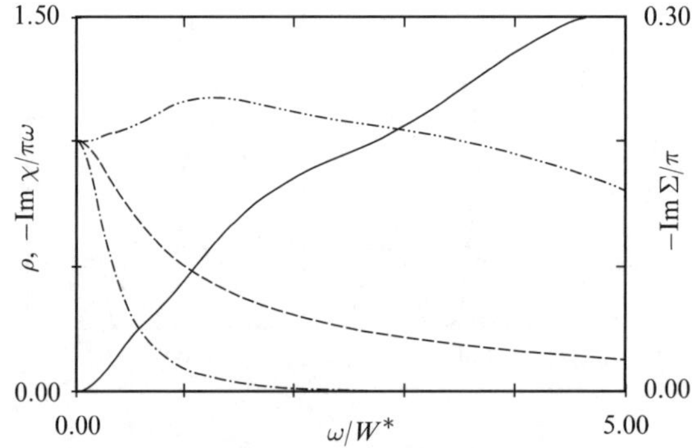

FIG. 3.21. Excitation spectra computed by the NRG method. The dashed line shows ρ with the half-width $W^* = 8.3 \times 10^{-3}$. The dot-dashed line shows Im $\chi_s(\omega)/\omega$ representing the local spin spectrum and double-dot-dashed line the charge spectrum. They are normalized to unity at $\omega = 0$, and the absolute magnitude of Im $\chi_s(\omega)/\omega$ is larger than Im $\chi_c(\omega)/\omega$ by 10^4. The solid line shows Im $\Sigma(\epsilon)$ with the scale on the right.

Suppose now that the Anderson lattice with a single conduction band and without orbital degeneracy of the f electrons has two electrons per unit cell. According to the energy-band theory, the ground state is insulating since the lower hybridized band is completely filled and the upper one is completely empty. One can ask if the situation remains the same under a large value of U. It may happen that the system orders magnetically. In such a case, the f electrons get localized and do not contribute to the band filling. For the simplest case of ferromagnetic order, the conduction band is half-filled and the system becomes metallic. In the case of antiferromagnetism with doubling of the unit cell, the ground state can be insulating. Another interesting case is when the system remains

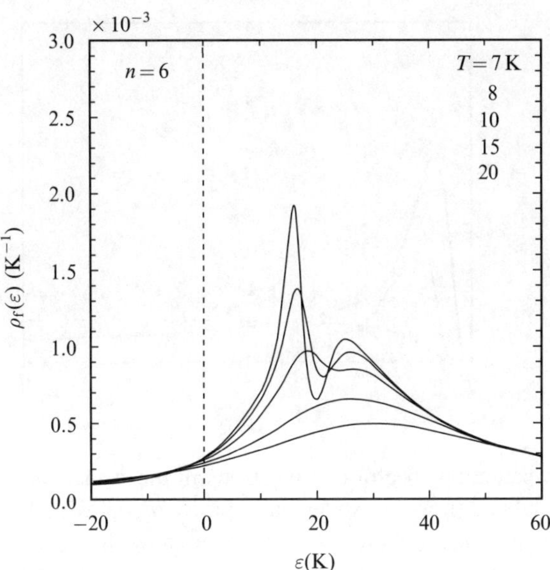

FIG. 3.22. The f-electron density of states in the Anderson lattice at finite temperatures [66]. The Kondo temperature T_K is 16 K according to eqn (3.37).

paramagnetic and can be adiabatically continued from the noninteracting limit. Then the band picture holds and the hybridization gap separates the filled and empty Bloch states.

This hybridization-gap picture is tested quantitatively by the quantum Monte Carlo simulation [83]. The result shows that although the Coulomb interaction reduces the energy gap, the hybridization-gap picture is supported by the result that the spin excitation gap is twice the gap in the density of states. With increasing temperature, however, the gap disappears, in strong contrast with the noninteracting system. Thus, the density of states is strongly dependent on temperature for large U.

Such an insulating ground state has also been studied by means of the NRG method [84]. The results for the density of states and dynamical susceptibilities of spin and charge are shown in Fig. 3.23. We remark that the thresholds for the spin and charge excitations are the same, in agreement with the band picture. However, the dominant spectral weight of the charge excitation lies in a much higher energy range than that of the spin excitation.

The Anderson lattice neglects the Coulomb interaction between the f and conduction electrons. In the insulating state, this interaction gives rise to an exciton-like correlation which is observed commonly in the optical property of semiconductors and insulators. In order to understand the experimental result on SmB$_6$ [30], it seems necessary to include this Coulomb interaction.

3.6.3 *Limitation of the infinite-dimensional model*

The infinite-dimensional model, which is powerful in deriving the dynamical properties of heavy electrons, is by no means a complete theory. Since the intersite correlation is

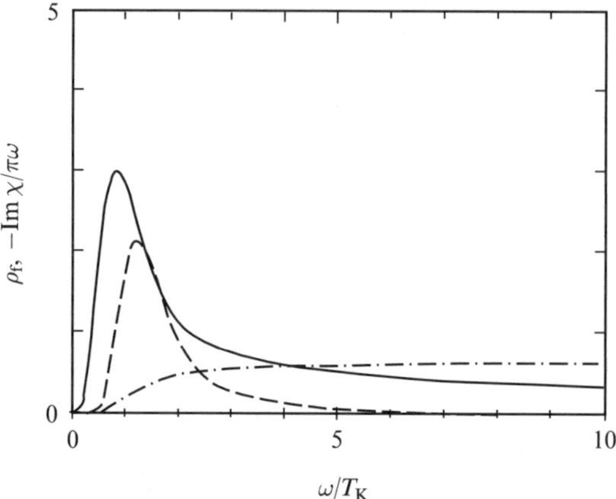

FIG. 3.23. Local excitation spectra of f electrons in the Anderson lattice. The unit of energy is T_K defined from the Anderson impurity model with the same parameters. The solid line shows $\rho_f(\epsilon)$, while the dashed line shows Im $\chi_s(\omega)/\omega$. The dot-dashed line shows Im $\chi_c(\omega)/\omega$ multiplied by 10^3.

treated only at the mean-field level, it cannot deal with phenomena which depend on more details of correlation. For example, the spin-pair singlet correlation is specific to finite dimensionality and can only be dealt with in higher order in the $1/d$ expansion. The pseudo-metamagnetic behaviour in some heavy electrons, as discussed in this chapter, also seems to require a more detailed account of the electron correlations than is possible in the infinite-dimensional model. We shall come back to this problem in Chapter 4.

The process of extension of the method beyond the large-d limit, however, encounters a violation of the analyticity in the Green functions [85]. The same difficulty was found in the effort to improve over the CPA [49]. On the other hand, the thermodynamic quantities, such as the static susceptibilities, do not have such problems and should be easier to treat by the $1/d$ expansion beyond the lowest order. We note that the susceptibility of the Ising model as explained in § 3.4.1 actually includes the $1/d$ correction.

In dealing with the insulating ground state in the Hubbard model, one must be careful about the residual entropy problem. The finite-entropy problem does not arise in the metallic state because the itinerant electron motion of the effective impurity removes the spin entropy as the temperature decreases. On the other hand, a localized spin has a finite entropy unless it is completely polarized or fluctuates quantum mechanically. In the infinite-dimensional theory, however, the quantum fluctuation of spins is not taken into account, although it does account for the quantum charge fluctuation. In other words, in the half-filled Hubbard model with charge fluctuations suppressed by large U, the entropy can vanish only with complete polarization at each site. Thus, it is not compatible with quantum magnetism where zero-point motion reduces more or less the magnitude of the ordered moments, or prevents the magnetic order as in one dimension. In order to deal with the quantum magnetism, an extension of the spherical model approximation has recently been proposed [86].

Bibliography

[1] D. Shoenberg, *Magnetic oscillation in metals* (Cambridge University Press, Cambridge, 1984).

[2] L. Taillefer, R. Newbury, G. G. Lonzarich, Z. Fisk, and J. L. Smith, *J. Magn. Magn. Mater.* **63 & 64**, 372 (1987).

[3] S. R. Julian, F. S. Tautz, G. J. McMillan, and G. G. Lonzarich, *Physica B* **199 & 200**, 63 (1994).

[4] G. G. Lonzarich, *J. Magn. Magn. Mater.* **76 & 77**, 1 (1988); H. Aoki, S. Uji, A. K. Albessard, and Y. Onuki, *J. Phys. Soc. Jpn.* **62**, 3157 (1993).

[5] P. H. P. Reinders, M. Springford, P. T. Coleridge, R. Boulet, and D. Ravot, *Phys. Rev. Lett.* **57**, 1631 (1986).

[6] T. Oguchi and A. J. Freeman, *J. Magn. Magn. Mater.* **61**, 233 (1986); M. R. Norman, R. C. Albers, A. M. Boring, and N. E. Christensen, *Solid State Commun.* **68**, 245 (1989).

[7] P. Haen, J. Flouquet, F. Lapierre, P. Lejay, and G. Remenyi, *J. Low. Tem. Phys.* **67**, 391 (1987).

[8] J. J. M. Franse, P. M. Fring, A de Visser, A. Menovsky, T. T. M. Palstra, P. H. Kess, and J. A. Mydosh, *Physica B* **126**, 116 (1984).

[9] H. Aoki, S. Uji, A. K. Albessard, and Y. Ōnuki, *Phys. Rev. Lett.* **71**, 2110 (1993).

[10] H. Yamagami and A. Hasegawa, *J. Phys. Soc. Jpn.* **61**, 3457 (1992); ibid., **62**, 592 (1993).

[11] I. Kouroudis *et al.*, *Phys. Rev. Lett.* **58**, 820 (1987).

[12] A. Lacerda *et al.*, *Phys. Rev. B* **40**, 11 429 (1990).

[13] Y. Kitaoka, K. Ueda, T. Kohara, Y. Kohori, and K. Asayama, *Nuclear magnetic resonance in heavy Fermion systems: Theoretical and experimental aspects of valence fluctuations and heavy Fermions* (Eds., L. C. Gupta and S. K. Malik, Plenum, 1987) p. 297.

[14] Y. Onuki and T. Komatsubara, *J. Magn. Magn. Mater.* **63–64**, 281 (1987).

[15] K. Ishida, Y. Kawasaki, Y. Kitaoka, K. Asayama, N. Nakamura, and J. Flouquet, *Phys. Rev. B* **57**, R11 050 (1998).

[16] J. Rossat-Mignod, L. P. Regnault, J. L. Jacoud, C. Vettier, P. Lejay, J. Flouquet, E. Walker, D. Jaccard, and A. Amato, *J. Magn. Magn. Mater.* **76 & 77**, 376 (1988).

[17] A de Visser, J. J. M. Franse, and J. Flouquet, *Physica B* **161**, 324 (1989).

[18] Z. Fisk *et al.*, *Science* **239**, 33 (1988).

[19] H. R. Ott, H. Rudigier, Z. Fisk, and J. L. Smith, *Phys. Rev. Lett.* **50**, 1595 (1983); H. R. Ott, *J. Magn. Magn. Mater.* **108**, 1 (1992).

[20] D. E. MacLaughlin, *J. Magn. Magn. Mater.* **47 & 48**, 121 (1985).

[21] A. I. Goldman, G. Shirane, G. Aeppli, E. Bucher, and J. Hufnagil, *Phys. Rev. B* **36**, 8523 (1987).

[22] N. R. Bernhoeft and G. G. Lonzarich, *J. Phys. Condens. Matter* **7**, 7325 (1995).

[23] R. Scherm, K. Guckelsberger, B. Fak, K. Skold, A. J. Dianoux, H. Godfrin, and W. G. Stirling, *Phys. Rev. Lett.* **59**, 217 (1987).

[24] G. Aeppli, E. Bocher, C. Broholm, J. K. Kjems, J. Baumann, and J. Hufnagl, *Phys. Rev. Lett.* **60**, 615 (1988).

[25] A. de Visser, A. Menovsky, and J. J. M. Franse, *Physica B* **147**, 81 (1987).

[26] P. H. Frings, J. J. M. Franse, F. R. de Boer, and A. Menovsky, *J. Magn. Magn. Mater.* **31–34**, 240 (1983).

[27] H. Tou, Y. Kitaoka, K. Asayama, N. Kimura, Y. Onuki, E. Yamamoto, and K. Maezawa, *Phys. Rev. Lett.* **77**, 1374 (1996).

[28] Y. Koike, N. Metoki, N. Kimura, E. Yamamoto, Y. Haga, Y. Onuki, and K. Maezawa, *J. Phys. Soc. Jpn.* **67**, 1142 (1998).

[29] E. A. Schuberth, B. Strickler, and K. Andres, *Phys. Rev. Lett.* **68**, 117 (1992).

[30] P. A. Alekseev, J.-M. Mignot, J. Rossat-Mignod, V. N. Lazukov, and I. P. Sadikov, *Physica B* **186–188**, 384 (1993); *J. Phys. Condens. Matter* **7**, 289 (1995).

[31] T. Takabatake *et al.*, *Physica B* **206 & 207**, 804 (1995).

[32] K. Nakamura, Y. Kitaoka, K. Asayama, T. Takabatake, H. Tanaka, and H. Fujii, *J. Phys. Soc. Jpn.* **63**, 433 (1994); ibid., *Phys. Rev. B* **53**, 6385 (1996).

[33] S. Tomonaga, *Prog. Theor. Phys.* **5**, 349 (1950).

[34] F. D. M. Haldane, *J. Phys. C* **14**, 2585 (1981).

[35] B. H. Brandow, *Phys. Rev. B* **33**, 215 (1986).

[36] H. Shiba, *J. Phys. Soc. Jpn.* **55**, 2765 (1986).

[37] G. Zwicknagl, *Adv. Phys.* **41**, 203 (1992).

[38] H. Tsunetsugu, M. Sigrist, and K. Ueda, *Rev. Mod. Phys.* **69**, 3042 (1997).

[39] P. Nozieres, *J. Low Temp. Phys.* **17**, 31 (1975).

[40] M. Yamanaka, M. Oshikawa, and Affleck, *Phys. Rev. Lett.* **79**, 1110 (1997).

[41] S. Watanabe and Y. Kuramoto, *Z. Phys B* **104**, 535 (1997).

[42] S. Watanabe, Y. Kuramoto, T. Nishino, and N. Shibata, *J. Phys. Soc. Jpn.* **68**, 159 (1999).

[43] T. Sakakibara, T. Tayama, K. Matsuhira, H. Mitamura, H. Amitsuka, K. Maezawa, and Y. Onuki, *Phys. Rev. B* **51**, 12 303 (1995).

[44] A. Georges, G. Kotliar, W. Krauth, and M. J. Rozenberg, *Rev. Mod. Phys.* **68**, 13 (1996).

[45] R. Brout, *Phys. Rev.* **122**, 469 (1961).

[46] K. Miyake, private communication (1990).

[47] E. Müller-Hartmann, *Z. Phys. B* **74**, 507 (1989).

[48] F. Yonezawa and K. Morigaki, *Suppl. Prog. Theor. Phys.* **76**, 621 (1973).

[49] R. J. Elliott, J. A. Krumhansl, and P. A. Leath, *Rev. Mod. Phys.* **46**, 465 (1974).

[50] U. Brandt and C. Mielsch, *Z. Phys. B* **75**, 365 (1989); ibid., **79**, 295 (1990); ibid., **82**, 37 (1991).

[51] W. Metzner and D. Vollhardt, *Phys. Rev. Lett.* **62**, 324 (1989).

[52] W. F. Brinkman and T. M. Rice, *Phys. Rev. B* **2**, 1324 (1970).

[53] B. Velicky, S. Kirkpatrick, and H. Ehrenreich, *Phys. Rev.* **175**, 747 (1968).

[54] J. Hubbard, *Proc. Roy. Soc. A* **281**, 401 (1964).

[55] J. Luttinger and J. Ward, *Phys. Rev.* **118**, 1417 (1960).

[56] G. Baym, *Phys. Rev.* **127**, 835 (1962).

[57] Y. Kuramoto and T. Watanabe, *Physica B* **148**, 80 (1987).

[58] A. Georges and G. Kotliar, *Phys. Rev. B* **45**, 6479 (1992).

[59] M. Jarrell, *Phys. Rev. Lett.* **69**, 168 (1992).

[60] V. Janis and D. Vollhardt, *Int. J. Mod. Phys. B* **6**, 283 (1992).

[61] F. J. Ohkawa, *Phys. Rev. B* **46**, 9016 (1992).

[62] K. Yamada, *Prog. Theor. Phys.* **53**, 970 (1975).

[63] E. Müller-Hartmann, *Z. Phys. B* **76**, 211 (1989).

[64] D. S. Hirashima, *Phys. Rev. B* **47**, 15 428 (1994).

[65] Y. Kuramoto, *Theory of heavy fermions and valence fluctuations* (Eds., T. Kasuya and T. Saso, Springer Verlag, 1985), p. 152.

[66] C.-I. Kim, Y. Kuramoto, and T. Kasuya, *J. Phys. Soc. Jpn.* **59**, 2414 (1990).

[67] O. Sakai, M. Motizuki, and T. Kasuya, *Core-level spectroscopy in condensed systems* (Eds., J. Kanamori and A. Kotani, Springer Verlag, 1988), p. 45.

[68] Th. Pruschke and N. Grewe, *Z. Phys. B* **74**, 439 (1989).

[69] N. E. Bickers, *Rev. Mod. Phys.* **54**, 845 (1987).

[70] J. Hirsch and R. Fye, *Phys. Rev. Lett.* **56**, 2521 (1986).

[71] J. E. Gubernatis *et al.*, *Phys. Rev. B* **44**, 6011 (1991).

[72] X. Y. Zhang, M. J. Rozenberg, and G. Kotliar, *Phys. Rev. Lett.* **70**, 1666 (1993).

[73] A. Georges and W. Kraut, *Phys. Rev. B* **48**, 7167 (1993).

[74] A. Georges, G. Kotliar and Q. Si, *Int. J. Mod. Phys. B* **6**, 257 (1992).

[75] M. J. Rozenberg, G. Kotliar, and X. Y. Zhang, *Phys. Rev. B* **49**, 10 181 (1994).

[76] M. Jarrell, H. Akhlaghpour, and Th. Pruschke, *Quantum Monte Carlo methods in condensed matter physics* (Ed., M. Suzuki, World Scientific, 1993), p. 221.

[77] L. N. Oliveira and J. W. Wilkins, *Phys. Rev. B* **23**, 1553 (1981).

[78] O. Sakai, Y. Shimizu, and T. Kasuya, *J. Phys. Soc. Jpn.* **58**, 3666 (1989).

[79] T. A. Costi and A. C. Hewson, *J. Phys. Cond. Matter* **5**, 361 (1993).

[80] O. Sakai and Y. Kuramoto, *Solid State Commun.* **89**, 307 (1994).

[81] Th. Pruschke, D. L. Cox, and M. Jarrell, *Phys. Rev. B* **47**, 3553 (1993).

[82] N. Grewe, *Z. Phys. B* **67**, 323 (1987).

[83] M. Jarrell, H. Akhlaghpour, and Th. Pruschke, *Phys. Rev. Lett.* **70**, 1670 (1993).

[84] Y. Shimizu and O. Sakai, *Computational physics as a new frontier in condensed matter research* (Eds., H. Takayama *et al.*, The Physical Society of Japan, 1995), p. 42.

[85] A. Schiller and K. Ingersent, *Phys. Rev. Lett.* **75**, 113 (1995).

[86] Y. Kuramoto and N. Fukushima, *J. Phys. Soc. Jpn.* **67**, 583 (1998).

4

ANOMALOUS MAGNETISM

4.1 Characteristics of heavy-electron magnetism

In this chapter, we discuss magnetically ordered or nearly ordered states of heavy electrons. In describing heavy-electron magnetism, there are two complementary pictures one can consider. First, one can regard a heavy-electron system as a lattice of Kondo centres coupled to each other by the RKKY interaction. Of central interest is the competition between the RKKY interaction toward magnetic order and the Kondo screening toward the nonmagnetic singlet state. The RKKY interaction is mediated by conduction band electrons which are only weakly renormalized. The resultant exchange interaction can be represented by $J(q)$ with no retardation. Let us assume that the system has strong antiferromagnetic (AF) correlation with the wavenumber Q. From eqns (1.84)–(1.89), the temperature dependence of $1/T_1$ is given by

$$\frac{1}{T_1} = 2\gamma_n^2 T |A_{hf}|^2 \left(\frac{\chi_L}{\Gamma}\right) \sum_q \frac{1}{[1 - J(q)\chi_L]^2}. \tag{4.1}$$

If magnetic order occurs close to the Kondo temperature T_K characterizing single-site spin fluctuations, both χ_L and Γ are weakly temperature-dependent near the transition temperature T_N. Then by expanding around $q = Q$ according to eqns (1.88) and (1.89), the temperature dependence of $1/T_1$ is expressed approximately by

$$\frac{1}{T_1 T} \propto \frac{\sqrt{\chi_Q(T)}}{|J''(Q)|^{3/2} \chi_L \Gamma}, \tag{4.2}$$

where $J''(Q)$ is the second derivative of $J(q)$ at $q = Q$. Provided that $\chi_Q(T)$ follows the Curie–Weiss law near T_N, $1/(T_1 T)$ behaves as $(T - T_N)^{-1/2}$. In some heavy-electron systems such as UNi_2Al_3, the above picture on the dynamical response function seems to hold qualitatively.

Second, in the alternative picture, one assumes that the magnetic degrees of freedom are described in terms of itinerant heavy quasi-particles. The dynamical properties reflect a strong residual interaction among quasi-particles. The itinerant picture is supported by the fact that even in the magnetically ordered state the density of states near the Fermi level remains finite. Then interesting diversity arises in electronic excitation spectra according to the band structure. For example, there are cases where the critical behaviour is not seen even around T_N, and where the Korringa law, $T_1 T = $ constant, is valid even at T far below T_N. On the contrary, if a nesting of the Fermi surface is significant, the excitation gap opens for a large part of the Fermi surface. The physical quantities then have nearly exponential T dependence below T_N, which saturates at lower T because of the residual density of states. In the following, we take up representative examples of anomalous magnetism of heavy electrons.

4.2 Weak antiferromagnetism in URu$_2$Si$_2$

We first study spin dynamics in URu$_2$Si$_2$. This compound undergoes two phase transitions: first at $T_N = 17.5$ K to an AF phase with a tiny moment of $\sim 0.04\mu_B$, and second to the superconducting state at $T_c = 1.2$ K. The AF order coexists with the superconducting order. Above 70 K, the resistivity decreases with increasing temperature as is usually observed in the high-temperature range of heavy-electron systems. The temperature independence of $1/T_1$ is also consistent with the picture that the system is described as an assembly of independently fluctuating local moments. Below 70 K, the resistivity drops rapidly upon cooling, and the Korringa law appears as shown in Fig. 4.1 [1]. The electronic specific heat just above T_N is approximately linear in T with $\gamma = 180$ mJ/(mol K^2). Below T_N, the heat capacity and T_1 indicate the presence of a gap $E_g \sim 100$ K in the spectrum. Nevertheless, a relatively large γ ($= 50$ mJ/(mol K^2)) still remains around T_c.

The bulk susceptibility of URu$_2$Si$_2$ shows a large anisotropy with an easy direction along the c-axis of the tetragonal symmetry. This is interpreted in terms of the CEF splitting of the J multiplet [2]. A few schemes of the CEF levels have been proposed

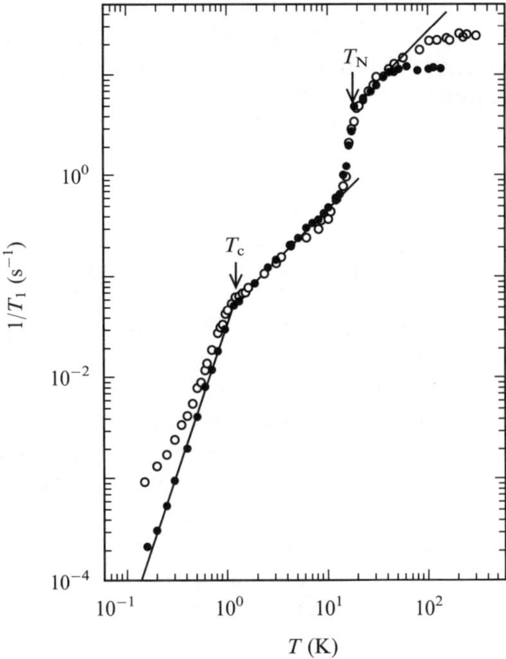

FIG. 4.1. Temperature dependence of $(1/T_1)$ of ^{29}Si (open circle) in external field and ^{105}Ru (solid circle) in zero field in URu$_2$Si$_2$ [1].

Table 4.1 *CEF states with* $J = 4$ *for the tetragonal symmetry with* α, β, γ, ϵ *being numerical coefficients.*

Representation	Basis functions
Γ_{t1}^{1}	$\epsilon[|4\rangle + |-4\rangle)] + \gamma|0\rangle$
Γ_{t1}^{2}	$\gamma[|4\rangle + |-4\rangle)] - \epsilon|0\rangle$
Γ_{t2}	$2^{-1/2}[|4\rangle - |-4\rangle]$
Γ_{t3}	$2^{-1/2}[|2\rangle + |-2\rangle]$
Γ_{t4}	$2^{-1/2}[|2\rangle - |-2\rangle]$
Γ_{t5}^{1}	$\alpha|\pm3\rangle + \beta|\mp1\rangle$
Γ_{t5}^{2}	$\beta|\pm3\rangle - \alpha|\mp1\rangle$

for the $5f^2$ configuration of U^{4+}. In terms of the basis set $|J_z\rangle$ with $|J_z| \leq 4$ for the $^3H_4$ configuration, the CEF eigenstates in tetragonal symmetry are given in Table 4.1. In the model of Nieuwenhuys [3], the energies of the levels are assumed to be

$$E(\Gamma_{t1}^{1}) = 0, \quad E(\Gamma_{t2}) = 46\,\text{K}, \quad E(\Gamma_{t1}^{2}) = 170\,\text{K}, \quad E(\Gamma_{t5}^{1}) = 550\,\text{K}. \tag{4.3}$$

The model calculation based on this scheme leads to an anisotropic susceptibility because the matrix element $\langle\Gamma_{t1}^{1}|J_z|\Gamma_{t2}\rangle$ is finite while $\langle\Gamma_{t1}^{1}|J_\alpha|\Gamma_{t2}\rangle$ with $\alpha = x$, y is 0. However, the Van Vleck term given by

$$\chi_{\text{VV}} = |\langle\Gamma_{t1}^{1}|J_z|\Gamma_{t1}^{2}\rangle|^2 \frac{2}{\Delta}\tanh\left(\frac{\Delta}{2T}\right)$$

does not decrease at low temperatures, in contrast with the experimental results [3]. In an alternative scheme [4], the CEF levels are assumed to be

$$E(\Gamma_{t3}) = 0, \quad E(\Gamma_{t1}^{1}) = 44\,\text{K}, \quad E(\Gamma_{t2}) = 111\,\text{K}, \quad E(\Gamma_{t5}^{1}) = 485\,\text{K}. \tag{4.4}$$

Now, the ground state does not have a finite off-diagonal matrix element of J_z with nearby levels, but the first and second excited states give rise to finite matrix elements. For this reason, the resultant χ_{zz} decreases as the population of Γ_{t1}^{1} becomes small with decreasing temperature, consistently with the experiments. However, the neutron scattering results to be presented below are difficult to explain in the scheme of eqn (4.4). We should mention that in the dilute alloy $U_{0.01}Th_{0.99}Ru_2Si_2$, $\chi_{zz}(T)$ increases as $\ln T$ with decreasing T down to 0.1 K (see Fig. 2.19). This non-Fermi-liquid behaviour has been interpreted by the assumption that the ground state CEF level consists of the doublet Γ_{t5}^{1} [5].

The antiferromagnetic order and fluctuations in URu_2Si_2 have been studied extensively by magnetic neutron scattering. The ordering pattern below 17.5 K is that of a type-I antiferromagnet with spins along the c-axis and antiparallel between adjacent planes. The ordered moment has magnitude $(0.04 \pm 0.01)\mu_B$, and is polarized along the c-axis [6]. A striking feature of the magnetic response is that both CEF and the itinerant aspects are manifested. Thus, at low energies, sharp CEF-like excitations propagate along the tetragonal basal plane, whereas at high energies and for fluctuations propagating along the c-axis the excitations constitute broad magnetic fluctuations as observed in other heavy-electron systems.

Figure 4.2 shows the spectrum of inelastic neutron scattering. Well-defined peaks are observed in energy scans at 1 K for various values of momentum transfer along $q_a = (1, \zeta, 0)$ [6]. These dispersive excitations have a gap of 2 meV at the AF zone centre, and are damped out above T_N. We note that the nuclear scattering is forbidden for $\zeta = 0$ because of the body-centred tetragonal structure. From a series of measurements with various energy and momentum transfers, the dispersion relation for the magnetic excitation is extracted (Fig. 4.3).

From the integrated intensity of magnetic excitations for polarization along the c-axis, it is found that the magnetic form factor is typical of 5f electrons like that of UO$_2$. The spin-wave-like excitation at $Q = (1, 0, 0)$ has a large transition-matrix element of $g\mu_B|\langle i|J^z|f\rangle| = 1.2\mu_B$. Furthermore, this longitudinal inelastic scattering exhibits no broadening beyond the resolution upon the application of the magnetic field along the c-axis. The last fact is consistent with the transitions between the two singlets and confirms that the dispersive state is not a part of continuum excitations.

We now turn to low-temperature magnetic excitations at energies well above the dispersive excitation. Figure 4.4 shows a perspective view of the neutron intensity vs energy and momentum transfer along $q_c = (1, 0, \zeta)$ at 5 K. This makes it possible to compare the dispersive excitation and the high-energy overdamped response. In the energy region higher than 5 meV, there emerges a continuous spectrum of dominantly AF spin fluctuations. As seen in the constant-energy scan data at 8 meV, the scattered intensity decreases when the momentum transfer varies from $Q = (1, 0, 0)$ to $q_F = (1, 0, \pm 1)$.

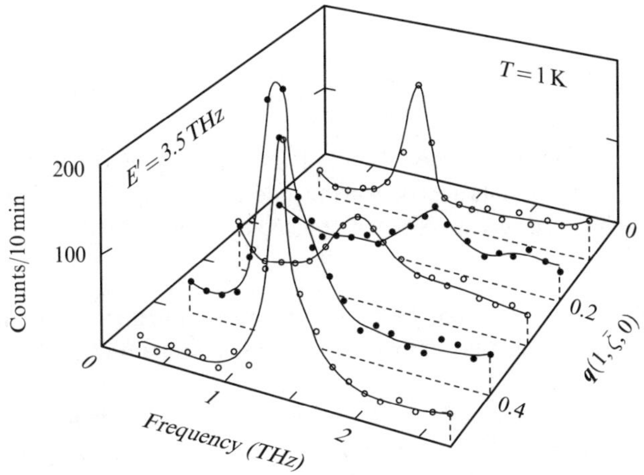

FIG. 4.2. Constant-Q scans in URu$_2$Si$_2$ along the $(1, -\zeta, 0)$ direction showing sharp magnetic excitations [6]. The energy corresponding to 1 THz is 4.1 meV.

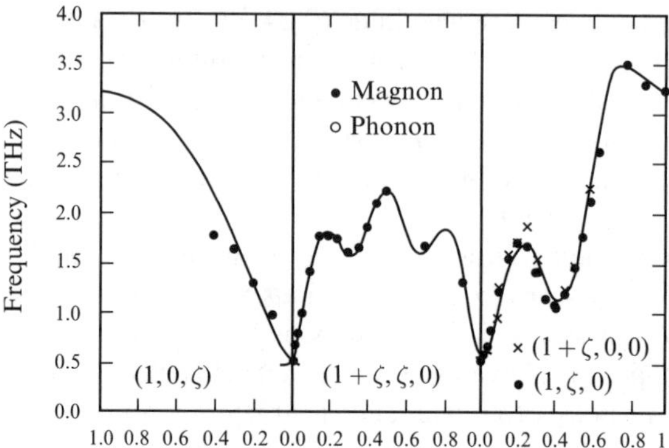

FIG. 4.3. Dispersion of excitations in URu_2Si_2 along the $(1, 0, \zeta)$, $(1+\zeta, \zeta, 0)$ and $(1, \zeta, 0)$ directions [6].

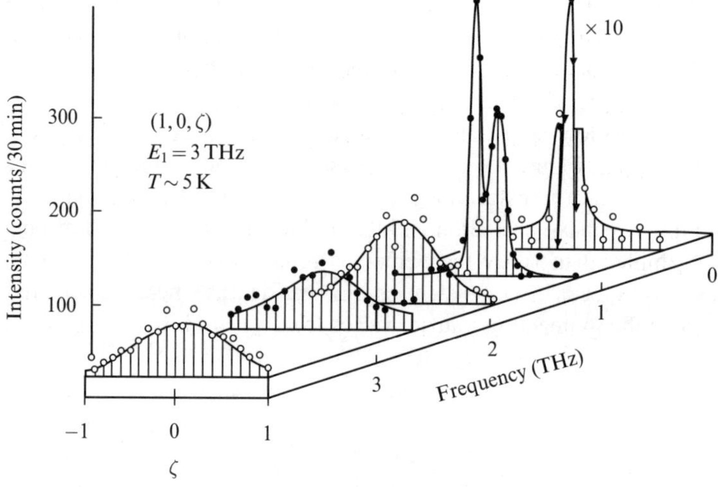

FIG. 4.4. Perspective view of scattered neutron intensity vs energy and momentum transfer along $(1, 0, \zeta)$. The data are taken in the ordered phase at $T = 5\,K$. At $\nu = 0.5\,THz$ (2 meV), the hatched region has been reduced by a factor of 10 [6].

The wavenumber q_F corresponds to the ferromagnetic fluctuation. From the width of the peak in this scan, it is deduced that the magnetic correlation length along the c-axis is only about one lattice unit.

Figure 4.5 shows the temperature dependence of the scattered intensity with energy transfer of 8 meV at $Q = (1, 0, 0)$ and $q_F = (1, 0, 1)$. The intensity of inelastic scattering above 6 meV remains almost unaffected by the magnetic transition at 17.5 K, and persists up to 40 K. Whereas the inelastic scattering intensity at $Q = (1, 0, 0)$ decreases slightly with increasing temperature, the intensity at $q_F = (1, 0, 1)$ increases, so that

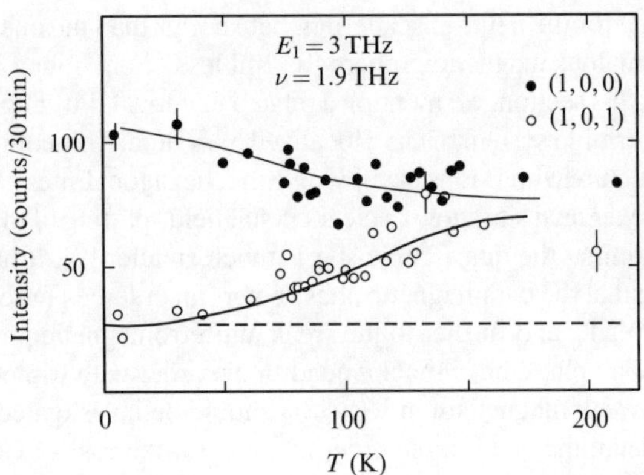

FIG. 4.5. Scattered neutron intensity vs temperature at energy transfer of 1.9 THz (8 meV) and scattering vectors (1, 0, 0) and (1, 0, 1). The dashed line is the background [6].

both become comparable above 100 K. Thus, the differences between ferromagnetic and AF correlations become significant only below 100 K. It should be noted that the development of AF correlations almost coincides with the temperature below which the resistivity of URu$_2$Si$_2$ begins to decrease.

The high-energy response may be regarded as due to itinerant particles which have a large f-electron weight. This view is supported by the variable amplitude of the magnetization and by the f-electron-like magnetic form factor. Thus, the response bears a resemblance to that observed in other heavy-electron systems. The formation of fermionic itinerant particles is also manifested by the onset of metallic T dependence in the resistivity, and the NMR result with the Korringa law. The itinerant particles become antiferromagnetically correlated below 100 K. We note that, since the temperature is high, these itinerant particles may have a character much different from that of the Landau quasi-particles.

The presence of sharp magnetic excitation in URu$_2$Si$_2$ is unique among heavy-electron systems. The dispersion relation and intensities of low-energy CEF excitations are described well by the singlet–singlet model. The model was also applied for explaining the T dependence and anisotropy of the magnetic susceptibility. With the assumption of a singlet ground state and with a quadrupole ordering [4], the CEF model of eqn (4.4) can approximately describe observed behaviours of linear and nonlinear susceptibilities and λ-type anomaly of the specific heat at T_N. However, the ordered moment predicted by this simple model is an order of magnitude larger than the experimental value.

The reason for the presence of the tiny ordered moment is an unresolved problem. Actually, no hyperfine broadening of the Si-NMR spectrum is detected at the onset of the long-range order. If the AF sublattice moment of $0.03\mu_B$ is really static, it should have led to observable hyperfine broadening. Thus, another possibility is that the magnetic order is not static but is slowly fluctuating. If this is the case, the time scale should be longer than the one for the neutron scattering, but shorter than the one for the NMR. We note that this anomalous magnetic property resembles the one found in UPt_3.

In concluding this section, we mention a related material UPd_3 although it is not classified as a heavy-fermion system but as a localized 5f system. This compound crystallizes in a hexagonal structure with U ions occupying either hexagonal sites or quasi-cubic sites. The U ion at the hexagonal site shows a clear crystal field splitting of the order of 10 meV, but the CEF splitting at the quasi-cubic site is much smaller [7]. It has been shown by neutron scattering that the paramagnetic phase enters an ordered phase with no magnetic moment at $T_1 = 6.5\,K$, and further to the weak antiferromagnetic phase at $T_2 = 4.5\,K$ [8]. The intermediate phase has in fact a quadrupole order with triple-Q structure. The relationship with weak magnetism in URu_2Si_2 should be investigated further. We shall explain in more detail the quadrupole order in § 4.6 for the case of CeB_6.

4.3   Antiferromagnetism in UPd_2Al_3 and UNi_2Al_3

UNi_2Al_3 and UPd_2Al_3 form a new series of antiferromagnetic heavy-electron superconductors with large uranium-derived moments $0.24\mu_B$ and $0.85\mu_B$, respectively [9]. The transition temperatures are $T_N = 4.6\,K$ and $T_c = 1\,K$ for UNi_2Al_3 and $T_N = 14.5\,K$ and $T_c = 2\,K$ for UPd_2Al_3. Of these, UPd_2Al_3 exhibits the highest T_c and the largest ordered moments as heavy electrons. The superconducting properties will be discussed in detail in Chapter 5.

The magnetic susceptibility $\chi(T)$ of single-crystal UPd_2Al_3 shows easy plane anisotropy, in contrast to the case of URu_2Si_2. A Curie–Weiss-type behaviour with effective moment $3.6\mu_B$/U-atom is observed for both directions at temperatures higher than 150 K. The susceptibility $\chi(T)$ shows a maximum at around 35 K and abruptly decreases below $T_N = 14.5\,K$. The magnetic order consists of ferromagnetic sheets in the basal plane, which are coupled antiferromagnetically along the c-axis with an ordering vector $Q = (0, 0, 1/2)$. The large λ-type anomaly of the specific heat at T_N for UPd_2Al_3 accompanies a large term linear in T, with $\gamma = 150\,mJ/(mol\,K^2)$. These features are similar to those in URu_2Si_2, but the size of the ordered moment and the nature of the ordered state are very different.

In contrast, the T dependence of $1/T_1$ of ^{27}Al in UPd_2Al_3 shows a behaviour common to that in URu_2Si_2 [10]. Below T_N, $1/T_1$ drops markedly, and is fitted by the following simple form:

$$\frac{1}{T_1 T} = A + B \exp\left(-\frac{E_g}{T}\right).$$

The first term originates from the particle–hole excitation in the Fermi-liquid state. When the superconductivity is suppressed by the magnetic field, the Korringa law remains valid down to $T < T_N$. This result means that the spin-wave excitation is not responsible for the relaxation process in the low-T regime. The second term should be related to the energy gap due to the magnetic ordering. According to a tentative estimate, half the density of states is lost below T_N.

The susceptibility $\chi(T)$ in UNi_2Al_3 does not exhibit a Curie–Weiss behaviour below 300 K. A small maximum of $\chi(T)$ vs T appears around 100 K, which is followed by a

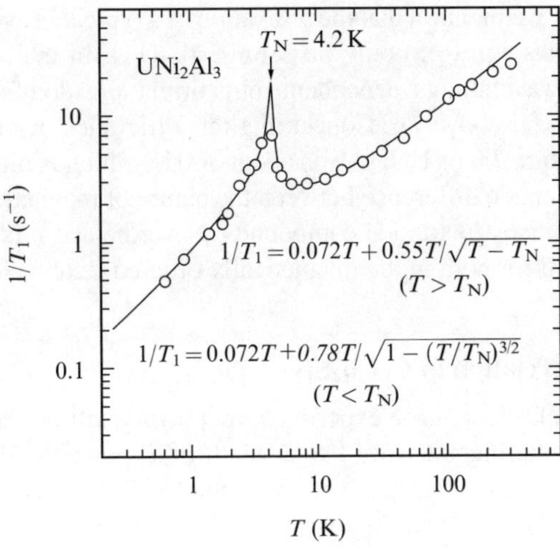

FIG. 4.6. Temperature dependence of $(1/T_1)$ of ^{27}Al in UNi$_2$Al$_3$ [10].

minimum around 30 K and a clear peak at $T_N = 4.6$ K [9]. Elastic neutron scattering measurements on single crystals have revealed that this compound has an incommensurate spin-density-wave-type order. The wave vector is $Q = (1/2 \pm \tau, 0, 1/2)$ with $\tau = 0.11$ and the size of the ordered moment is $0.24\mu_B$/U. As shown in Fig. 4.6, $1/T_1$ of ^{27}Al does not follow the Korringa law in the paramagnetic state. It tends to saturate above the room temperature [10]. $1/T_1$ approaches a linear T dependence well below $T_N = 4.6$ K when the superconducting transition is suppressed by the magnetic field. These features are different from those commonly observed in other heavy-electron systems. Namely, $1/T_1$ usually undergoes a smooth crossover from the behaviour $1/T_1 = $ constant at the T region higher than the Kondo temperature to the Korringa law at low T (see Fig. 4.1).

The T dependence of the relaxation in UNi$_2$Al$_3$ is described well by the following expression:

$$\frac{1}{T_1 T} = a + b \frac{1}{\sqrt{(T - T_N)}}. \tag{4.5}$$

Here the first term is due to the Al-3p orbital relaxation and the second due to the transferred hyperfine interactions of ^{27}Al nuclei with uranium 5f-electron spins. The characteristic T dependence of the second term coincides with the expression in eqn (4.2) of the staggered susceptibility, which follows the Curie–Weiss law at the wave vector $q = Q$ above T_N. From the T dependence of $^{27}(1/T_1)$, it is concluded that the spin fluctuation in UNi$_2$Al$_3$ possesses a large q dependence, in contrast to the standard behaviour of heavy-electron compounds. In the latter case, the magnetic response is described as arising from an assembly of local moments at high T, whereas the response at low T is that of the Fermi liquid. Even in the latter case, the q dependence is weak.

For comparison, we mention that the relaxation of a typical heavy-electron antiferro-magnet UCu_5 follows eqn (4.5) only just above T_N [11]. In the case of UNi_2Al_3, on the other hand, the wavenumber-dependent spin correlations dominate in the wide temperature range from far above T_N. Concerning this difference, we remark that the spin fluctuation temperature T^* of UNi_2Al_3 is about 300 K, which is much larger than T_N.

We have thus seen the difference between the nature of magnetic fluctuation and the spin structure of two isostructural U compounds. Nevertheless, it is remarkable that the Korringa law is valid for both in the magnetically ordered state well below T_N.

4.4   Magnetic correlation in $CeCu_2Si_2$

According to the inelastic neutron experiment on polycrystalline samples of $CeCu_2Si_2$, the quasi-elastic scattering intensity has a width of $\Gamma \sim 10$ K [12], which suggests that $T_K \sim 10$ K. This value lies between the value for $CeCu_6$ with $T_K = 6$ K and that for $CeRu_2Si_2$ with $T_K = 23$ K. As expected, the value of the coefficient of specific heat γ, linear in T, is correspondingly large, $\gamma = 800$ mJ/(mol K^2) and we obtain $\gamma T_K/k_B \sim 6 \times 10^{23}$/mol. Note that we have reinstated the Boltzmann constant here. The latter value, close to the Avogadro's number, is comparable to those in $CeCu_6$ and $CeRu_2Si_2$, and is a measure of the number of heavy electrons.

$CeCu_2Si_2$ shows anomalous magnetism near the superconducting transition at T_c=0.7 K. The NQR intensity observed around 3.435 MHz decreases upon cooling below 1 K without any broadening associated with the spontaneous magnetic moments [13,14]. This anomalous state is called the phase A, which does not have a static magnetic order, but has a very slow magnetic fluctuation, with frequencies comparable to the NQR frequency ω_N [14]. This dynamical character of the phase A is consistent with recent μSR experiments [15]. Neutron scattering experiments detect neither magnetic Bragg peaks nor any superstructure which indicates an order parameter of the spin-density-wave (SDW) or the charge-density-wave (CDW) for the phase A. We mention that the phase A was first suggested by magnetoresistance [16], and its magnetic nature was confirmed by NMR [13,17,18] and muon spin rotation (μSR) [19].

A series of polycrystalline $Ce_xCu_{2+y}Si_2$ in the vicinity of stoichiometric composition together with a high-quality single crystal have been studied by measurements of elastic constant and thermal expansion [15,20–23]. In this process, another phase called the phase B was identified above 7 T. Figure 4.7 shows the phase diagram obtained on a high-quality single crystal of $CeCu_2Si_2$ [20]. The phase A and the superconducting phase are extremely sensitive to sample preparation, especially to the nominal content x [22] of Ce. The phase A under zero field seems to be expelled below T_c by the onset of superconductivity in both high-quality single-crystalline [20] and polycrystalline samples [22]. From these macroscopic measurements, it is suggested that $CeCu_2Si_2$ is close to a quantum critical point [23], where different ground states meet each other.

Figure 4.8 shows the temperature dependence of $1/T_1$ of ^{63}Cu under zero field for a series of $Ce_xCu_{2+y}Si_2$ compounds [24]. For $x = 0.975$ below 1 K, there appear long (T_{1L}) and short (T_{1S}) components in the relaxation. In this case, both components are estimated as shown in Fig. 4.8. In other cases of $Ce_xCu_{2+y}Si_2$, there is only a single component in T_1.

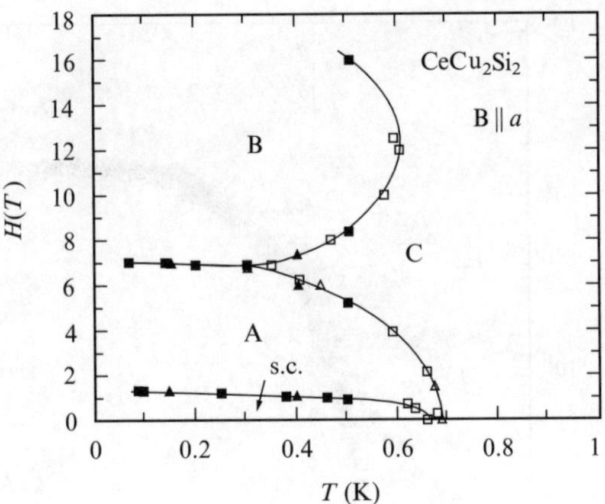

FIG. 4.7. Magnetic field H vs temperature phase diagram of CeCu$_2$Si$_2$ with H in the tetragonal basal plane ($H \perp c$-axis). The phase boundaries are determined from elastic constant, magnetostriction and thermal expansion anomalies [20].

Above 3 K, $1/T_1$ values for all the samples fall on the same curve, which shows that $T_K \sim 10$ K is nearly independent of the Ce concentration x. Below 2 K, the T dependence of $1/T_1$ reflects the difference in the ground state of each concentration. We note that $1/T_1$ below 1 K has both short (T_{1S}) and long (T_{1L}) components [24]. A cusp in $1/T_{1S}$ for $x = 0.975$ is observed at 0.6 K. This cusp is associated with the onset of a static magnetic order. On the other hand, $1/(TT_1)$ for $x = 1.025$ is nearly constant between 1.2 K and T_c. This behaviour reflects the formation of a Fermi-liquid state before the superconductivity appears. For the state with $x = 0.99$, which seems to have the phase A as the ground state, magnetic fluctuations comparable to $\omega_N \sim 3.4$ MHz dominate down to 0.012 K. In this sense, the phase A is characterized as a 'critically magnetic phase'. Note that the $1/T_1$ values in the superconducting state for $x = 1.025$ and $x = 1.00$ follow the T^3 dependence in the range $T = 0.6$–0.1 K, and fall on the single line. This result supports the conclusion based on elastic measurements on a high-quality single crystal [20] that the phase A is expelled below T_c by the onset of the superconducting phase.

Figure 4.9 shows the phase diagram of Ce$_{1-x}$Th$_x$Cu$_2$Si$_2$ with the Th content x as the abscissa. The phase diagram of CeCu$_{2.02}$Si$_2$, with the magnetic field taken as the abscissa, is also shown [14]. The solid line in the H–T plane in CeCu$_{2.02}$Si$_2$ above 4 T corresponds to those temperatures and fields at which anomalies appear in the magneto-resistance and the dHvA signal [25]. For CeCu$_2$Si$_2$, T_M, the transition temperature to the phase A, is determined as the temperature below which the NQR intensity decreases without broadening. On the other hand, the (static) magnetic ordering temperature T_N is determined as the temperature below which the NQR linewidth starts to increase. From the measurements of thermal expansion and the elastic constants, a lattice anomaly has been found at T_M [26]. It is seen that antiferromagnetic order is induced by substitution of Ce for Th. This seems to be due to the decrease of an effective f-electron number.

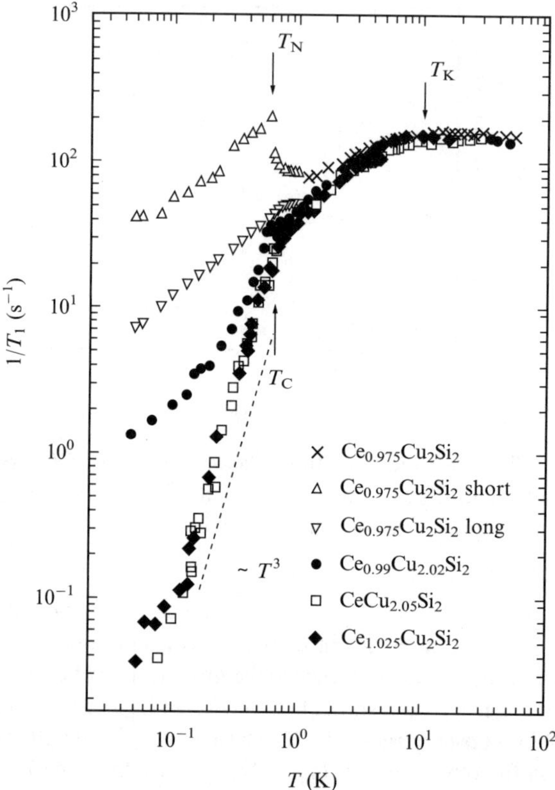

FIG. 4.8. T dependence of NQR $1/T_1$ in $Ce_xCu_{2+y}Si_2$. In the particular case of $Ce_{0.975}Cu_2Si_2$ below 1 K, short (T_{1S}) and long (T_{1L}) components of $1/T_1$ appear [24].

As shown by the dashed line in Fig. 4.9, T_c of $CeCu_2Si_2$ decreases gradually with increasing magnetic field or content of Th. On the contrary, T_c is enhanced by applying pressure from 0.7 K at ambient pressure to 2.2 K at 30 kbar [27]. This is understood if one assumes that $CeCu_2Si_2$ is on the border of antiferromagnetism. Then the magnetic fluctuations above T_c should have extremely low frequencies. The application of pressure increases hybridization and hence T_K, which means an increase of the bandwidth of the heavy quasi-particle [18]. The superconducting ground state is favoured under such a condition, whereas the magnetically ordered state is favoured by the opposite condition; decrease of either T_K or the effective f-electron number.

4.5 Non-Fermi-liquid behaviour near the antiferromagnetic phase boundary

When the Kondo exchange interaction J between the f and conduction electrons is strong enough, the ground state is the paramagnetic Fermi liquid. Then weakening the exchange interaction, i.e., reducing the c–f hybridization can lead to long-range magnetic order. For example, in nonmagnetic $CeCu_6$, Au substitution of Cu expands the lattice parameter and,

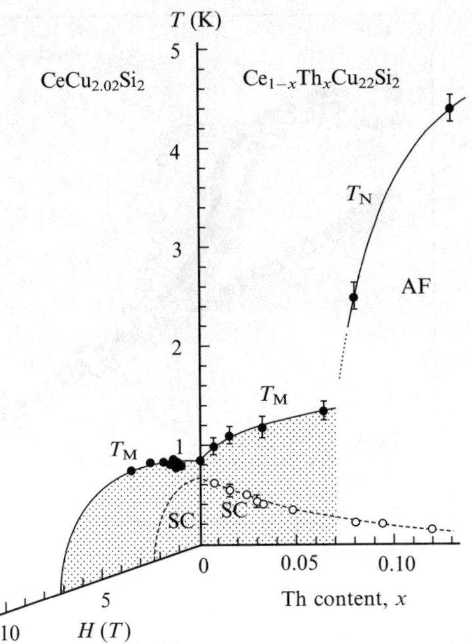

FIG. 4.9. Magnetic and superconducting phase diagrams for CeCu$_2$Si$_2$ as functions of the Th content and the magnetic field [14].

as a result, reduces J_K. Actually, the AF order is observed in CeCu$_{6-x}$Au$_x$ above a critical concentration $x_c \sim 0.1$ [28]. Conversely, in a CeCu$_{5.7}$Au$_{0.3}$ alloy with $T_N = 0.49$ K at zero pressure, the breakdown of the AF order occurs at a critical value $p_c \sim 8$ kbar. In the vicinity of such a magnetic–nonmagnetic transition, strong deviations from Fermi-liquid behaviour are seen. Typical anomalies in the physical quantities are: $C/T \sim -\ln(T/T_0)$, $\chi \sim (1 - \alpha T^{1/2})$, and $\rho \sim \rho_0 + AT$. Figure 4.10 shows the temperature dependence of C/T at various pressure values. A strong deviation from the canonical behaviour, $C/T = $ constant is evident from its logarithmic dependence $C/T \sim -\ln(T/T_0)$ in a wide T range [28]. This non-Fermi-liquid behaviour has a different microscopic origin from that in the single-ion case discussed in Chapter 2. In the present case, the anomalies originate from the dominant collective magnetic excitations caused by the incipient AF order near the magnetic–nonmagnetic phase boundary.

4.6   Quadrupolar and magnetic orderings in CeB$_6$

CeB$_6$ crystallizes in a cubic CsCl-type structure with a B$_6$ octahedron in the body centre of the cube. The cubic crystal field around each Ce ion lifts the degeneracy of the six-fold multiplet with $J = 5/2$ of the trivalent 4f^1 configuration into two CEF levels: the ground-state quartet Γ_8 and the excited doublet Γ_7 lying above 530 K from Γ_8. The phase diagram of CeB$_6$ exhibits anomalous features with two kinds of ordered phases as shown in Fig. 4.11 [29]. In the paramagnetic phase (called phase I) at $T > T_Q = 3.2$ K

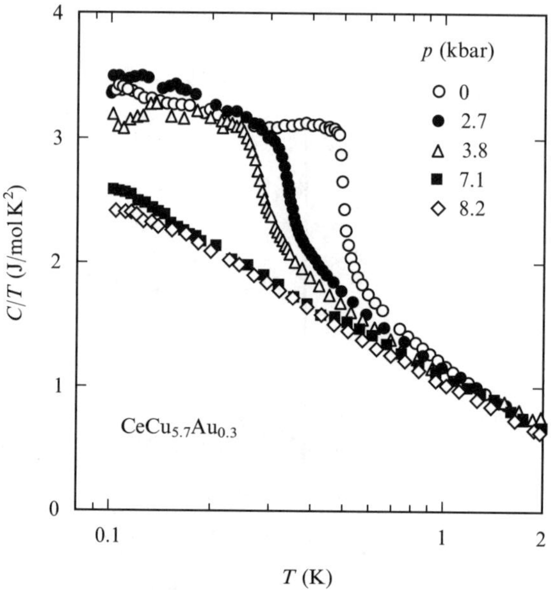

FIG. 4.10. Specific heat C of $CeCu_{5.7}Au_{0.3}$ plotted as C/T vs ln T for various values of pressure p [28].

in zero field, the resistivity shows a Kondo-type behaviour with $T_K = 1$ K. In the temperature range $T_N = 2.4$ K $< T < T_Q$, a new phase (denoted II) appears, which is characterized by the antiferro-quadrupolar (AFQ) ordering with the ordering vector $Q = [1/2, 1/2, 1/2]$ in units of the reciprocal lattice vectors. With the quadrupole order, the Γ_8 ground state should be split into two doublets. Although the AFQ order itself cannot be observed directly by neutron scattering experiments, the external magnetic field induces an AF order with the same wave vector as that of the AFQ ordering. The induction is due to different local susceptibilities associated with two inequivalent Ce ions. In phase III, which appears in the lowest-T range, a magnetic order develops with the ordered moment $0.28\mu_B$ and with a double-k commensurate structure of wave vectors $k_\pm = [1/4, \pm 1/4, 1/2]$.

The anomalous magnetism apparently originates from the interplay of a few relevant effects such as the single-site Kondo-type fluctuation and intersite quadrupolar and RKKY interactions. We note that T_K and T_Q are of the same order of magnitude in CeB_6.

The Kondo effect appears also as the reduced magnetic moment and a large residual value of electronic specific heat with $\gamma = 240$ mJ/(mol K^2) below T_N. Furthermore, the applied magnetic field enhances T_Q and the specific heat anomaly markedly. An interpretation for this is that the application of the magnetic field progressively suppresses the Kondo state; as a result, T_Q shifts to higher temperatures. It is also suggested [30,31] that the actual T_Q has been reduced appreciably from the mean-field value by the fluctuations

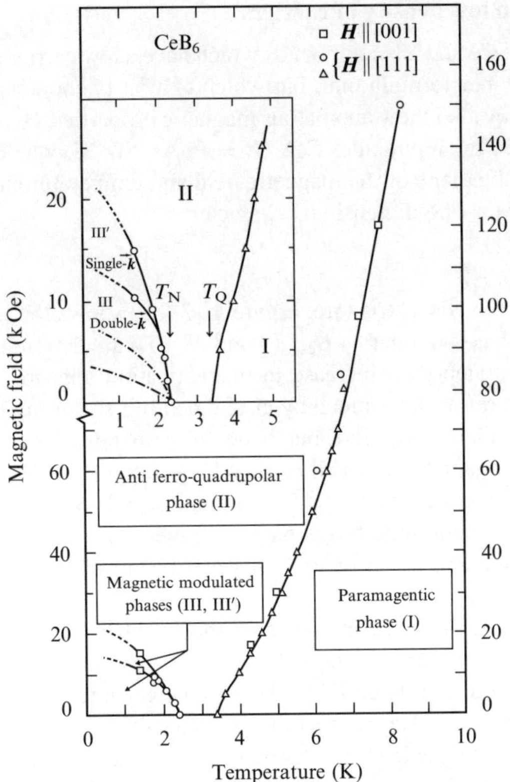

FIG. 4.11. Phase diagram of CeB$_6$ obtained for a magnetic field along [001] and [111] direction of the cubic structure [29].

of the AFQ order. The number of fluctuating components is an important parameter for the reduction. Since the applied magnetic field suppresses the fluctuation by lowering the number of equivalent components, T_Q is enhanced. For the same reason, the specific heat anomaly is also enhanced. There is also an interpretation which assumes that the octupolar intersite interaction is enhanced by the spin polarization [32,33]. The octupole degrees of freedom play an important role in reconciling the NMR results [34] with the order parameter in phase II [35].

Upon alloying with La, the system Ce$_x$La$_{1-x}$B$_6$ shows a complex phase diagram with intriguing dependence on x. A new phase (phase IV) appears with $x < 0.75$ in the low-field region. On entering this phase from the phase I, the susceptibility shows a cusp as if it entered a Néel state [36]. However, phase IV has very small magnetic anisotropy and small magnetoresistance [37], in contrast with phase III. It also shows a prominent elastic anomaly [38]. It seems that the orbital degrees of freedom play an important role in realizing these unusual properties.

4.7 Systems with low density of carriers

There are classes of Ce and Yb compounds which have a low carrier density, of the order of only 10^{-3}–10^{-2} per formula unit, but which exhibit phenomena like those seen in Kondo systems. They also show anomalous magnetic properties. Here we present typical examples such as Ce-monopnictides CeX (X = P, As, Sb, Bi) which show complicated phase diagrams in the plane of the magnetic field and temperature, and Yb$_4$As$_3$, where spin excitations have a one-dimensional character.

4.7.1 *CeP*

CeP crystallizes in the NaCl structure. Figure 4.12 [39] shows the magnetic phase diagram of CeP. The phase boundaries $H_{c1}(T)$ and $H_{c2}(T)$ are determined from the values of the field where a step-wise increase in magnetization appears. Figure 4.13 shows the magnetic structures as determined by elastic neutron scattering [39]. In phase I, Ce spins in each (001) plane align ferromagnetically perpendicular to the plane. What is unusual is that the magnetic unit cell is very long along the [001] direction; the ferromagnetic double layers containing a Ce moment of $\sim 2\mu_B$ are stacked periodically. The Ce ions in intervening nine layers have a moment of $0.7\mu_B$. The values $2\mu_B$ and $0.7\mu_B$ of ordered moments are close to those of CEF states labelled as Γ_8 and Γ_7 in the cubic symmetry.

In phase II, the AF ordered moments of the Γ_7 layers change their direction as shown in Fig. 4.13, while there is no change in the ordering of the Γ_8 double layers. In phase III, the period of ordered Γ_8 Ce layers changes into *ten* layers, and the Γ_7 Ce layers sandwiched by the Γ_8 layers become paramagnetic.

These anomalous magnetic structures seem to originate from the low carrier density of this compound [40]. At low temperatures, holes tend to localize so as to obtain the energy gain through the p–f mixing interaction. This results in the polarization of 4fΓ_8 state. It is not fully understood why the Γ_8 double layers have such long periodicity as *eleven* layers.

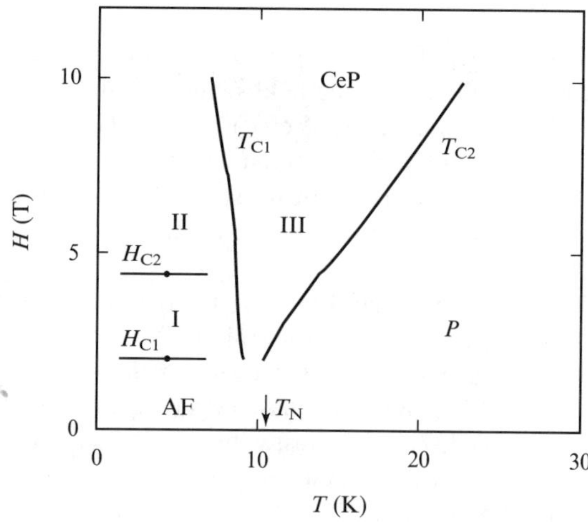

FIG. 4.12. Phase diagram of CeP for magnetic field along [001] direction [39].

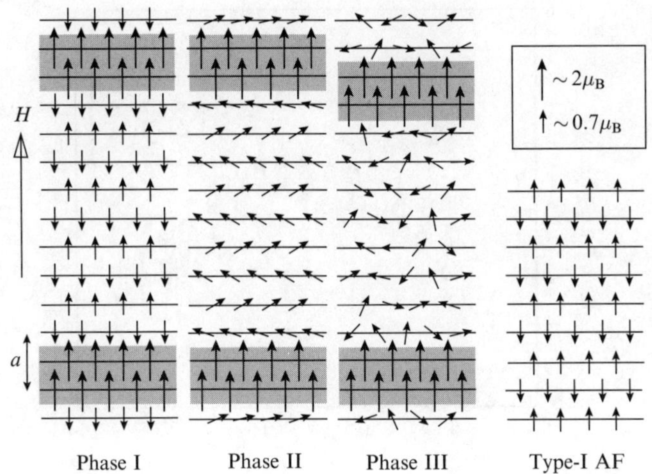

FIG. 4.13. Magnetic structure of CeP under magnetic field [39].

4.7.2 Yb_4As_3

Yb_4As_3 exhibits a structural phase transition at 290 K from the mixed-valent state in the cubic phase to the charge ordered state in the trigonal phase. In the mixed-valent state, the proportion of two valences is about $Yb^{+3} : Yb^{+2} \sim 1 : 3$, and the average $4f$ hole number is ~ 0.25 per Yb. This leads to metallic character for the system. On the other hand, the carrier density at low temperatures is extremely low (10^{-3} per formula). Nevertheless, the low-temperature properties are similar to those of the typical heavy-electron materials. It shows a large T-linear term in specific heat with $\gamma = 205\,\mathrm{mJ/mol\,K^2}$, and the resistivity has a T^2 dependence at low temperatures, followed by a $- \log T$ behaviour at higher T [41].

In the charge ordered state, the distance between Yb atoms becomes shorter in the chain along one of the [111] axes and longer in the other directions. This trigonal distortion makes the linear chains of $Yb^{+3}(4f^{13})$, which are magnetically active. The chains seem to interact only weakly with each other in the nonmagnetic background consisting of divalent $Yb^{+2}(4f^{14})$. As a matter of fact, inelastic neutron scattering experiments revealed that the magnetic excitation shows the characteristics of a one-dimensional (1D) Heisenberg antiferromagnet [42,43]. The spectrum indicated in Fig. 4.14 is close to the des Cloiseaux–Pearson mode with $E_1(q) = \pi J \sin(dq)$, where d is the atomic distance in the Yb^{3+} along the chain. Here q is the projection of the wave vector along the Yb^{3+} chain direction. In the 1D Heisenberg model, the specific heat at low T is linear in T with $\gamma = 190\,\mathrm{mJ/(mol\,K^2)}$ at $\pi J = 3.5\,\mathrm{meV}$. Then, with this value of J, the temperature at which the susceptibility is maximum is predicted to be 17 K, which is close to the experimental observation. Hence, the heavy-electron behaviour in Yb_4As_3 seems to be due to the 1D-like spin excitation caused by the charge ordering, rather than due to the Kondo effect.

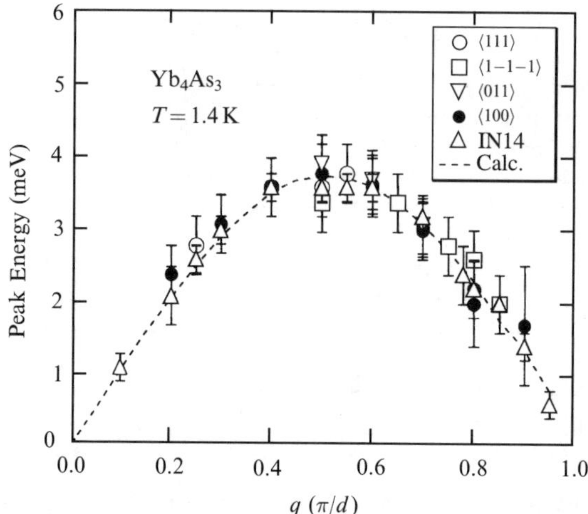

FIG. 4.14. Dispersion relation of inelastic peaks of Yb$_4$As$_3$ in the 1D representation. The open triangles show the newest results, and other symbols indicate previous results for different directions from the (002) reciprocal lattice point. The broken line represents the calculated peak position in the 1D Heisenberg model [42].

4.8 Quantum phenomenology for the dual character

4.8.1 *Coexistence of itinerant and localized characters*

At present, there is no microscopic theory to explain the small ordered moment and anomalous magnetism in heavy electrons. The difficulty lies in the fact that these properties depend rather strongly on the details of the individual system. Some heavy electrons show neither magnetic order nor any metamagnetism, while others do show such behaviour. Because the magnetic fluctuation depends strongly on momentum in the case of anomalous magnetism, the renormalization theory utilizing $1/n$ or $1/d$ as small parameters is not applied straightforwardly. Hence, the description in the present section becomes necessarily qualitative; we merely present our ideas concerning the possible physics without confirmation of their relevance.

In considering the anomalous magnetism of heavy-electron systems, it is instructive to see the effect of magnetic field on the f-electron density of states. Figure 4.15 shows sample results for the single-impurity Anderson model calculated by the NCA [44]. The parameters in the calculation are $\epsilon_f = -1500$ K, $\pi W_0 = 500$ K, and the density of states for the conduction band is constant for $|\epsilon| < D = 10^4$ K and 0 otherwise. Then the Kondo temperature T_K, given by

$$T_K = D \left(\frac{2W_0}{D} \right)^{1/2} \exp \left(\frac{\epsilon_f}{2W_0} \right),$$

amounts to 16 K. As shown in Fig. 4.15(a), the magnetic field splits the Kondo resonance into up- and down-spin resonances, which is analogous but not identical to the Zeeman splitting of a fermion level. The first difference from the simple level splitting is that the

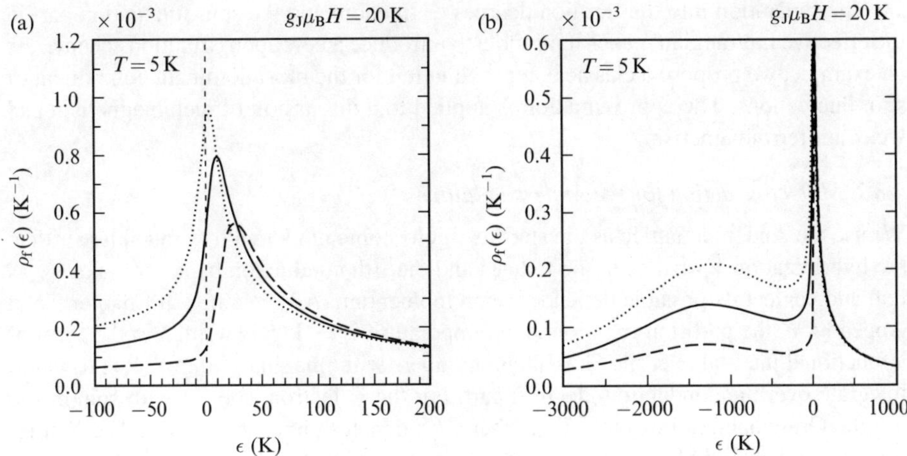

FIG. 4.15. The density of states of the Anderson model in a magnetic field calculated in the NCA. The solid lines show the zero-field results, while the dotted lines correspond to majority spins in a magnetic field, and the dashed lines to the minority spins.

up- and down-spin resonances have different weights, and the second is that the splitting is not symmetric about the zero-field peak. The strong correlation is responsible for the deviation. A dramatic effect of the correlation appears in the energy range deep below the Fermi level. As Fig. 4.15(b) shows, the minority spin component loses much of the spectral weight, although the magnetic energy is very small as compared with $|\epsilon_f|$. This demonstrates that the magnetic polarization of f electrons has little to do with the scale ϵ_f, but is controlled by the scale T_K.

In the case of the Anderson lattice, as we have seen in Chapter 3, the momentum distribution of the f electrons has only a small discontinuity at the Fermi surface, and in general it is rather similar to the localized case. This shows that the f electrons possess an itinerant character close to the Fermi level, but show a dominantly localized character for the magnetization. At temperatures above T_K, many features of heavy-electron systems are common to dilute Kondo systems. This means that the excitation spectrum, except for the low-energy limit, is similar in both systems. The essential idea for constructing a quantum phenomenology is that in the intermediate stage of the renormalization toward the Fermi-liquid fixed point, there should be local variables which are common in the single-site Anderson model and the Anderson lattice. It is reasonable to assume that the Kondo spin compensation is already substantial at this stage.

We set up an effective Lagrangian which includes the itinerant fermions and local spin fluctuations. These variables represent the dual nature of strongly correlated f electrons. Hence, we call the present scheme the duality model [45]. In this model, in order to understand heavy electrons, we make maximum use of the known results for the single-site system. Fortunately the single-site Kondo system is now well understood both for static and dynamic properties using various theoretical methods. A renormalization flow which goes ultimately off the paramagnetic Fermi liquid toward the superconducting phase, for example, can also be treated by the duality model.

The separation into the fermion degrees of freedom and the spin-fluctuation part in the effective Lagrangian makes it possible to introduce a new approximation scheme. As an example, we propose a classical approximation for the most dominant component of spin fluctuations. The approximation is applied to a discussion of metamagnetism and weak antiferromagnetism.

4.8.2 *Effective action for the Anderson lattice*

We use the Anderson lattice as the starting microscopic model with U much larger than the hybridization V_μ between the f states and the μth conduction band. We take V_μ as real and neglect its possible dependence on momentum. As explained in Chapter 3 and Appendix F, the partition function Z at temperature $T = 1/\beta$ is written in the form of a functional integral over the Grassmann numbers with imaginary time τ. It is trivial to integrate over the conduction-electron part. For the f electron at site i with spin σ, we use the Grassmann numbers $f_{i\sigma}^{\pm}(\tau)$, where $f_{i\sigma}^{\pm}$ denotes either $f_{i\sigma}^{\dagger}$ or $f_{i\sigma}$. The Fourier transforms with odd Matsubara frequency $\epsilon_n = (2n+1)\pi T$, with n integer, are given by

$$f_{i\sigma}^{\pm}(\tau) = \sum_n f_{i\sigma}^{\pm}(i\epsilon_n) \exp(-i\epsilon_n \tau). \tag{4.6}$$

For low-energy fermion excitations, only those components with $|\epsilon_n|$ of $O(T_{\mathrm{K}})$ are relevant. However, spin excitations with low energy are dominated by $f_{i\sigma}^{\pm}(i\epsilon_n)$ with $|\epsilon_n|$ much larger than T_{K}. This is because the low-energy spin excitations are represented by a bilinear form of the Grassmann numbers each of which can have high frequencies.

In order to realize the idea, we introduce the auxiliary field via the Hubbard–Stratonovich identity as explained in Appendix F. The partition function is represented by

$$Z_{\mathrm{f}} \equiv Z/Z_{\mathrm{c}} = \int \mathcal{D}f^{\dagger}\mathcal{D}f\mathcal{D}\boldsymbol{\phi}\exp(-A), \tag{4.7}$$

where Z_{c} is the conduction-electron part, and $\boldsymbol{\phi}$ is an auxiliary variable representing the fluctuating magnetic field. The action $A = A_0 + A_1$ is given by

$$A_0 = -\sum_{ij\sigma n} f_{i\sigma}^{\dagger}(-i\epsilon_n)[g_{\mathrm{f}}(i\epsilon_n)^{-1}]_{ij} f_{j\sigma}(i\epsilon_n) - \frac{1}{2U}\sum_{im}|\boldsymbol{\phi}_i(i\nu_m)|^2, \tag{4.8}$$

$$A_1 = -\sum_{i\alpha\beta}\sum_{mn} f_{i\alpha}^{\dagger}(-i\epsilon_n - i\nu_m)f_{i\beta}(i\epsilon_n)\boldsymbol{\sigma}_{\alpha\beta} \cdot \boldsymbol{\phi}_i(i\nu_m), \tag{4.9}$$

where $g_{\mathrm{f}}(i\epsilon_n)$ is the Green function matrix of f electrons with no Coulomb interaction. We integrate over high-frequency variables in the action. The remaining ones are $f_{i\sigma}^{\pm}(i\epsilon_n)$ and $\boldsymbol{\phi}_i(i\nu_m)$ with $|\epsilon_n|$ and $|\nu_m|$ smaller than the cutoffs of $O(T_{\mathrm{K}})$. Furthermore, we change variables from $\boldsymbol{\phi}_i(i\nu_m)$ to magnetization $\partial A_1/\partial\boldsymbol{\phi}_i(i\nu_m)$. Then we need to incorporate the Jacobian associated with this change of variables. Since the correlation function of magnetization coincides with that of the f-electron spin at low frequencies, we choose the notation $\boldsymbol{S}(i\nu_m)$ to represent the slow part of magnetization. Note that the quantity $\boldsymbol{S}(i\nu_m)$ is a c-number.

With this formal procedure, we are left with an effective Lagrangian or its integral, the action A_{eff}, which has only low-frequency variables:

$$Z_{\text{f}} = \int \mathcal{D}f^{\dagger}\mathcal{D}f\mathcal{D}S \exp(-\beta A_{\text{eff}}), \qquad (4.10)$$

where all variables have now a cutoff energy of $O(T_{\text{K}})$. It is not possible to obtain the explicit form of the action A_{eff} by microscopic calculation, since important terms are spread over many orders in perturbation theory. We thus construct the action phenomenologically, following the 'minimal coupling' principle. Namely, we assume a local and instantaneous interaction between the fermions and the 'spins' represented by $S(i\nu_m)$. The action, neglecting the shift of the ground-state energy, consists of three parts [45]:

$$A_{\text{eff}} = A_{\text{f}} + A_{\text{s}} + A_{\text{int}}, \qquad (4.11)$$

$$A_{\text{f}} = -\sum_{ij\sigma n} f_{i\sigma}^{\dagger}(-i\epsilon_n)[G_{\sigma}^{(0)}(i\epsilon_n)^{-1}]_{ij} f_{j\sigma}(i\epsilon_n), \qquad (4.12)$$

$$A_{\text{s}} = \sum_{ijm} S_i(-i\nu_m) \cdot S_j(i\nu_m)[\chi_0(i\nu_m)^{-1}\delta_{ij} - J_{ij}]$$

$$- \sum_i h_i S_{iz}(0), \qquad (4.13)$$

$$A_{\text{int}} = -\lambda_0 \sum_{i\alpha\beta} \sum_{mn} f_{i\alpha}^{\dagger}(-i\epsilon_n - i\nu_m) f_{i\beta}(i\epsilon_n)\boldsymbol{\sigma}_{\alpha\beta} \cdot S_i(i\nu_m). \qquad (4.14)$$

In the fermion part A_{f}, $G_{\sigma}^{(0)}(i\epsilon_n)^{-1}$ is the Green function matrix of the f electrons. This involves the site-diagonal self-energy $\Sigma_{\sigma}^{(0)}(i\epsilon_n)$ caused by many-body interactions other than those by spin fluctuations. The matrix element is given by

$$\left[G_{\sigma}^{(0)}(i\epsilon_n)^{-1}\right]_{ij} = \left[i\epsilon_n - \epsilon_{\text{f}} - \Sigma_{\sigma}^{(0)}(i\epsilon_n) - \frac{1}{2}\sigma h_i\right]\delta_{ij} - \sum_{\mu} V_{\mu}^2 g_{ij}^{\mu}(i\epsilon_n), \qquad (4.15)$$

where ϵ_{f} denotes the f-electron level, $g_{ij}^{\mu}(i\epsilon_n)$ is the bare propagator of μth conduction band and h_i is a magnetic field at site i. The Green function $g_{\text{f}}(i\epsilon_n)$ in eqn (4.8) corresponds to the one without $\Sigma_{\sigma}^{(0)}(i\epsilon_n)$ in eqn (4.15).

In the spin part A_{s}, $\chi_0(i\nu_m)$ is a partially renormalized spin susceptibility before the inclusion of the RKKY interaction and the coupling with fermions. For the RKKY interaction J_{ij}, we neglect the dependence on energy since the characteristic energy in J_{ij} is usually much larger than T_{K}. We deal with an exceptional case where the energy dependence cannot be neglected later. In the lowest order in hybridization, J_{ij} is written as

$$J_{ij} = 2\sum_{\mu}\left(\frac{V_{\mu}^2}{\epsilon_{\text{f}}}\right)^2 T \sum_n g_{ij}^{\mu}(i\epsilon_n)g_{ji}^{\mu}(i\epsilon_n). \qquad (4.16)$$

If we integrate over the spin degrees of freedom in eqn (4.10), we are left with the standard Fermi-liquid theory. In order to discuss the magnetic properties, it is convenient to integrate first over the fermion variables. The result is given by

$$Z = \det G^{(0)} \int \mathcal{D}S \exp(-\beta A_m), \tag{4.17}$$

$$A_m = A_s - T \operatorname{Tr} \ln(1 + \lambda_0 G^{(0)} \boldsymbol{\sigma} \cdot \boldsymbol{S}), \tag{4.18}$$

where the partially renormalized Green function $G^{(0)}$ is a matrix in the space of site, spin, and frequency. Let us first derive the dynamical susceptibility. By taking the second derivative of A_m with respect to $S_q(i\nu_m)$, which is the Fourier transform of $S_i(i\nu_m)$, we obtain the Gaussian approximation for the dynamical susceptibility. We improve the result by renormalizing the coupling constant λ_0 and the Green function in the polarization bubble. The RKKY interaction is treated in the mean-field approximation, and intersite effects on the renormalized spin–fermion interaction vertex λ are neglected. Then we obtain

$$\chi(\boldsymbol{q}, i\nu_m)^{-1} = \chi_0(i\nu_m)^{-1} - 2\lambda^2 \Pi(\boldsymbol{q}, i\nu_m) - J(\boldsymbol{q}), \tag{4.19}$$

where $J(\boldsymbol{q})$ is the Fourier transform of J_{ij} and

$$\Pi(\boldsymbol{q}, i\nu_m) = -\frac{1}{N} \sum_{k} T \sum_{n} G_f(\boldsymbol{k}, i\epsilon_n) G_f(\boldsymbol{k} + \boldsymbol{q}, i\epsilon_n + i\nu_m). \tag{4.20}$$

4.8.3 *Duality picture applied to the Anderson impurity*

We assume that the local susceptibility $\chi_0(i\nu_m)$ and the coupling constant λ are determined mainly by the local correlation. Then these magnitudes can be estimated by applying the duality model to the single-site Anderson model. For this purpose, we first reformulate the local Fermi-liquid theory at zero temperature in terms of spin fluctuations and fermion excitations. For simplicity, we assume a single conduction band with constant density of states ρ_c near the Fermi level. The vertex parts Λ_3 and λ_3 are introduced as follows:

$$\langle T_\tau f_\uparrow^\dagger(\tau_1) f_\downarrow(\tau_2) f_\downarrow(\tau) f_\uparrow^\dagger(\tau) \rangle$$
$$= -\int_{-\infty}^{\infty} \frac{d\epsilon_1}{2\pi} \int_{-\infty}^{\infty} \frac{d\epsilon_2}{2\pi} \exp[-i\epsilon_1\tau_1 + i\epsilon_2\tau_2$$
$$-i(\epsilon_1 - \epsilon_2)\tau] G_\uparrow(i\epsilon_1) G_\downarrow(i\epsilon_2) \Lambda_3(i\epsilon_1, i\epsilon_2; i\epsilon_2 - i\epsilon_1), \tag{4.21}$$
$$\Lambda_3(i\epsilon_1, i\epsilon_2; i\epsilon_2 - i\epsilon_1) = 1 + 2\lambda_3(i\epsilon_1, i\epsilon_2; i\epsilon_2 - i\epsilon_1)\chi(i\epsilon_2 - i\epsilon_1), \tag{4.22}$$

where $G_\sigma(i\epsilon)$ is the renormalized f-electron Green function. Figure 4.16 shows the relation diagramatically. It is obvious that λ_3 can be interpreted as a coupling strength between fermions and spin fluctuations. In terms of the self-energy $\Sigma_\sigma(i\epsilon)$, the Green function is given by

$$G_\sigma(i\epsilon) = [i\epsilon - \epsilon_f - \tfrac{1}{2}h\sigma - \Sigma_\sigma(i\epsilon) + i\Delta \operatorname{sgn}\epsilon]^{-1}, \tag{4.23}$$

where $\Delta = \pi V^2 \rho_c$ is assumed to be a constant in the relevant energy range.

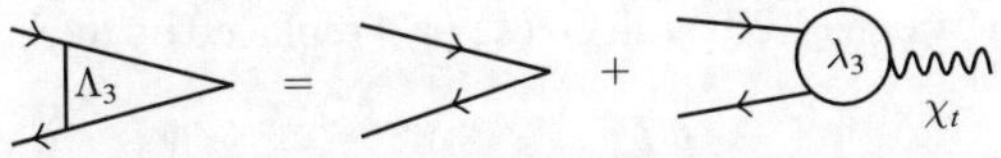

FIG. 4.16. Feynman diagrams for eqn (4.22).

We invoke an important relation which is an example of the Ward–Takahashi identity [46]:

$$\Sigma_\uparrow(i\epsilon) - \Sigma_\downarrow(i\epsilon) = -2M\lambda_3(i\epsilon, i\epsilon; 0), \tag{4.24}$$

where $M = \langle S_z \rangle$ with $S_z = [f_\uparrow^\dagger f_\uparrow - f_\downarrow^\dagger f_\downarrow]/2$. The identity represents the many-body correction to the Zeeman splitting. It will be shown later that in the Kondo regime where charge fluctuations are suppressed, the many-body correction overwhelms the bare Zeeman energy.

It is possible to derive $\lambda_3(0, 0; 0) \equiv \lambda$ exactly in the case of $h = 0$. We make use of the Friedel sum rule for the spin polarization as explained in Appendix G. The result is given by

$$M = \frac{1}{2\pi} \sum_\sigma \sigma \, \mathrm{Im} \ln G_\sigma(0), \tag{4.25}$$

where (and in the following) we omit writing the positive infinitesimal imaginary part in the argument of the Green function. Then the static susceptibility χ is given by

$$\chi = \frac{\partial M}{\partial h} = \frac{1}{2}\rho_f(0)[1 + 2\lambda\chi], \tag{4.26}$$

where $\rho_f(0) = -\mathrm{Im}\, G_\sigma(0)/\pi$. Thus, λ is given in terms of $\rho_f(0)$ and χ. In the Kondo limit where $\chi \gg \rho_f(0)$, we obtain

$$\lambda = \pi\Delta. \tag{4.27}$$

The full self-energy $\Sigma_\sigma(\epsilon)$ near the Fermi level depends so strongly on ϵ that the wave function renormalization factor is much smaller than unity. Then the fermion degrees of freedom alone do not contribute to charge and spin susceptibilities. However, the magnetic response is influenced by fermions through the coupling λ. It follows that for small magnetic field the effective Zeeman splitting of the fermion levels is *twice* of that without hybridization. This is seen if we recognize that the effective Zeeman splitting is given by $a_f[\Sigma_\uparrow(0) - \Sigma_\downarrow(0)]$ and $M = h/(4T_K)$.

In the duality picture, the contribution of the local part of the f electrons to the dynamical susceptibility is written as $\chi_0(\omega)$. This is fixed by the condition that the full susceptibility $\chi(\omega)$ given by the duality model must be identical to that given by the Fermi-liquid formula, eqn (2.69). In the duality model, we regard the spin–fermion coupling constant λ as independent of ω. Then we obtain

$$\chi(\omega) = \frac{\chi_0(\omega)}{1 - 2\lambda^2\Pi(\omega)\chi_0(\omega)}, \tag{4.28}$$

where $\Pi(\omega)$ is given by eqn (4.20) with $G_f(k, i\epsilon_n)$ replaced by the impurity counterpart:

$$G_f(i\epsilon_n) = a_f \bigg/ \left(i\epsilon_n - \tilde{\epsilon}_f - a_f \sum_k \frac{V^2}{i\epsilon_n - \epsilon_k} \right). \qquad (4.29)$$

Equation (4.28) is exact in the static limit since the coupling constant $\lambda = \pi\Delta$ takes precise account of the vertex correction there. By comparing eqns (4.28) and eqn (2.69), we obtain

$$\chi_0(\omega)^{-1} = 2\lambda^2\Pi(\omega) + \frac{2}{\Lambda_0^2\Pi(\omega)} - 4T_K, \qquad (4.30)$$

There is no ω-linear imaginary part in $\chi_0(\omega)$ because of the cancellation between the first and second terms. This quasi-gap behaviour is consistent with our definition of the spin degrees of freedom, which do not involve the low-energy fermion excitations as their constituents. Thus, the spin fluctuations do not contribute to the T-linear specific heat without coupling to the fermion degrees of freedom. The specific heat $C = \gamma T$ of the Anderson model is determined by the renormalized Green function $G_f(\epsilon)$ as

$$\gamma = \frac{4\pi}{3}\mathrm{Im}\frac{\partial}{\partial\epsilon}\ln G_\sigma(\epsilon)\bigg|_{\epsilon=0},$$

just as in the Fermi-liquid theory.

We can rely on eqn (2.69) up to frequencies of the order of T_K. Then eqn (4.30) fixes $\chi_0(\omega)$ for these frequencies. Figure 4.17 shows the real and imaginary parts of the susceptibilities $\chi(\omega)$, $\chi_0(\omega)$, and $\Lambda_0^2\Pi(\omega)$. As is evident in Fig. 4.17(a), Im $\chi_0(\omega)/\omega$ displays a structure which is interpreted as a pseudo-gap of the order of T_K for spin excitations. There is a corresponding structure in Re $\chi_0(\omega)$ as shown in Fig. 4.17(b). Note that $\chi_0(0) = \chi(0)/3$. We remark that $\Lambda_0^2\Pi(\omega)/2$ corresponds to the dynamical susceptibility of ref. [47] if one regards $2T_K$ as the Kondo temperature. At high frequencies, the imaginary parts converge to the same asymptotic behaviour proportional to ω^{-3}. The odd power is caused by the logarithmic singularity at $\omega = \infty$ in eqn (4.30). By comparison with a numerical result obtained by the Wilson-type renormalization theory [45,48], it is seen that the quasi-particle RPA given by eqn (2.69) is an excellent approximation for all frequencies. Hence, we expect that eqn (4.30) is also a reasonably accurate representation of $\chi_0(\omega)$ for all ω.

4.8.4 *Metamagnetism in the duality model*

We study the magnetic equation of state on the basis of eqns (4.17) and (4.18). The basic observation is that the coupling between spins and fermions causes a large nonlinear effect as the spin polarization increases. The experimental results on $CeRu_2Si_2$ have been discussed in § 3.1.1. Now we outline a theoretical aspect of the metamagnetic behaviour of heavy electrons. It should be remarked that there is no established result on the mechanism of the phenomena. Therefore, the discussion below is presented just to show a consequence of the duality model.

We assume the presence of a static magnetic field h_q which is modulated with wavenumber q. Then the spin variable $S_q \equiv S_q(i\nu_m = 0)$ becomes macroscopic. We

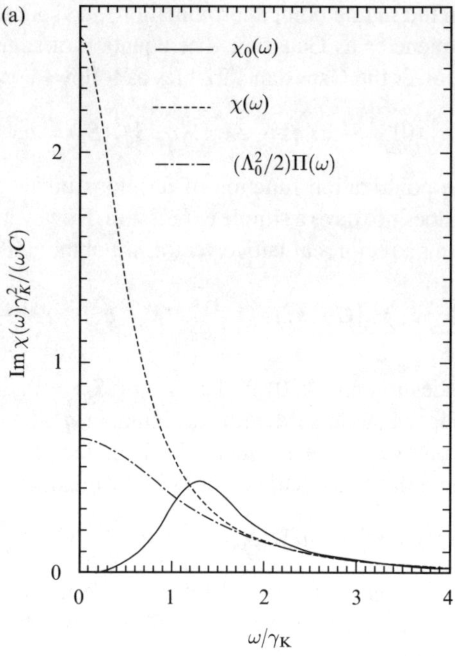

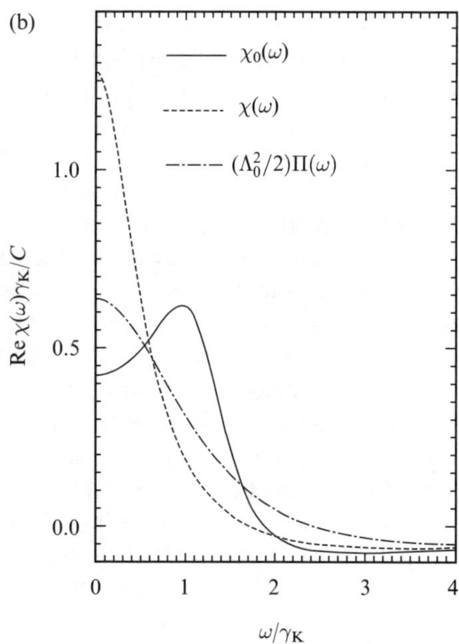

FIG. 4.17. Three kinds of susceptibilities: $\chi_0(\omega)$ (solid line), $\chi(\omega)$ (dashed line), and $a_{\mathrm{f}}^{-2}\Pi(\omega)$ (dash-dotted line). Imaginary parts are shown in (a) and real parts in (b).

treat this component in the saddle point approximation, and spin fluctuations with other wavenumbers and frequencies as Gaussian. The equation of state is given by the zero average of $\partial A_m / \partial S_{-q}$ over the Gaussian variables as follows:

$$[\chi_0(0)^{-1} - J(q) - 2\lambda^2 \Pi(q; S_q)]S_q = h_q, \tag{4.31}$$

where $\Pi(q; S_q)$ is the polarization function of fermions under finite S_q. For general values of q, $\Pi(q; S_q)$ does not have a simple expression. However, for the case of $q = 0$ or $q = Q$, with Q being a reciprocal lattice vector, we obtain the compact formula

$$\Pi(q; S_q) = -\frac{1}{N}\sum_k T \sum_n [G_f(k, i\epsilon_n)^{-1} G_f(k+q, i\epsilon_n)^{-1} - |\lambda S_q|^2]^{-1}, \tag{4.32}$$

which, of course, reduces to eqn (4.20) in the case of $S_q = 0$. The itinerant fermions 'feel' λS_q as a fictitious magnetic field with wavenumber q.

Figure 4.18 shows an example of a numerical calculation [49]. The parameters in the calculation are taken in units of the half-width D of the conduction band as follows:

$$\rho_c = 1/2, \quad a_f\lambda = 0.01\pi^2/2, \quad a_f V^2 \rho_c = 5 \times 10^{-3}, \quad \chi_0(0)^{-1} - J(0) = 2a_f\lambda.$$

The parameter ξ_F is the energy (measured from the Fermi level) of the conduction electron with the Fermi wavenumber. The magnetization is enhanced for large enough coupling between fermions and spins. The jump corresponds to the first-order metamagnetic transition. With smaller coupling, there appears a smooth but nonlinear behaviour of the magnetization, which seems to be related to the pseudo-metamagnetic transition discussed in § 3.1.1. This behaviour in the duality model comes from nonmonotonic dependence of $\Pi(0; S_0)$ on the magnetization $M = |S_0|$. The itinerant band has a peak in the density of states near the Fermi level due to hybridization, and the Zeeman splitting pushes the peak of either up or down spin toward the Fermi level. Then, $\Pi(0; S_0)$ increases up to certain magnitude of M. With further increase of M, the Zeeman-split peaks of both spins go off the Fermi level, so that $\Pi(0; S_0)$ turns to decrease. We note that the mechanism in the present model calculation is not qualitatively different from a rigid-band picture of f electrons. The many-body effect is taken into account only in reducing the energy scale. It is difficult in the duality model to incorporate more details of the interaction between the quasi-particles and magnetization.

A very interesting feature in heavy-electron systems is that the external magnetic field causes a nonlinear effect in the itinerant state of f electrons. Experimentally, this appears as a change of the Fermi surface probed by the dHvA effect. In the case of $CeRu_2Si_2$ the Fermi surface below $H = 7.7\,\mathrm{T} = H_M$ is roughly consistent with the one predicted by the energy band theory. As the magnetic field becomes close to H_M, the signal shows a strange temperature dependence, suggesting a dependence of the effective mass and/or the Zeeman splitting on T. Above H_M, the shape of the observed Fermi surface is consistent with the localized picture [50]. We note that $CeRu_2Si_2$ has a strong Ising-type anisotropy. Similar magnetization is observed in UPt_3, which has a strong XY-type (easy-plane) anisotropy [51].

There are also many heavy-electron systems without the metamagnetic behaviour. The characteristics in magnetization thus depend on the details of the system. Such

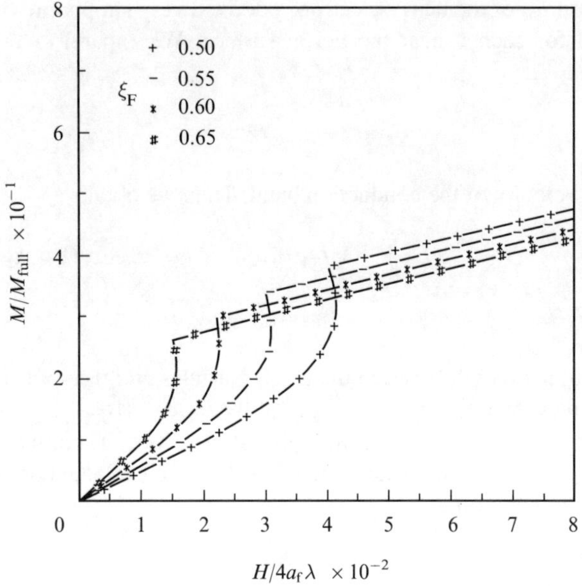

FIG. 4.18. Magnetization vs applied magnetic field calculated in the duality model. The parameters in the calculation are explained in the text.

details are best dealt with by the energy band theory. In the purely itinerant picture, the metamagnetic behaviour is ascribed to characteristic band structures such as a high peak in the density of states near the Fermi level. Since the Zeeman splitting of the energy band causes the peak to cross the Fermi level, the differential susceptibility should be enhanced at the crossing. In this picture, however, the structure of the density of states should be extremely sharp, of the order of 0.1 K in $CeRu_2Si_2$. It is impossible to reproduce such a tiny structure by the ordinary band calculation. We remark that the energy scale of 0.1 K is much smaller than the Kondo temperature of $CeRu_2Si_2$, which is of the order of 10 K. In the duality model, the tiny structure comes from a combination of the Kondo effect and the coherent hybridization in the periodic lattice. It is an open problem whether such tiny structure reflects an intersite spin–spin correlation in addition to the Kondo effect.

4.8.5 *Weak antiferromagnetism*

On the basis of the magnetic equation of state in the duality model, we now proceed to discuss weak antiferromagnetism. In this section, we take a one-dimensional model to demonstrate the effect of nesting in the simplest way. In the duality model, the weak antiferromagnetism results from the strong sensitivity of the itinerant exchange to the magnitude of magnetization. Let us assume that there is only one heavy-electron band which crosses the Fermi level. We evaluate $\Pi(Q; S_Q)$ using the quasi-particle form given by eqn (4.29). We note that the renormalization factor a_f is of the order of $T_K/(\pi^2 V^2 \rho_c)$.

The spectrum E_k of the heavy electron is derived very simply provided that $|E_k| \ll |\epsilon_k|$ is satisfied for each k near the Fermi surface. We expand ϵ_k around the Fermi wavenumber $\pm k_F$ as

$$\epsilon_k = \epsilon_F \pm v_c(k \mp k_F), \tag{4.33}$$

where v_c is the velocity of the conduction band. Then we obtain

$$E_k \cong \pm a_f \sum_\mu (V/\epsilon_F)^2 \, v_c(k \mp k_F) \equiv \pm v_f(k \mp k_F), \tag{4.34}$$

$$G_f(k, i\epsilon_n) \cong a_f/(i\epsilon_n - E_k). \tag{4.35}$$

The interaction strength between the quasi-particles and the spins is determined by $\lambda_f \equiv a_f \lambda$. We introduce the misfit energy δ_m by $\delta_m \equiv v_f|Q - 2k_F|$ which simulates the deviation from complete nesting in higher dimensions. Then, using eqn (4.34) up to a cutoff energy $\pm D_f$, which should be of $O(T_K)$, we evaluate eqn (4.32) at $T = 0$. Straightforward integration gives

$$\Pi(Q; S_Q) = \begin{cases} \dfrac{1}{2}a_f^2 \rho_f \ln \dfrac{D_f}{[(\delta_m)^2 - |\lambda_f S_Q|^2]^{1/2} + \delta_m}, & (\delta_m > |\lambda_f S_Q|), \\[2ex] \dfrac{1}{2}a_f^2 \rho_f \ln \left| \dfrac{D_f}{\lambda_f S_Q} \right|, & (\delta_m < |\lambda_f S_Q|), \end{cases} \tag{4.36}$$

where we have defined the quasi-particle density of states $\rho_f \equiv (\pi v_f)^{-1}$ with the lattice constant being unity. In the case of complete nesting, the spontaneous magnetization at $T = 0$ is given by the solution of

$$\frac{1}{\chi_0(0)} - J(Q) - \lambda_f^2 \rho_f \ln \left| \frac{D_f}{\lambda_f S_Q} \right| = 0. \tag{4.37}$$

In eqn (4.37), the smaller magnetization causes larger exchange field from the itinerant part. Since we have $D_f \sim \lambda_f \sim \lambda_f^2 \rho_f \sim T_K$ as order of magnitudes, the condition for weak antiferromagnetism under very good nesting is that $\chi_0(0)^{-1} - J(Q)$ should be much larger than T_K. Namely, the relevant exchange interaction responsible for the weak antiferromagnetism is not the usual RKKY interaction. Instead, the heavy electrons themselves mediate the coupling of the spins. In this respect, the weak antiferromagnetism of heavy electrons is very similar to that of the band electrons in transition metals [52].

In contrast to transition metals, however, the presence of the RKKY interaction leads to specific dynamical and temperature effects. Suppose that the RKKY interaction does not favour the wavenumber Q so that $J(Q)$ is negative. Then, $\chi_0(0)^{-1} - J(Q)$ can be much larger than T_K and, hence, than $\lambda_f^2 \rho_f$. In the present model, the dominant exchange interaction changes from the RKKY type at high T to the anomalous one mediated by the itinerant heavy electrons with decreasing T. In this connection, we note that the wavenumbers of most dominant magnetic fluctuations in UPt$_3$ depend on the temperature and the excitation energy. The relevant Q for the Néel order can be seen only at low temperature and with low excitation energy [53].

The Néel temperature T_N can be derived by evaluating $\Pi(Q; 0)$ at finite temperatures. The result at T larger than δ_m is given by

$$\Pi(Q; S_Q) \cong \frac{1}{2} a_f^2 \rho_f \ln\left(\frac{D_f}{T}\right). \tag{4.38}$$

Comparison with eqn (4.37) gives the relation between T_N and the zero-temperature magnetization S_Q as

$$T_N \cong \lambda_f |S_Q|. \tag{4.39}$$

Thus, in the present model the weak antiferromagnetism requires T_N to be much smaller than λ_f and, hence, than T_K. The small energy scale T_N is generated by the nesting property of the heavy-electron band. This is again similar to the weak antiferromagnetism in itinerant electrons, where the role of T_K is played by the Fermi energy of the conduction band.

Bibliography

[1] T. Kohra, Y. Kohori, K. Asayama, Y. Kitaoka, M. B. Maple, and M. S. Torikachvilli, *Jpn. J. Appl. Phys.* **26**, 1247 (1987); Y. Kohori *et al.*, *J. Phys. Soc. Jpn.* **65**, 679 (1996).

[2] T. T. M. Palstra, A. A. Menovsky, J. van den Berg, A. J. Dirkmaat, P. H. Kes, G. J. Nieuwenhuys, and J. A. Mydosh, *Phys. Rev. Lett.* **55**, 2727 (1985).

[3] G. J. Nieuwenhuys, *Phys. Rev. B* **35**, 5260 (1987).

[4] P. Santini and G. Amoretti, *Phys. Rev. Lett.* **73**, 1027 (1994).

[5] H. Amitsuka and T. Sakakibara, *J. Phys. Soc. Jpn.* **63**, 736 (1994).

[6] C. Broholm, H. Lin, P. T. Matthews, T. E. Mason, W. J. L. Buyers, M. F. Collins, A. A. Menovsky, J. A. Mydosh, and J. K. Kjems, *Phys. Rev B* **43**, 12 809 (1991).

[7] W. J. L. Buyers and T. M. Holden, *Handbook on the physics and chemistry of actinides* (Eds, G. H. Lander and A. J. Freeman, North-Holland, 1985) Vol. 2, p. 239.

[8] M. B. Walker, C. Kappler, K. A. McEwen, U. Steigenberger, and K. N. Clausen, *J. Phys. Cond. Matter* **6**, 7365 (1994).

[9] C. Geibel, S. Thies, D. Kaczorowski, A. Mehner, A. Grauel, B. Seidel, U. Ahlheim, R. Helfrich, K. Peterson, C. D. Bredl, and F. Steglich, *Z. Phys. B* **83**, 305 (1991).

[10] M. Kyogaku, Y. Kitaoka, K. Asayama, C. Geibel, C. Schank, and F. Steglich, *J. Phys. Soc. Jpn.* **62**, 4016 (1993).

[11] S.Takagi, T. Homma, and T. Kasuya, *J. Phys. Soc. Jpn.* **58**, 4610 (1989).

[12] S. Horn, E. Holland-Moritz, M. Loewenhaupt, F. Steglich, H. Scheuer, A. Benoit, and J. Flouquet, *Phys. Rev. B* **23**, 3171 (1981).

[13] H. Nakamura, Y. Kitaoka, T. Iwai, H. Yamada, and K. Asayama, *J. Phys. Cond. Matter* **4**, 473 (1992).

[14] Y. Kitaoka, H. Nakamura, T. Iwai, K. Asayama, U. Ahlheim, C. Geibel, C. Schank, and F. Steglich. *J. Phys. Soc. Jpn.* **60**, 2122 (1992).

[15] R. Feyerherm *et al.*, *Physica B* **206 & 207**, 596 (1995); *Phys. Rev. B* **56**, 699 (1997).

[16] U. Rauchschwalbe *et al.*, *J. Magn. Magn. Mater.* **63 & 64**, 347 (1987).

[17] H. Nakamura, Y. Kitaoka, H. Yamada, and K. Asayama, *J. Magn. Magn. Mater.* **76 & 77**, 517 (1988).

[18] Y. Kitaoka, H. Tou, G.-q. Zheng, K. Ishida, K. Asayama, T. C. Kobayashi, A. Kohda, N. Takeshita, K. Amaya, Y. Onuki, G. Geibel, C. Schank, and F. Steglich, *Physica B* **206 & 207**, 55 (1995).

[19] Y. J. Uemura *et al.*, *Phys. Rev. B* **39**, 4726 (1989); G. M. Luke *et al.*, *Phys. Rev. Lett.* **59**, 1853 (1994).

[20] G. Bruls, B. Wolf, D. Finsterbusch, P. Thalmeier, I. Kouroudis, W. Sun, W. Assmus, and B. Luthi, *Phys. Rev. Lett.* **72**, 1754 (1994).

[21] F. Steglich *et al.*, *J. Phys. Cond. Matter* **8**, 9909 (1996).

[22] R. Modler *et al.*, *Physica B* **206 & 207**, 586 (1995).

[23] P. Gegenwart *et al.*, *Phys. Rev. Lett.* **81**, 1501 (1998).

[24] K. Ishida *et al.*, *Phys. Rev. Lett.* **82**, 5353 (1999).

[25] M. Hunt, P. Messon, P. A. Probst, P. Reinders, M. Springford, W. Assmus, and W. Sun, *J. Phys. Cond. Matter* **2**, 6859 (1990).

[26] M. Lang, R. Molder, U. Ahlheim, R. Helfrich, P. H. P. Reinders, and F. Steglich, *Phys. Scripta T* **39**, 135 (1991).

[27] F. Thomas, J. Thomasson, C. Ayache, C. Geibel, and F. Steglich, *Physica B* **186 & 188**, 303 (1993).

[28] B. Bogenberger and H. v. Lohneysen, *Phys. Rev. Lett.* **74**, 1016 (1995).

[29] J. M. Effantin, J. Rossat-Mignod, P. Burlet, H. Bartholin, S. Kunii, and T. Kasuya, *J. Magn. Magn. Mater.* **47 & 48**, 145 (1985).

[30] G. Uimin, Y. Kuramoto, and N. Fukushima, *Solid State Commun.* **97**, 595 (1996).

[31] N. Fukushima and Y. Kuramoto, *J. Phys. Soc. Jpn.* **67**, 2460 (1998).

[32] F. Ohkawa, *J. Phys. Soc. Jpn.* **52**, 3897 (1983).

[33] R. Shiina, H. Shiba, and P. Thalmeier, *J. Phys. Soc. Jpn.* **67**, 1741 (1997).

[34] M. Takigawa, H. Yasuoka, T. Tanaka, and Y. Ishizawa, *J. Phys. Soc. Jpn.* **52**, 728 (1983).

[35] O. Sakai, R. Shiina, H. Shiba, and P. Thalmeier, *J. Phys. Soc. Jpn.* **66**, 3005 (1997).

[36] T. Tayama, T. Sakakibara, K. Tenya, H. Amitsuka, and S. Kunii, *J. Phys. Soc. Jpn.* **66**, 2268 (1997).

[37] M. Hiroi, M. Sera, N. Kobayashi, and S. Kunii, *Phys. Rev. B* **55**, 8339 (1997).

[38] S. Nakamura, O. Suzuki, T. Goto, S. Sakatsume, T. Matsumura, and S. Kunii, *J. Phys. Soc. Jpn.* **66**, 552 (1997).

[39] M. Kohgi, T. Osakabe, K. Ohyama, and T. Suzuki, *Physica B* **213 & 214**, 110 (1995).

[40] T. Suzuki, *Physica B* **186–189**, 347 (1993).

[41] A. Ochiai *et al.*, *J. Phys. Soc. Jpn.* **59**, 4129 (1990).

[42] M. Kohgi *et al.*, *Physica B* **259–261**, 269 (1999).

[43] P. Fulde, B. Schmidt, and P. Thalmeier, *Europhys. Lett.* **31**, 323 (1995).

[44] Y. Kuramoto, *Physica B* **156 & 157**, 789 (1989).

[45] Y. Kuramoto and K. Miyake, *J. Phys. Soc. Jpn.* **59**, 2831 (1990).

[46] T. Koyama and M. Tachiki, *Prog. Theor. Phys.* **Suppl. 80**, 108 (1984).

[47] Y. Kuramoto and E. Müller-Hartmann, *J. Magn. Magn. Mater.* **52**, 122 (1985).

[48] O. Sakai, Y. Shimizu, and T. Kasuya, *J. Phys. Soc. Jpn.* **58**, 3666 (1989).

[49] K. Miyake and Y. Kuramoto, *J. Magn. Magn. Mater.* **90 & 91**, 438 (1990).

[50] M. Takashima *et al.*, *J. Phys. Soc. Jpn.* **65**, 515 (1996).

[51] A. de Visser *et al.*, *Physica B* **171**, 190 (1991).

[52] T. Moriya, *Spin fluctuations in itinerant electron magnetism* (Springer Verlag, Berlin, 1985).

[53] G. Aeppli *et al.*, *Phys. Rev. Lett.* **60**, 615 (1988).

5

SUPERCONDUCTING STATES

5.1 Historical overview

Superconductivity, one of the best understood many-body problems in physics [1], again became a challenging problem when a new kind of superconducting phenomenon was discovered in $CeCu_2Si_2$ by Steglich *et al.* [2] in 1979. The system behaves as a heavy-electron material close to magnetic instability. In the subsequent decade, intensive investigations of a class of uranium compounds established a new field of heavy-electron superconductivity by successive discoveries of superconductivity in UBe_{13}, UPt_3, URu_2Si_2, UPd_2Al_3 and UNi_2Al_3 [3,4]. The most important characteristics for a series of uranium heavy-electron superconductors, except for UBe_{13}, are that superconductivity coexists with antiferromagnetism and that the specific heat coefficient γ lies in a broad range from $700\,mJ/(mol\,K^2)$ (UBe_{13}) to $60\,mJ/(mol\,K^2)$ (URu_2Si_2). The f-shell electrons, which are strongly correlated by Coulomb repulsive interaction, determine the properties of heavy quasi-particles at the Fermi level. This gives rise to a large γ value as well as an enhanced spin susceptibility. Hence, the Fermi energy is also quite small: $T_F = 10$–$100\,K$ and, as a result, the transition temperature T_c is also small, 0.5–$2\,K$. The magnetic ordering temperature $T_N = 5$–$20\,K$ is by one order of magnitude higher than T_c. A jump $(C_s - C_n)/C_n$ of the specific heat normalized by the value C_n just above T_c is of $O(1)$ in all compounds. This result demonstrates that the superconductivity is produced mainly by the heavy quasi-particles. Due to the strong Coulomb repulsion among the f electrons, it seems difficult for the heavy quasi-particles to form the ordinary s-wave Cooper pairs with large amplitude at zero separation of the pair. In order to avoid the Coulomb repulsion, the system would favour an anisotropic pairing channel like spin triplet p-wave or spin singlet d-wave. These types of paired states are actually realized in superfluid ^{3}He where pairings with p-wave spin triplet are formed with a few phases of different symmetries such as A, A_1 and B phases [5].

Ferromagnetic spin fluctuations (paramagnons) play a major part in the effective potential which produces the anisotropic pairing. For example, the pairing in the A-phase of ^{3}He was identified as an anisotropic type, the Anderson–Brinkman–Morel (ABM) state, with the energy gap vanishing on points at the Fermi surface [6]. In many respects, the analogy with ^{3}He is a good guide for the interpretation of the experimental results in heavy-electron superconductors. As a low-energy excitation, the on-site Coulomb repulsive interaction produces AF spin fluctuations, which play a role similar to ferromagnetic spin fluctuations in liquid ^{3}He. There are, on the other hand, important differences between these two systems. The strong correlation effect, the spin–orbit interaction, and the CEF characterize the heavy-electron systems. Thus, the problem of the heavy-electron superconductivity turns out to be much more complicated than in ^{3}He.

The heavy-electron superconductivity also has intimate connection with another new kind of superconductivity; the high-temperature superconductivity discovered in a copper oxide $La_{2-x}Ba_xCuO_4$ by Bednorz and Müller in 1986 [7]. Intensive studies of the class of copper oxides led to successive discoveries of new superconducting systems, with higher and higher T_c values: Y–Ba–Cu–O system with $T_c \sim 90$ K, Bi–Sr–Ca–Cu–O system with $T_c \sim 110$ K, Tl–Ba–Ca–Cu–O system with $T_c \sim 120$ K and Hg–Ba–Cu–O system with $T_c \sim 130$ K [8]. These new superconductors exhibit unconventional properties, and there is increasing evidence that the order parameter symmetry could be of the anisotropic $d_{x^2-y^2}$ type. In copper oxides, the AF spin fluctuation is considered to play a vital role in producing the anisotropic pairing [9,10]. Even if there is a difference of more than two orders of magnitude between the T_c's in heavy-electron and high-T_c copper oxide superconductors, they share the common feature of strongly correlated electrons in partially filled f or d shells. It is a common property of the anisotropic superconductivity that the low-lying excitations in the quasi-particle spectrum originate from the presence of zero gap either on points or along a line at the Fermi surface. These zeroes give rise to a power-law temperature dependence of various physical quantities instead of the exponential dependence observed in conventional s-wave superconductors. We resume a more detailed comparison in Chapter 6.

5.2   Fundamentals of anisotropic pairing

5.2.1   *Symmetry of the pairing*

Spherical symmetry   Let us consider the symmetry property of the pairing. For simplicity, we first assume spherical symmetry in the system and neglect the spin–orbit interaction. The situation is approximately realized in superfluid ^{3}He. In the case of heavy electrons, the effects of CEF and spin–orbit interaction are very important and will be discussed later. For a description of the symmetry, we draw an analogy with the one-body density matrix given by $\rho_{\alpha\beta}(k) = \langle a_\beta(k)^\dagger a_\alpha(k) \rangle$ in k-space. This can be parametrized as

$$2\rho_{\alpha\beta}(k) = n(k)\delta_{\alpha\beta} + m(k) \cdot \sigma_{\alpha\beta}(k), \tag{5.1}$$

where $n(k)$ is the occupation number, $m(k)$ the spin polarization, and σ a vector composed of the Pauli matrices. The average magnetization $\langle M(k) \rangle$ is given by

$$\langle M(k) \rangle = -\tfrac{1}{4}g\mu_B \mathrm{Tr}[\sigma\rho(k)] = -\tfrac{1}{2}g\mu_B m(k), \tag{5.2}$$

where g is the g-factor.

We now turn to the pairing amplitude $\Psi_{\alpha\beta}(k)$ defined by

$$\Psi_{\alpha\beta}(\hat{k}) = \langle a_\alpha(k)a_\beta(-k) \rangle, \tag{5.3}$$

where we have assumed zero total momentum of the pair. Here $\hat{k}$ denotes a unit vector in the direction of k which is near the Fermi surface. Because of the small energy scale of superconductivity as compared with the Fermi energy, we have neglected slight dependence of $\Psi_{\alpha\beta}(\hat{k})$ on $|k|$. In order to make analogy with the symmetry of the density matrix, we introduce the time reversal operator K acting on the single-particle state ϕ_α with a spin component α as

$$\langle r\beta \mid K\phi_\alpha \rangle \equiv \phi_\beta(r)^*(i\sigma_y)_{\beta\alpha}. \tag{5.4}$$

If $\phi_\alpha(r)$ is an eigenstate of the single-particle Hamiltonian with a magnetic field, $K\phi_\alpha$ represents a solution with reversed magnetic field. Equation (5.4) for spin 1/2 is an example of the more general property of K as the product of a unitary operator and taking the complex conjugate of the wave function. The appearance of the real unitary operator $i\sigma_y$ here is connected with the property

$$i\sigma_y\boldsymbol{\sigma}(i\sigma_y)^{-1} = -\boldsymbol{\sigma}^*,$$

which means that the spin operator $\boldsymbol{\sigma}/2$ is odd under time reversal.

The convenience of using $i\sigma_y$ is clear from the identity for the singlet pair creator:

$$a_\uparrow^\dagger(k)a_\downarrow^\dagger(-k) - a_\downarrow^\dagger(k)a_\uparrow^\dagger(-k) = \sum_{\alpha\beta} a_\alpha^\dagger(k)(i\sigma_y)_{\alpha\beta}a_\beta^\dagger(-k),$$

where both sides transform as a scalar. On the other hand, the triplet pair with $S_z = 1$ is given by

$$a_\uparrow^\dagger(k)a_\uparrow^\dagger(-k) = \frac{1}{2}\sum_{\alpha\beta} a_\alpha^\dagger(k)[(\sigma_x + i\sigma_y)i\sigma_y]_{\alpha\beta}a_\beta^\dagger(-k).$$

It can easily be checked that the triplet pair with $S_z = -1$ obtains $(\sigma_x - i\sigma_y)$ instead of $(\sigma_x + i\sigma_y)$ in the above, and the pair with $S_z = 0$ obtains $2\sigma_z$ instead. Thus, we parametrize the pairing amplitude as

$$\Psi_{\alpha\beta}(\hat{k}) = \{[\Psi_s(\hat{k}) + \boldsymbol{\Psi}_t(\hat{k}) \cdot \boldsymbol{\sigma}]i\sigma_y\}_{\alpha\beta}, \tag{5.5}$$

where $\Psi_s(\hat{k})$ describes the singlet part and the vector $\boldsymbol{\Psi}_t(\hat{k})$ the triplet one. In the case of s-wave pairing, $\Psi_s(\hat{k})$ is just a constant. With finite angular momentum l, it is expanded in terms of spherical harmonics as

$$\Psi_s(\hat{k}) = \sum_m c_m Y_{lm}(\hat{k}).$$

On the other hand, for the triplet pair with $S_z = 0$, for example, we obtain

$$\boldsymbol{\Psi}_t(\hat{k}) = \hat{z}f(\hat{k}), \tag{5.6}$$

where $\hat{z}$ is the unit vector along the z-axis and $f(\hat{k})$ specifies the orbital degrees of freedom. The latter is expanded in terms of spherical harmonics as the singlet pairing. In the case of $S_z = \pm 1$, $\hat{z}$ is replaced by $\hat{x} \pm i\hat{y}$. The Fermi statistics require that $\Psi_s(\hat{k})$ is an even function of $\hat{k}$ and $\boldsymbol{\Psi}_t(\hat{k})$ odd. These types of pairings do not occur at the same time.

Let us introduce the gap function $\Delta_{\alpha\beta}(\hat{k})$ which is related to the pairing amplitude by

$$\Delta_{\alpha\beta}(\hat{k}) = \sum_{\mu\nu} \sum_{p} \langle \alpha\beta | V(k, p) | \nu\mu \rangle \Psi_{\mu\nu}(\hat{p}).$$

Here $\langle \alpha\beta | V(k, p) | \nu\mu \rangle$ is the matrix element of the pairing interaction. As before, the convention $|\alpha\beta\rangle^{\dagger} = \langle \beta\alpha |$ is adopted in ordering the one-particle states. Then we parametrize

$$\Delta_{\alpha\beta}(\hat{k}) = \{[D(\hat{k}) + d(\hat{k}) \cdot \sigma] i \sigma_y\}_{\alpha\beta}. \tag{5.7}$$

The quantities $D(\hat{k})$ and $d(\hat{k})$ have the same transformation properties as $\Psi_{\mathrm{s}}(\hat{k})$ and $\Psi_{\mathrm{t}}(\hat{k})$, respectively. In contrast to $n(\hat{k})$ and $m(\hat{k})$ in the density matrix, the gap function need not be real. This is associated with the gauge symmetry broken spontaneously in the superconducting phase.

In addition to gauge symmetry, other symmetries like spherical symmetry and time reversal can be broken. In order to deal with a general case, we introduce [11] the four-component field $\psi_i(k)$ ($i = 1, 2, 3, 4$) by the vector

$$\psi(k) = (a_{\uparrow}(k), a_{\downarrow}(k), a_{\uparrow}^{\dagger}(-k), a_{\downarrow}^{\dagger}(-k))^{\mathrm{t}}, \tag{5.8}$$

where the superscript t means the transpose. The first and second components are said to be conjugate to each other. The third and the fourth components are also conjugate partners, while the first and the fourth, as well as the second and the third, are time-reversal partners. The mean-field Hamiltonian is written as

$$H = \sum_{k} \sum_{ij} h_{ij}(k) \psi_i^{\dagger}(k) \psi_j(k) = \sum_{k} \psi^{\dagger}(k) \hat{h}(k) \psi(k),$$

where the 4×4 matrix $\hat{h} = \{h_{ij}\}$ is given by

$$\hat{h}(k) = \begin{pmatrix} \epsilon_k & \Delta(\hat{k}) \\ \Delta(\hat{k})^{\dagger} & -\epsilon_{-k} \end{pmatrix}. \tag{5.9}$$

Here the entries on the right-hand side are themselves 2×2 matrices.

The eigenvalues of $\hat{h}(k)$ are most easily obtained by taking the square of it. Then we get

$$(\hat{h}(k))^2 = \begin{pmatrix} \epsilon_k^2 + \Delta\Delta^{\dagger} & 0 \\ 0 & \epsilon_k^2 + \Delta^{\dagger}\Delta \end{pmatrix}, \tag{5.10}$$

where

$$\Delta\Delta^{\dagger} = |D|^2 + |d|^2 + w \cdot \sigma, \qquad \Delta^{\dagger}\Delta = |D|^2 + |d|^2 - w \cdot \sigma,$$

with $w = i d \times d^* = w^*$. One calls the pairing with $w = 0$ unitary and the others non unitary.

We diagonalize the 2×2 submatrices in eqn (5.10) and obtain the quasi-particle spectrum as

$$E_{\tau\sigma}(\boldsymbol{k}) = \tau \sqrt{\epsilon(\boldsymbol{k})^2 + |D(\hat{k})|^2 + |\boldsymbol{d}(\hat{k})|^2 + \sigma|\boldsymbol{w}(\hat{k})|},$$

where $\tau = \pm 1$ specifies the orbital (or the particle–hole) branch and $\sigma = \pm 1$ the spin branch. In the nonunitary case, the spin degeneracy is removed, which means that time-reversal symmetry is spontaneously broken. The superconducting gap vanishes at $\boldsymbol{k}$ such that $|\boldsymbol{d}(\hat{k})|^2 - |\boldsymbol{w}(\hat{k})| = 0$ in the triplet case, and $|D(\hat{k})|^2 = 0$ in the singlet case. In the nonunitary case, the gap may open only in one spin branch.

In the case of superfluid ^{3}He, the following types of p-wave pairings have been considered:

$$\boldsymbol{d}(\hat{k}) \propto \begin{cases} \hat{x}k_x + \hat{y}k_y + \hat{z}k_z & \text{(BW)}, \\ \hat{z}(k_x + ik_y) & \text{(ABM)}, \\ \hat{z}k_z & \text{(polar)}. \end{cases} \tag{5.11}$$

Here and in what follows, we write $\hat{k}_\alpha$ ($\alpha = x, y, z$) simply as k_α for notational simplicity. The first pairing type corresponds to vanishing total angular momentum, namely $J = L + S = 0$, and is called the Balian–Werthamer (BW) state [11]. The BW state has no node. The second one, with $S_z = 0$ and $L_z = 1$, is realized in a narrow range of the phase diagram and is called the ABM state [12]. The ABM state has a point node at $k_x = k_y = 0$. The third one with $S_z = L_z = 0$ is called the polar state which has a line node at $k_z = 0$. It is noted that the spherical symmetry is spontaneously broken in the ABM and polar states.

Singlet pairing under tetragonal symmetry In the presence of a periodic potential, the degeneracy associated with a finite angular momentum of the pair is lifted. The pair should instead be labelled by an irreducible representation of the point group [13–15]. As long as the deviation from spherical symmetry is small, one can still use a terminology such as the s-wave or p-wave pairing. Under strong crystal potential, however, wave functions with different angular momenta are mixed. Then, labelling by the angular momentum loses its significance. Let us consider as an example the tetragonal symmetry. The simplest pairing has Γ_1^+ (A_{1g}) symmetry represented by

$$D_0(\hat{k}) = a + b(k_x^2 + k_y^2) + ck_z^2 + \cdots, \tag{5.12}$$

where $a, b, c, \ldots$ are parameters to be determined on the basis of an analysis of energies. Here a originates from the s wave, b and c from the d wave, and so on. One has $|a| \gg |b|, |c|$ if the pairing predominantly has an s-wave character. The function $D_0(\hat{k})$ transforms as a scalar under the point-group operation, and obviously there are no nodes.

Another type of the singlet pairing has the Γ_3^+ (B_{1g}) symmetry and is represented by

$$D_1(\hat{k}) = (k_x^2 - k_y^2)f_0(\hat{k}) = \text{Re}\, k_+^2 f_0(\hat{k}),$$

where $k_+ = k_x + ik_y$ and $f_0(\hat{k})$ is a scalar function which is, in general, different from $D_0(\hat{k})$ in eqn (5.12). The function $D_1(\hat{k})$ vanishes on the planes $k_x = \pm k_y$. Therefore,

Table 5.1 *Pairing basis functions in tetragonal symmetry* D_{4h}. *The functions separated by a semicolon are to be taken as linear combinations to minimize the energy. The functions with the suffix $\pm$ or $\mp$ denote degenerate partners.*

Representation	Basis functions
(a) Singlet pairing	
$\Gamma_1^+\ (A_{1g})$	1
$\Gamma_2^+\ (A_{2g})$	$\mathrm{Im}\,k_+^4$
$\Gamma_3^+\ (B_{1g})$	$\mathrm{Re}\,k_+^2$
$\Gamma_4^+\ (B_{2g})$	$\mathrm{Im}\,k_+^2$
$\Gamma_5^+\ (E_g)$	$k_z k_\pm;\ k_z k_\pm^3$
(b) Triplet pairing	
$\Gamma_1^-\ (A_{1u})$	$\hat{z}k_z;\ \mathrm{Re}\,k_+\hat{r}_-;\ \mathrm{Re}\,k_+^3\hat{r}_+$
$\Gamma_2^-\ (A_{2u})$	$\mathrm{Im}\,k_+\hat{r}_-;\ \mathrm{Im}\,k_+^3\hat{r}_+;\ k_z\mathrm{Im}\,k_+^4\hat{z}$
$\Gamma_3^-\ (B_{1u})$	$\mathrm{Re}\,k_+\hat{r}_+;\ k_z\mathrm{Re}\,k_+^2\hat{z};\ \mathrm{Re}\,k_+^3\hat{r}_-$
$\Gamma_4^-\ (B_{2u})$	$\mathrm{Im}\,k_+\hat{r}_+;\ k_z\mathrm{Im}\,k_+^2\hat{z};\ \mathrm{Im}\,k_+^3\hat{r}_-$
$\Gamma_5^-\ (E_u)$	$k_\pm^{n+1}\hat{z};\ k_z k_\pm^n\hat{r}_\pm;\ k_z k_\pm^{n+2}\hat{r}_\mp\ (n=0;2)$

along the intersection of the plane with the Fermi surface, which makes a line, the gap vanishes.

The pairing symmetry $\Gamma_5^+\ (E_g)$ belongs to a two-dimensional representation where the Cooper pair has an internal degree of freedom, e.g. a finite angular momentum. The gap function is given by

$$D_2(\hat{k}) = (\eta_1 k_x + \eta_2 k_y)k_z[f_0(\hat{k}) + k_z^2 f_1(\hat{k})].$$

The coefficients η_1, η_2 are arbitrary, in contrast to the case of eqn (5.12). On the other hand, the scalar functions $f_0(\hat{k})$ and $f_1(\hat{k})$ are completely determined by minimizing the energy. Again, the gap vanishes on the line $k_z = 0$.

The list including other representations is given in Table 5.1(a). We notice that all the basis functions in these tables should be multiplied by a scalar function which depends on each irreducible representation.

Triplet pairing under tetragonal symmetry We now turn to triplet pairing. The simplest pairing has $\Gamma_1^-\ (A_{1u})$ symmetry represented by

$$\boldsymbol{d}(\hat{k}) = \hat{z}k_z f_0(\hat{k}) + (\hat{x}k_x + \hat{y}k_y)f_1(\hat{k}) + (\hat{x}k_x - \hat{y}k_y)(k_x^2 - k_y^2)f_2(\hat{k})$$

$$= \hat{z}k_z f_0(\hat{k}) + (\mathrm{Re}\,k_+\hat{r}_-)f_1(\hat{k}) + (\mathrm{Re}\,k_+^3\hat{r}_+)f_2(\hat{k}), \tag{5.13}$$

where $\hat{x}$ denotes the unit vector along the x-axis and $\hat{r}_\pm = \hat{x} \pm i\hat{y}$. The scalar functions $f_i(\hat{k})$ ($i = 0, 1, 2$) are to be determined by minimizing the energy. There is no node because, in general, $f_0 f_1 f_2 \neq 0$. If, however, there is a spontaneous breakdown of

Table 5.2 *Basis functions for the pair in the hexagonal symmetry D_{6h}.*

Representation	Basis functions
(a) Singlet pairing	
Γ_1^+ (A_{1g})	1
Γ_2^+ (A_{2g})	$\mathrm{Im}\, k_+^6$
Γ_3^+ (B_{1g})	$k_z \mathrm{Im}\, k_+ d^3$
Γ_4^+ (B_{2g})	$k_z \mathrm{Re}\, k_+^3$
Γ_5^+ (E_{1g})	$k_z k_\pm;\ k_z k_\mp^5$
Γ_6^+ (E_{2g})	$k_\pm^2;\ k_\mp^4$
(b) Triplet pairing	
Γ_1^- (A_{1u})	$\hat{z} k_z;\ \mathrm{Re}\, k_+ \hat{r}_-;\ \mathrm{Re}\, k_+^5 \hat{r}_-$
Γ_2^- (A_{2u})	$\mathrm{Im}\, k_+ \hat{r}_-;\ \mathrm{Im}\, k_+^5 \hat{r}_-;\ k_z \mathrm{Im}\, k_+^6 \hat{z}$
Γ_3^- (B_{1u})	$\mathrm{Im}\, k_+^3 \hat{r}_-;\ k_z \mathrm{Im}\, k_+^2 \hat{r}_+;\ k_z \mathrm{Im}\, k_+^4 \hat{z}$
Γ_4^- (B_{2u})	$\mathrm{Re}\, k_+^3 \hat{r}_-;\ k_z \mathrm{Re}\, k_+^2 \hat{r}_+;\ k_z \mathrm{Re}\, k_+^4 \hat{z}$
Γ_5^- (E_{1u})	$k_\pm \hat{z};\ k_z \hat{r}_\pm;\ k_z k_\pm^2 \hat{r}_\mp;\ k_\mp^5 \hat{z};\ k_z k_\pm^4 \hat{r}_\mp;\ k_z k_\mp^6 \hat{r}_\pm$
Γ_6^- (E_{2u})	$k_\pm \hat{r}_\pm;\ k_z k_\pm^2 \hat{z};\ k_\pm^3 \hat{r}_\mp;\ k_\mp^3 \hat{r}_\mp;\ k_\mp^5 \hat{r}_\pm;\ k_z k_\mp^4 \hat{z}$

symmetry to realize $S_z = 0$, for example, we have $f_1 = f_2 = 0$ and a line node appears at $k_z = 0$.

The other simple symmetry is Γ_3^- (B_{1u}) with

$$\boldsymbol{d}(\boldsymbol{k}) = (\hat{x} k_x - \hat{y} k_y) f_0(\hat{k}) = (\mathrm{Re}\, k_+ \hat{r}_+) f_0(\hat{k}),$$

which has zeroes at $k_x = k_y = 0$. The list including other representations is given in Table 5.1(b). We also quote all basis functions for hexagonal and cubic symmetries [15] in Tables 5.2 and 5.3.

5.2.2 *Effect of spin–orbit interaction*

In the presence of the spin–orbit interaction, the Bloch function is specified by a pair of indices: the crystal momentum $\boldsymbol{k}$ and the quasi-spin index σ, which reduces to the pure spin as the spin–orbit interaction decreases. The energy level for each $\boldsymbol{k}$ is doubly degenerate as long as the time-reversal symmetry is not broken. We call the degenerate states a conjugate pair. Since the choice of the quasi-spin is not unique, it is convenient to formulate the theory such that the symmetry appears independently of the representation [16]. We shall illustrate this for the case with the Zeeman splitting. We introduce the g-tensor $\hat{g}(\boldsymbol{k}) = \{g_{ij}(\boldsymbol{k})\}$ and a vector $\boldsymbol{\gamma}(\boldsymbol{k})$, which describes the magnetic moment, by the relation

$$\gamma_i(\boldsymbol{k}) = \frac{1}{2} \sum_j g_{ij}(\boldsymbol{k}) \sigma_j, \tag{5.14}$$

Table 5.3 *Basis functions for the pair in the cubic symmetry O_h. Functions separated by a comma represent degenerate partners, and $\hat{d}_{\mathrm{BW}} = k_x\hat{x} + k_y\hat{y} + k_z\hat{z}$. The semicolons mean that linear combinations of terms with different n's indicated in each parenthesis are to be taken.*

Representation	Basis functions
(a) Singlet pairing	
Γ_1^+ (A_{1g})	1
Γ_2^+ (A_{2g})	$(k_x^2 - k_y^2)(k_y^2 - k_z^2)(k_z^2 - k_x^2)$
Γ_3^+ (E_g)	$2k_z^n - k_x^n - k_y^n, k_x^n - k_y^n$ $(n = 2; 4)$
Γ_4^+ (T_{1g})	$k_y k_z(k_y^n - k_z^n), k_z k_x(k_z^n - k_x^n), k_x k_y(k_x^n - k_y^n)$ $(n = 2; 4; 6)$
Γ_5^+ (T_{2g})	$k_y k_z k_x^n, k_z k_x k_y^n, k_x k_y k_z^n$ $(n = 0; 2; 4)$
(b) Triplet pairing	
Γ_1^- (A_{1u})	$\hat{x}k_x^n + \hat{y}k_y^n + \hat{z}k_z^n$ $(n = 1; 3; 5)$
Γ_2^- (A_{2u})	$\hat{x}k_x^n(k_y^2 - k_z^2) + \hat{y}k_y^n(k_z^2 - k_x^2) + \hat{z}k_z^n(k_x^2 - k_y^2)$ $(n = 1; 3; 5)$
Γ_3^- (E_u)	$\hat{x}k_x^n + \hat{y}k_y^n - 2\hat{z}k_z^n, \hat{x}k_x^n - \hat{y}k_y^n$ $(n = 1; 3; 5)$
	$(k_x^n - k_y^n)\hat{d}_{BW}, (2k_z^n - k_x^n - k_y^n)\hat{d}_{BW}$ $(n = 1; 3; 5)$
	$(k_x^5 k_z^2 \hat{x} + k_x^2 k_z^5 \hat{z}) + (x \leftrightarrow y), [k_x^5(k_z^2 - 2k_y^2)\hat{x} + k_x^2 k_z^5 \hat{z}] + (x \leftrightarrow y)$
Γ_4^- (T_{1u})	$k_x^m(\hat{y}k_z^n - \hat{z}k_y^n), k_y^m(\hat{z}k_x^n - \hat{x}k_z^n), k_z^m(\hat{x}k_y^n - \hat{y}k_x^n)$
	$(m = 0; 2, n = 1; 3; 5)$
	$k_x k_y(k_x^2 - k_y^2)k_z^n\hat{z}, k_y k_z(k_y^2 - k_z^2)k_x^n\hat{x}, k_z k_x(k_z^2 - k_x^2)k_y^n\hat{y}$
	$(n = 1; 3; 5)$

where each component $\gamma_i(k)$ is a matrix. Then the magnetic moment operator $M(k)$ is given by

$$M(k) = -\mu_{\mathrm{B}} \sum_{\alpha\beta} a_\alpha(k)^\dagger \gamma_{\alpha\beta}(k) a_\beta(k). \tag{5.15}$$

Under the crystal symmetry operation, $M(k)$ transforms like a vector and is independent of the choice of basis. With the one-body density matrix parametrized by eqn (5.1), we obtain the magnetization

$$\langle M(k)\rangle = -\mu_{\mathrm{B}}\mathrm{Tr}[\gamma(k)\rho(k)] = -\tfrac{1}{2}\mu_{\mathrm{B}}\hat{g}(k)m(k).$$

The Zeeman splitting $2E_{\mathrm{Z}}(k)$ is given by the eigenvalues $\pm E_{\mathrm{Z}}(k)$ of the Hermitian matrix

$$-M(k) \cdot H.$$

We obtain, from eqns (5.14) and (5.15),

$$E_{\mathrm{Z}}(k)^2 = \tfrac{1}{4}\mu_{\mathrm{B}}^2 H \cdot \hat{G}(k) H,$$

where $\hat{G}(k) = \hat{g}(k)\hat{g}(k)^{\mathrm{t}}$. Note that this result is independent of the choice of basis.

The gap function is now parametrized by

$$\Delta_{\alpha\beta}(\hat{k}) = \{[D(\hat{k}) + \boldsymbol{d}(\hat{k}) \cdot \boldsymbol{\gamma}(\hat{k})]i\sigma_y\}_{\alpha\beta}. \qquad (5.16)$$

Then we get

$$\Delta\Delta^{\dagger} = |D(\hat{k})|^2 + (\boldsymbol{d} \cdot \boldsymbol{\gamma})(\boldsymbol{\gamma} \cdot \boldsymbol{d}^*) = |D(\hat{k})|^2 + |\boldsymbol{d}_g|^2 + \boldsymbol{w} \cdot \boldsymbol{\sigma},$$

where $\boldsymbol{d}_g = \hat{g}^{\mathrm{t}}\boldsymbol{d}$ and $\boldsymbol{w} = i\boldsymbol{d}_g \times \boldsymbol{d}_g^* = \boldsymbol{w}^*$. The symmetry property of $\boldsymbol{d}_g(\hat{k})$ is the same as that of $\boldsymbol{d}(\hat{k})$. Thus, the energy spectrum in the presence of the spin–orbit interaction is obtained just by using $\boldsymbol{d}_g(\hat{k})$ in place of $\boldsymbol{d}(\hat{k})$ in the previous section.

It has been argued that the triplet in general cannot have line nodes [16]. The main line of argument for this is the following: the gap vanishes if an eigenvalue of the matrix $\{h_{ij}(\boldsymbol{k})\}$ is given by $\pm\epsilon(\boldsymbol{k})$; namely, we require

$$\det\{\Delta_{\alpha\beta}(\hat{k})\} = D(\hat{k})^2 - \boldsymbol{d}_g(\hat{k}) \cdot \boldsymbol{d}_g(\hat{k}) = 0. \qquad (5.17)$$

In the singlet case, $D(\hat{k})$ can be made real. Then the zero condition specifies a plane in the $\boldsymbol{k}$-space, and the intersection with the Fermi surface constitutes the line node. For the unitary triplet case, $\boldsymbol{d}_g(\hat{k})$ can also be made real. Then the left-hand side of eqn (5.17) is the sum of three non-negative quantities. If these quantities are independent, there is no solution on the Fermi surface. If, by symmetry, the number of independent equations is reduced to two, then a point zero is realized. In order to have a line node, there must be only one independent equation. However, this cannot be expected from the symmetry alone. In the nonunitary case, $\boldsymbol{d}_g(\hat{k})$ is a complex vector, and eqn (5.17) leads to two real equations. Therefore, some special situation is also necessary in order to have line nodes in the nonunitary triplet pairing.

A candidate realizing such a situation is the pair with a definite direction of $\boldsymbol{d}_g(\hat{k})$ as in the case of the polar state of eqn (5.11). Under hexagonal symmetry, for example, the E_{1u} pair with $S_z = \pm 1$ of the quasi-spin always contains a factor k_z (see Table 5.2). The same remark applies also to the E_{2u} pair with $S_z = 0$. These pairs with line nodes are suggested for pairing in UPt$_3$, as will be explained later.

5.2.3 *Density of states of quasi-particles*

In the superconducting state, the energy of quasi-particles is given as the eigenvalues of $\hat{h} = \{h_{ij}\}$ given by eqn (5.9). The density of states $N(E)$ of quasi-particles is then

$$N(E) = \frac{1}{2}\sum_{\boldsymbol{k}\tau\sigma}\delta(E - E_{\tau\sigma}(\boldsymbol{k})) = |E|\sum_{\boldsymbol{k}\sigma}\delta(E^2 - E_{\boldsymbol{k}\sigma}^2), \qquad (5.18)$$

where $E_{\boldsymbol{k}\sigma} = |E_{\tau\sigma}(\boldsymbol{k})|$. By putting the eigenvalues of $(\hat{h}(\boldsymbol{k}))^2$, we can derive the density of states. For simplicity, we first treat cases with spherical symmetry.

s-Wave pairing In BCS superconductivity mediated by the electron–phonon interaction [1], the energy gap opens over the entire Fermi surface. With the notation $D(\hat{k}) = \Delta$, the density of states $N_{\text{BCS}}(E)$ in the superconducting state is given by

$$N_{\text{BCS}}(E) = N_0 \frac{|E|}{\sqrt{E^2 - \Delta^2}} \quad (|E| > \Delta) \tag{5.19}$$

and $N_{\text{BCS}}(E) = 0$ for $|E| < \Delta$. Here, $N_0 = 2\sum_k \delta(\epsilon_k)$ is the density of states at the Fermi level in the normal state. Therefore, various physical quantities obey an exponential law well below T_c.

p-Wave pairing   In the case of a p-wave triplet pairing, we obtain

$$N(E) = N_0 \int \frac{d\Omega}{8\pi} \sum_{\sigma=\pm 1} \text{Re} \frac{|E|}{\sqrt{E^2 - |\boldsymbol{d}(\hat{k})|^2 - \sigma|\boldsymbol{w}(\hat{k})|}}, \tag{5.20}$$

where the integral is over the solid angle of $\hat{k}$. In the ABM state in the superfluid ^{3}He-A, the Cooper pair consists of parallel spin pairing as given by eqn (5.11). Then, we have $|\boldsymbol{d}(\hat{k})|^2 = \Delta^2 \sin^2\theta$ and $\boldsymbol{w}(\hat{k}) = 0$. Correspondingly, the quasi-particle density of states in the ABM state is given by

$$N_{\text{ABM}}(E) = \frac{N_0|E|}{4\pi} \int_0^{2\pi} d\phi \int_0^{\pi} d\theta \, \text{Re} \frac{\sin\theta}{\sqrt{E^2 - \Delta^2 \sin^2\theta}}$$

$$= \frac{N_0 E}{2\Delta} \ln\left|\frac{E + \Delta}{E - \Delta}\right|. \tag{5.21}$$

This result is obtained by choosing $x = \Delta\cos\theta/\sqrt{E^2 - \Delta^2}$ as the integration variable for the case of $|E| > \Delta$. The result in the case of $|E| < \Delta$ is obtained by analytic continuation of E. Hence, eqn (5.21) is valid for both cases. We note that $N_{\text{ABM}}(E)$ exhibits a divergence at the gap edge, which is weaker than the BCS case, and is proportional to E^2 at low energy as shown in Fig. 5.1.

In the case of the p-wave polar type with $\boldsymbol{d}(\hat{k}) = \Delta\hat{z}k_z$, the gap vanishes along a line at the Fermi surface. Accordingly, we have

$$N_{\text{polar}}(E) = \frac{1}{2} N_0|E| \int_0^{\pi} d\theta \, \text{Re} \frac{\sin\theta}{\sqrt{E^2 - \Delta^2 \cos^2\theta}}$$

$$= \begin{cases} \pi N_0|E|/(2\Delta) & (|E| < \Delta), \\ N_0(E/\Delta)\arcsin(\Delta/E) & (|E| > \Delta). \end{cases} \tag{5.22}$$

This result is obtained with $x = \Delta\cos\theta/|E|$ taken as the integration variable. In this case, $N_{\text{polar}}(E)$ is finite at the gap edge and is proportional to $|E|$ at low energy. In the triplet BW [11] state realized in the superfluid ^{3}He-B phase, the density of states is the same as the BCS state, since the energy gap is isotropic. These results are summarized in Fig. 5.1.

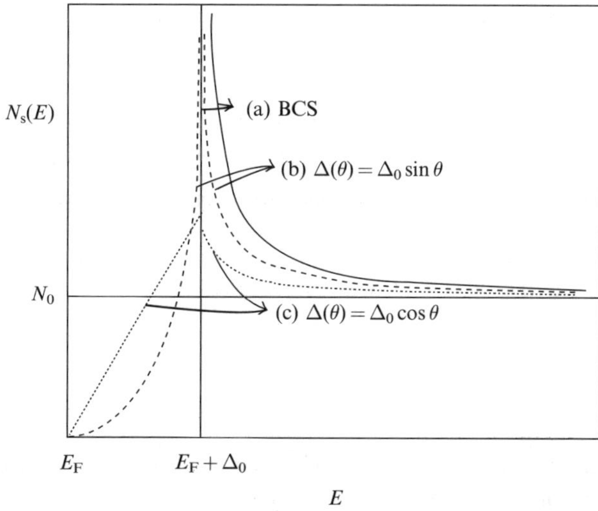

FIG. 5.1. Quasi-particle density of states in the superconducting phase: (a) s-wave, (b) axial (ABM) p wave with vanishing gap at points, and (c) polar p wave with vanishing gap along a line.

5.3   NMR as a probe of superconducting states

5.3.1   *Knight shift*

The Knight shift is the only convenient measure of the local spin susceptibility reflecting the symmetry of the order parameter, since the diamagnetic shielding by supercurrents overwhelms all other contributions to the bulk susceptibility. In the presence of a weak magnetic field H along the z-axis, the itinerant part of the magnetization M_z is given by

$$M_z = -\frac{1}{2}g\mu_\mathrm{B}\sum_{k}[f(E_{k\uparrow}) - f(E_{k\downarrow})], \tag{5.23}$$

where $f(E_{k\sigma})$ is the Fermi distribution function for a quasi-particle with a spin-dependent energy $E_{k\sigma}$. We derive the spin susceptibility χ_s for various cases.

s-Wave pairing   In the case of s-wave singlet pairing, we obtain $E_{k\sigma} = E_k + \sigma h$, with $h = g\mu_\mathrm{B}H/2$. Then χ_s for the s wave is derived [17] using $N_\mathrm{BCS}(E)$ in eqn (5.19) as

$$\chi_\mathrm{s} = \frac{1}{4}(g\mu_\mathrm{B})^2 \int_{-\infty}^{\infty} N_\mathrm{BCS}(E)\left(-\frac{df}{dE}\right)dE. \tag{5.24}$$

This result is plotted in Fig. 5.2. At low-T, χ_s decreases exponentially as $\exp(-\Delta/T)$. In the d wave, since $N_\mathrm{d}(E)$ is proportional to the energy E, or its square (E^2), χ_s becomes zero in proportion to T or T^2, respectively, at low T.

The above argument is valid only in the clean limit where the transport mean free path, l_tr is larger than the superconducting coherence length ξ, i.e. $\Delta > 1/\tau$. Here the scattering time τ is defined by $l_\mathrm{tr} = v_\mathrm{F}\tau$. In the case where a strong spin–orbit scattering

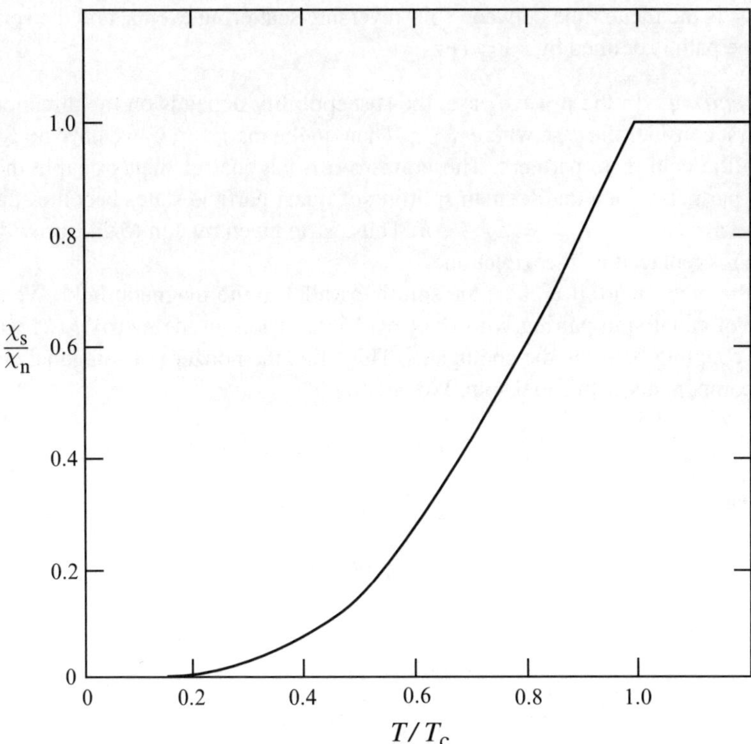

FIG. 5.2. Temperature dependence of the Knight shift in s-wave superconductors as given by eqn (5.24). In actual materials there remains a residual shift due to the spin–orbit interaction.

($\lambda \boldsymbol{L} \cdot \boldsymbol{S}$) with $\lambda \gg \Delta$ occurs due to the presence of imperfections and the boundary in small particles, those states which would be different eigenstates in the absence of the spin–orbit interaction can be admixed. This is time-reversal-invariant, so that the new eigenstates can retain the same pairing in the superconducting state as in the absence of spin–orbit scattering. However, these eigenstates contain admixtures of both up and down spin directions. As a result, a conventional picture of spin-singlet pairing no longer holds. In the limit of large spin–orbit interaction, the opening of a finite energy gap below T_c does not produce a large perturbation on the spin susceptibility. For the two limits of strong and weak spin–orbit scattering, Anderson derived the fraction of the residual spin susceptibility at $T = 0$ as [18]

$$\chi_\mathrm{s}/\chi_\mathrm{n} \simeq 1 - 2\Delta\tau_\mathrm{so}, \quad \xi \gg l_\mathrm{so},$$
$$\chi_\mathrm{s}/\chi_\mathrm{n} \simeq (6\Delta\tau_\mathrm{so})^{-1}, \quad \xi \ll l_\mathrm{so},$$

where τ_{so} is the mean time between spin-reversing scattering events, and the spin–orbit mean free path is defined by $l_{\text{so}} = v_{\text{F}}\tau_{\text{so}}$.

p-Wave pairing   In the p-wave case, the susceptibility depends on the direction of $\boldsymbol{d}$. Let us first consider the case where $\boldsymbol{d} \parallel \hat{z}$. Then, in the matrix $\hat{h}(\boldsymbol{k})$ we have no coupling between the conjugate partners. The nonzero off-diagonal elements couple the time-reversal partners. Then the Zeeman splitting of quasi-particle states becomes the same as the s-wave case i.e. $E_{\boldsymbol{k}\sigma} = E_{\boldsymbol{k}} + \sigma h$. Thus, χ_{s} is given by eqn (5.24), provided that $N_{\text{BCS}}(E)$ is replaced by the triplet one.

On the other hand, if $\boldsymbol{d} \perp \hat{z}$, the spin is parallel to the magnetic field. We assume the case of equal-spin pairing with $\boldsymbol{d}(\hat{k}) = \Delta \hat{x} k_{+}$. Then, in the matrix $\hat{h}(\boldsymbol{k})$ we again have no coupling between the conjugates. This time, the nonzero off-diagonal elements couple components with equal spin. We obtain

$$E_{\boldsymbol{k}\sigma}^2 = (\epsilon_{\boldsymbol{k}} + \sigma h)^2 - \Delta^2 \sin^2 \theta,$$

which leads to

$$\frac{\partial E_{\boldsymbol{k}\sigma}}{\partial H} = \frac{1}{2}\mu_{\text{B}}\sigma\frac{\partial E_{\boldsymbol{k}\sigma}}{\partial \epsilon_{\boldsymbol{k}}}.$$

Then χ_{s} is given by

$$\begin{aligned}
\chi_{\text{s}} &= \sum_{\boldsymbol{k}\sigma} \frac{\partial}{\partial H}\left[f(E_{\boldsymbol{k}\sigma})\frac{\partial E_{\boldsymbol{k}\sigma}}{\partial H}\right] \\
&= -\frac{1}{4}(g\mu_{\text{B}})^2 N_0 \int_{-\infty}^{\infty} d\epsilon \frac{\partial}{\partial \epsilon}\left[f(E_{\epsilon})\frac{\partial E_{\epsilon}}{\partial \epsilon}\right] \\
&= \frac{1}{4}(g\mu_{\text{B}})^2 N_0.
\end{aligned} \tag{5.25}$$

Thus, the spin susceptibility remains the same as in the normal state.

If one can neglect the anisotropy in the system as in liquid ^{3}He, $\boldsymbol{d}$ can rotate so that $\boldsymbol{d} \perp \hat{z}$ is satisfied. Then χ_{s} stays constant in any direction in the parallel-spin pairing (ABM) state. On the contrary, provided that the spin is pinned along an easy axis or an easy plane due to the anisotropy below T_{c}, χ_{s} is reduced if H is perpendicular to such an axis or plane. In the isotropic triplet state (BW state), χ_{s} is decreased down to 2/3 of the value in the normal state. This is interpreted as occurring because the average $|d_z|^2$ of the component parallel to H is 1/3 of $|\boldsymbol{d}|^2$ in the BW state. Thus, the Knight shift measurement provides an important clue for the determination of the symmetry of the order parameter.

5.3.2 *Nuclear spin–lattice relaxation rate*

A modification of the excited state spectrum by the onset of superconductivity has a substantial effect on the temperature dependence of the NMR relaxation rate $1/T_1$. The dynamical response of the excited states is also affected by the presence of the mixed state, or magnetic and nonmagnetic impurities. All these effects can be probed by the nuclear relaxation study.

At finite temperature, some pairs are broken up into single-particle excitations, while the presence of the coherent condensed state modifies the matrix element of the external perturbations. The latter effect is called a coherence effect. Since the nuclear relaxation in the superconducting state is significantly affected by the coherence effect, NMR serves as a good probe of coherence. We consider a case where the magnetic field is along the z-axis and we have axial symmetry in the x–y plane. Then, by eqns. (1.129) and (1.130), we have

$$\frac{1}{T_1 T} = A_\perp^2 (g_J \mu_B)^2 \sum_q \frac{1}{\omega_n} \text{Im}\, \chi_\perp(q, \omega_n), \qquad (5.26)$$

where $A_\perp$ is the momentum average of the hyperfine interaction. The transverse dynamical susceptibility is derived in the mean-field approximation in the superconducting state. For χ we consider the y-component χ_{yy}. We note

$$\sum_{k, p} \sum_{\alpha\beta} a_{k\alpha}^\dagger (\sigma_y)_{\alpha\beta} a_{p\beta} = \frac{1}{2} \sum_{k, p} \psi(k)^\dagger \begin{pmatrix} \sigma_y & 0 \\ 0 & \sigma_y \end{pmatrix} \psi(p), \qquad (5.27)$$

where $\psi(p)$ is the four-component field given by eqn (5.8). The 4×4 matrix in eqn (5.27) is written as $\hat\sigma_y = \sigma_y \otimes \tau_0$, where τ_0 is the 2×2 unit matrix. By evaluating the zeroth order polarization part in the Green function formalism, we obtain

$$\text{Im}\, \chi_{yy}(q, \omega) = \sum_{k, p} \int_{-\infty}^{\infty} d\epsilon [f(\epsilon) - f(\epsilon + \omega)]$$

$$\times \text{Tr}\,[\hat\sigma_y \delta(\epsilon - \hat h(k)) \hat\sigma_y \delta(\epsilon + \omega - \hat h(p))], \qquad (5.28)$$

where the delta function corresponds to the imaginary part of the retarded Green function of the four-component field. In the unitary case, to which we restrict ourselves from now on, the latter is given by

$$-\frac{1}{\pi} \text{Im}\,[\epsilon - \hat h(k) + i\delta]^{-1} = \frac{1}{2}\left(1 + \frac{\hat h(k)}{E_k}\right) \delta(\epsilon - E_k)$$

$$+ \frac{1}{2}\left(1 - \frac{\hat h(k)}{E_k}\right) \delta(\epsilon + E_k), \qquad (5.29)$$

where we have used the fact that $(\hat h(k))^2$ is proportional to a unit matrix given by $E_k^2 \sigma_0 \otimes \tau_0$.

We divide $\hat h(k)$ into the diagonal part $\hat h_0(k) = \epsilon_k \sigma_0 \otimes \tau_z$ and the rest $\hat h_1(k)$, which depends on the pairing type. Each case is discussed in the following.

s-Wave pairing In this case, we have $\hat\sigma_y \hat h(k) = \hat h(k) \hat\sigma_y$. Then the trace in eqn (5.28) is calculated as

$$\left(1 + \frac{\epsilon_k \epsilon_p + D(\hat k) D(\hat p)}{E_k E_p}\right)$$

$$\times [\delta(\epsilon - E_k)\delta(\epsilon + \omega - E_p) + \delta(\epsilon + E_k)\delta(\epsilon + \omega + E_p)], \qquad (5.30)$$

where the factor in front of the delta functions is an example of the coherence factor. We note that the summation over k makes the term with ϵ_k vanish because it is an odd function of ϵ_k. Then, for the s-wave pairing we obtain

$$\frac{1}{T_1 T} = \pi A^2 \int_0^\infty dE \left(1 + \frac{\Delta^2}{E^2}\right) N_{\text{BCS}}(E)^2 \left(-\frac{\partial f}{\partial E}\right), \tag{5.31}$$

where it is assumed that $\omega \ll T$. The density of states $N_{\text{BCS}}(E)$ diverges at $E = \Delta$, as shown in Fig. 5.1. Near T_c, where $|f'(\Delta)|$ is still large, the divergence of $N_{\text{BCS}}(E)$ gives rise to a divergence of $1/T_1$. However, the lifetime effect of quasi-particles by the electron–phonon and/or the electron–electron interactions, and the anisotropy of the energy gap due to the crystal structure broadens the quasi-particle density of states. This results in the suppression of the divergence of $1/T_1$. Instead, a peak of $1/T_1$ is observed just below T_c. This peak, characteristic of singlet pairing, was observed first by Hebel and Slichter [19] in the nuclear relaxation rate of Al, and is called the coherence peak. At low temperatures, $1/T_1$ decreases exponentially due to the uniform gap.

The coherence peak is not always observed in the case of strong-coupling super-conductors. As an example of the relaxation behaviour for weak and strong coupling s-wave superconductors, Fig. 5.3 shows $1/T_1$ of ^{119}Sn and ^{205}Tl in the Chevrel phase superconductors, $\text{Sn}_{1.1}\text{Mo}_6\text{Se}_{7.5}$ and $\text{TlMo}_6\text{Se}_{7.5}$ with $T_c = 4.2$ K and 12.2 K, respectively [20]. In the normal state, the law $T_1 T = $ constant holds for both compounds. In the superconducting state, $1/T_1$ of ^{119}Sn in $\text{Sn}_{1.1}\text{Mo}_6\text{Se}_{7.5}$ has a coherence peak just below T_c and decreases exponentially with $2\Delta = 3.6 k_B T_c$, while $1/T_1$ of ^{205}Tl in the strong-coupling superconductor $\text{TlMo}_6\text{Se}_{7.5}$ has no coherence peak just below T_c and decreases exponentially over five orders of magnitude below $0.8 T_c$ (10 K) with $2\Delta = 4.5 k_B T_c$. Even though the coherence peak is depressed, the s-wave picture is evidenced by the exponential decrease of $1/T_1$ below T_c. Figure 5.3 also shows another important example, the high-T_c cuprates, which are discussed in the next chapter.

p-Wave pairing   In the p-wave case, the coherence factor for Im χ_{yy} is calculated from eqn (5.30) as [11]

$$1 + \frac{1}{E_k E_p} \text{Re} \left(\epsilon_k \epsilon_p + \boldsymbol{d}(\hat{k}) \cdot \boldsymbol{d}(\hat{p})^* - 2 d_y(\hat{k}) d_y(\hat{p})^*\right). \tag{5.32}$$

Because ϵ_k and $\boldsymbol{d}(\hat{k})$ are odd functions of $\hat{k}$, the bilinear terms in them vanish by summation over k. Then the coherence factor is reduced to unity. The same situation occurs in the case of Im χ_{xx}. Therefore, the relaxation rate is obtained by replacing the quasi-particle density of states in eqn (5.31) by $N_{\text{ABM}}(E)$ as follows:

$$\frac{1}{T_1 T} = \frac{\pi A^2 N_0^2}{4 \Delta^2} \int_0^\infty E^2 \left(\ln \left|\frac{E + \Delta}{E - \Delta}\right|\right)^2 \left(-\frac{\partial f}{\partial E}\right) dE. \tag{5.33}$$

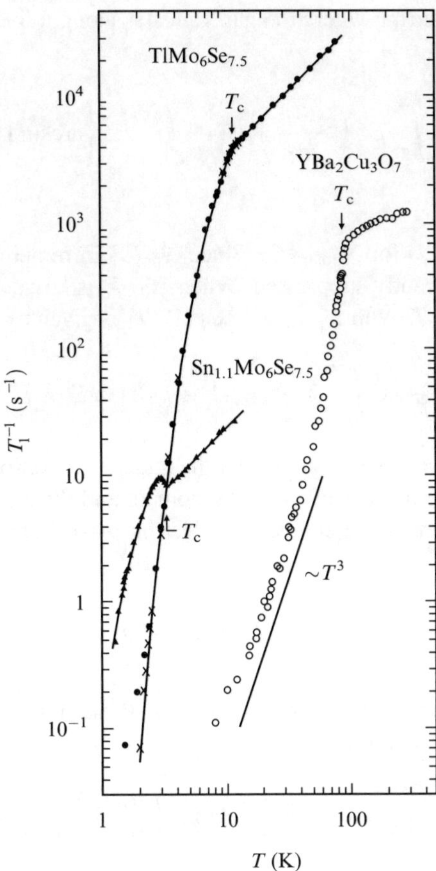

FIG. 5.3. Temperature dependence of $1/T_1$ of ^{205}Tl and ^{119}Sn in strong-coupling (TlMo$_6$Se$_{7.5}$) and weak-coupling (Sn$_{1.1}$Mo$_6$Se$_{7.5}$) s-wave superconductors, and of ^{63}Cu in YBa$_2$Cu$_3$O$_7$ at zero field. Solid lines above T_c represent the law $T_1 T$ = constant, and below T_c the exponential law with $1/T_1 = A \exp(-\Delta/k_B T)$. For TlMo$_6Se_8$, the ratio $2\Delta/k_B T_c = 4.5$ is obtained and for SnMo$_6$Se$_8$ the ratio is 3.6 [20].

Since the divergence of $N_{ABM}(E)$ at $E = \Delta$ is weak, the peak of $1/T_1$ just below T_c is much smaller than in the BCS case. The larger Δ_0/T_c is, where Δ_0 is the gap at zero temperature, the more suppressed is the peak of $1/T_1$. At low temperatures $T \ll T_c$, we obtain

$$\frac{1}{T_1} \propto \int_0^\infty E^4 \exp\left(-\frac{E}{T}\right) dE = T^5 \Gamma(5) \tag{5.34}$$

which gives the T^5 dependence at low T.

In the polar-type pairing where the gap vanishes along a line on the Fermi surface, $1/T_1$ is calculated as

$$\frac{1}{T_1 T} = \frac{\pi A^2 N_0^2}{\Delta^2} \left[\frac{\pi^2}{4} \int_0^\Delta E^2 \left(-\frac{\partial f}{\partial E} \right) dE + \int_\Delta^\infty E^2 \left\{ \arcsin \left(\frac{\Delta}{E} \right) \right\}^2 \left(-\frac{\partial f}{\partial E} \right) dE \right],$$

(5.35)

where we use eqn (5.22) for $N_{\text{polar}}(E)$. Since $N_{\text{polar}}(E)$ remains finite at $E = \Delta$, the peak of $1/T_1$ is significantly suppressed. When Δ_0/T_c is large, $1/T_1$ decreases rapidly just below T_c. At $T \ll T_c$ with $N_{\text{polar}}(E) \propto E$, $1/T_1$ is given by

$$\frac{1}{T_1} \propto \int_0^\infty E^2 f(E)(1 - f(E)) \, dE \propto T^3.$$

(5.36)

Quadrupolar relaxation We now consider the nuclear relaxation brought about by the interaction between the nuclear quadrupole moment and the electric field gradient due to the nonspherical part of the conduction-electron wave function. The interaction is expressed by

$$H_Q = \sum_{k p \sigma} B_{kp} a_{k\sigma}^\dagger a_{p\sigma},$$

(5.37)

which does not break the time-reversal symmetry. If one replaces B_{kp} by an average, the relevant coupling is written as

$$\sum_{k,p} \sum_\alpha a_{k\alpha}^\dagger a_{p\alpha} = \frac{1}{2} \sum_{k,p} \psi(k)^\dagger \begin{pmatrix} \sigma_0 & 0 \\ 0 & -\sigma_0 \end{pmatrix} \psi(p),$$

(5.38)

where we have neglected a contribution coming from anticommutation of fermion operators with $k = p$ because it is smaller by $O(1/N)$. Then we obtain

$$\frac{1}{T_1 T} = B^2 \sum_q \frac{1}{\omega} \text{Im} \, \chi_c(q, \omega),$$

(5.39)

where the charge susceptibility is given by

$$\text{Im} \, \chi_c(q, \omega) = \sum_{k,p} \int_{-\infty}^\infty d\epsilon \, [f(\epsilon) - f(\epsilon + \omega)]$$

$$\times \text{Tr}\left[\hat{\tau}_z \delta(\epsilon - \hat{h}(k)) \hat{\tau}_z \delta(\epsilon + \omega - \hat{h}(p)) \right],$$

(5.40)

with $\hat{\tau}_z = \sigma_0 \otimes \tau_z$. The coherence factor for singlet pairing is now given by

$$1 + \frac{\epsilon_k \epsilon_p - \Delta(\hat{k})\Delta(\hat{p})}{E_k E_p},$$

(5.41)

where the minus sign originates from $\hat{\tau}_z \hat{h}_1(\boldsymbol{k}) = -\hat{h}_1(\boldsymbol{k})\hat{\tau}_z$. We note that the coherence factor $(1 - \Delta^2/E^2)$ cancels the divergence of $N_{\mathrm{BCS}}(E)$. Explicitly, we have

$$\frac{1}{T_1 T} = \pi B^2 \int_{-\infty}^{\infty} dE \left(1 - \frac{\Delta^2}{E^2}\right) N_{\mathrm{BCS}}(E)^2 \left(-\frac{\partial f}{\partial E}\right) = \frac{\pi B^2 N_0}{1 + \exp(\beta \Delta)}. \tag{5.42}$$

In this case, the coherence peak just below T_c disappears. We note that the temperature dependence of $1/(T_1 T)$ is the same as that of ultrasonic attenuation. These contrasting relaxation behaviours between the magnetic and quadrupole processes were reported in the ^{119}Sn and ^{181}Ta relaxation in TaSn$_3$ [21].

As for the impurity effect on the relaxation behaviour in conventional s-wave super-conductors, we note that nonmagnetic impurities smear the anisotropy due to the crystal structure. Since the anisotropy has an effect of suppressing the coherence peak just below T_c, such impurities *enhance* the peak. On the other hand, magnetic impurities broaden the density of states by the pair-breaking effect, and suppress the peak of $1/T_1$ substantially. With increasing number of impurities, the superconductivity becomes gapless, leading to the $T_1 T = $ constant behaviour below T_c. The magnetic field, which lowers T_c, also depresses the peak.

5.4   Characteristic features of heavy-electron superconductivity

The heavy-electron superconductors discovered to date may be classified into two groups, depending on their different magnetic behaviours and the extent of quasi-particle renormalization. The first group comprises CeCu$_2$Si$_2$, UBe$_{13}$, and UPt$_3$, which exhibit weak or no magnetic order. The quasi-particle masses as derived from the specific-heat coefficient $\gamma = C(T)/T$ are as large as $(\gamma \geq 0.4\,\mathrm{J/(mol\ K^2})$.

The second group includes URu$_2$Si$_2$, UNi$_2$Al$_3$, and UPd$_2$Al$_3$, which show antiferromagnetic order. The effective masses are not so large with $\gamma \sim 0.1\,\mathrm{J/(mol\ K^2)}$, but T_c of 1–2 K are higher than that in the first group. The highest T_c is 2 K in UPd$_2$Al$_3$. Most remarkably, UNi$_2$Al$_3$ and UPd$_2$Al$_3$ possess sizable magnetic moments of 0.24 and $0.85\mu_B/$(U atom), respectively. In both groups, heavy quasi-particles are responsible for the superconductivity. This follows from the large jump of the specific heat ΔC at T_c with $\Delta C/\gamma T_c \sim 1$, which is close to the value 1.43 for conventional superconductors, and from the large slope of the upper critical field $(dH_{c2}/dT)_{T=T_c}$. The multiplicity of the superconducting phases in UPt$_3$ is analogous to the complex phase diagram of superfluid ^{3}He. This suggests the unconventional nature of the superconductivity. We focus on selected properties studied by the NMR and the neutron scattering experiments. We should mention that enormous amount of work has also been done on the thermodynamic and transport properties, which cannot be covered in the following.

5.4.1   *CeCu$_2$Si$_2$*

Phase diagram under pressure   As discussed in Chapter 4, superconductivity in CeCu$_2$Si$_2$ occurs close to a mysterious magnetic phase called the phase A. In order to gain further insight into the relationship between the magnetism and superconductivity, we show the pressure and temperature (P–T) phase diagram in Fig. 5.4 for CeCu$_2$Si$_2$ and

$CeCu_2Ge_2$ [22–24]. The heavy-electron antiferromagnet $CeCu_2Ge_2$ exhibits a pressure-induced superconducting transition around $P = 7\,GPa$ [22]. The phase diagrams show nearly identical pressure dependence if they are transposed so that $P = 0$ for $CeCu_2Si_2$ coincides with $P = 7.6\,GPa$ for $CeCu_2Ge_2$ [24]. The electronic state of $CeCu_2Ge_2$ at $P = 7.6\,GPa$ seems nearly identical to that of $CeCu_2Si_2$ at $P = 0$. The enhancement of T_c observed for $CeCu_2Si_2$ in Fig. 5.4 in the range of $P = 2$–$3\,GPa$ [23] is observed for $CeCu_2Ge_2$ as well. Since the application of pressure increases the hybridization between the 4f and the conduction electrons, there exists an optimum strength of hybridization for the superconductivity. Namely, as the hybridization increases, the heavy-electron superconductivity emerges immediately after the disappearance of antiferromagnetic order.

The pressure-induced superconductivity is often observed in heavy-electron systems when the AF order is suppressed under pressure. For example, close to the border of AF order, an unconventional normal-state property is observed in the heavy-electron antiferromagnet $CePd_2Si_2$ ($T_N = 10\,K$), which crystallizes in the same tetragonal structure as $CeCu_2Si_2$. Near the critical pressure $p_c \sim 30\,kbar$, at which the AF ordering temperature is extrapolated to zero, the electrical resistivity $\rho(T)$ exhibits a quasi-linear variation over two orders of magnitude. This non-Fermi-liquid form of $\rho(T)$ extends down to the onset of the superconducting transition below $0.43\,K$ [25].

Besides the series of CeM_2X_2 tetragonal compounds, the cubic stoichiometric AF compound $CeIn_3$ displays an unusual normal-state resistivity followed by superconductivity around $0.2\,K$ near $p_c \sim 25\,kbar$ [26]. For the onset of superconductivity near the border of AF order, a possible scenario is that the critical low-lying magnetic excitations caused by the incipient AF order contribute to the formation of anisotropic even-parity Cooper pairs.

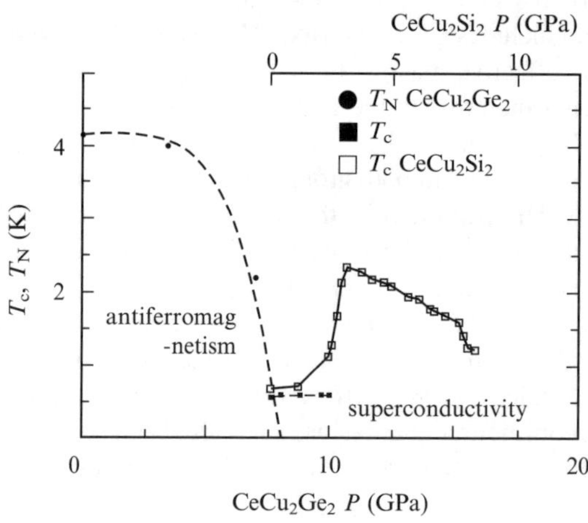

FIG. 5.4. Combined phase diagram for superconductivity and antiferromagnetism in $CeCu_2Si_2$ and $CeCu_2Ge_2$ as a function of pressure. Normal and superconducting states for $CeCu_2Ge_2$ at $7\,GPa$ bear resemblance to those in $CeCu_2Si_2$ at ambient pressure [22–24].

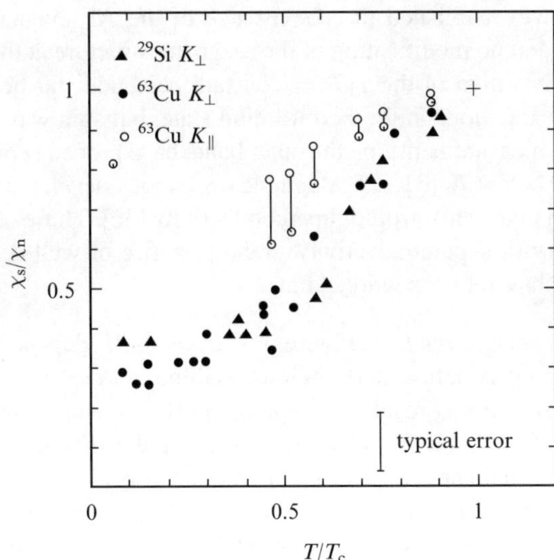

FIG. 5.5. Temperature dependence of spin susceptibility $\chi_s(T)$ below T_c. This is deduced from the Knight shift of ^{63}Cu and ^{29}Si in CeCu$_2$Si$_2$, and is normalized by the value at T_c [27].

Symmetry of the order parameter   The temperature dependence of the Knight shift provides a clue to the parity of the order parameter. From the Knight shift measurements of ^{29}Si and ^{63}Cu, the spin susceptibilities parallel and perpendicular to the tetragonal c-axis have been extracted using the oriented powder. From the T dependence of the Knight shift, an appreciable reduction of the spin susceptibility is found. This suggests strongly the singlet nature of the order parameter, hence the even-parity superconductivity. The normalized T dependence of the spin susceptibility in the superconducting state is plotted in Fig. 5.5 against (T/T_c) [27]. The residual shift at $T=0$ is attributed to the spin–orbit scattering and/or the T-independent Van Vleck shift. Because of the residual shift, the detailed structure of the gap function is difficult to deduce from the T dependence of the spin susceptibility.

As shown in Fig. 4.8 [28], the T^3 law for $x = 1.00$ and 1.025 reveals that the gap function of the order parameter vanishes along lines on the Fermi surface. Application of a simple model with $\Delta(\theta) = \Delta_0 \cos(\theta)$ yields the best fit to the experiment with $2\Delta_0 = 5T_c$. Both the Knight shift and T_1 results for CeCu$_2$Si$_2$ are consistent with a d-wave superconductor with vanishing gap along lines on the Fermi surface. The specific heat $C(T)$, however, does not confirm the line node: although $C(T)$ follows a T^2 dependence near T_c, which is consistent with a line node, $C(T)$ goes like $\sim T^3$ at low temperatures [29].

5.4.2   UPd$_2$Al$_3$

Coexistence of AF order and superconductivity   The Néel temperature T_N of UPd$_2$Al$_3$ is 14.5 K and the superconducting transition temperature T_c is 1.98 K [30–32]. Various measurements such as Al-NQR [33], Pd-NQR [34], specific heat [35], and neutron diffraction [31] have established the coexistence of the AF magnetic order with the superconductivity and no modification of the magnetic structure at the superconducting transition. The observation of the $T_1 T = $ constant behaviour far below T_N confirmed the quasi-particle excitations in the Fermi-liquid state, but spin waves were not found. Neutron scattering measurements, on the other hand, have probed spin-wave excitations, which are reduced below T_c [31,32]. Magnetic excitations develop a gap feature which seems to be related to the anisotropic superconductivity [36]. These observations should constrain the theory of superconductivity in the presence of well-developed antiferro-magnetism, which has not been worked out.

Symmetry of the order parameter   Figure 5.6 shows the T dependence of the Knight shift below T_c for polycrystalline and single-crystalline samples of UPd$_2$Al$_3$ [37]. In the μSR experiment also, similar results are reported [38]. The Knight shift is reduced only by about 0.08–0.11% at 0.4 K for all directions. Since the sample is in the clean limit, impurity scattering should not be the origin of the anisotropy in the residual Knight shift. Therefore, the anisotropy in the residual Knight shift should be ascribed to that of the AF susceptibility which does not change below T_c. This contribution is shown as K_{AF} in Fig. 5.6. If one subtracts K_{AF} from the total, the rest decreases nearly isotropically. The latter part is written as K_i and is ascribed to the itinerant part forming the Cooper pairs. The division of the f-electron degrees of freedom into itinerant and localized parts has been discussed in detail in § 4.8 .

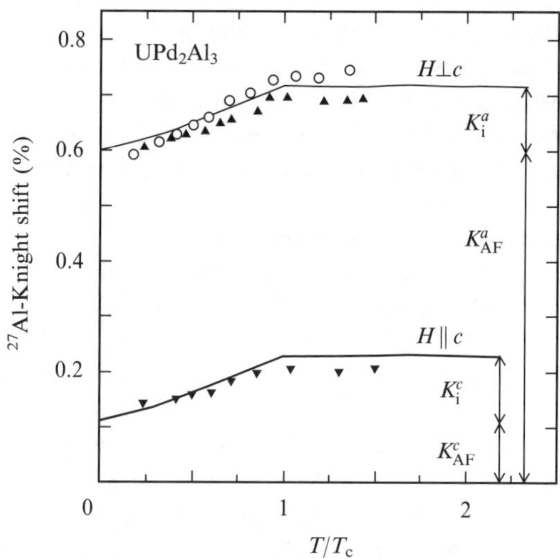

FIG. 5.6. Temperature dependence of Knight shifts. Open circles show Knight shift perpendicular to the field in an oriented powder of UPd$_2$Al$_3$. Solid up-triangles denote the shifts with c-axis perpendicular to the magnetic field, and down-triangles those parallel to the field. Solid lines represent the calculation based on the d-wave model [37]. See text for the meaning of K_i and K_{AF}.

According to the Fermi-liquid picture, a simple estimate of the itinerant part χ_i of the susceptibility yields $\chi_i \sim 1.6 \times 10^{-3}$ emu/mol with $\gamma = 150$ mJ/(mol K^2) in the normal state. On the other hand, a polarized neutron experiment extracted a value $\chi_i \sim 1.8 \times 10^{-3}$ emu/mol as the itinerant part [31], which is nearly the same as the above value. By using these values of χ_i, and the ^{27}Al hyperfine coupling constant $H_{hf} = 3.5$ kOe/μ_B, the itinerant part K_i of the Knight shift is estimated from the formula

$$K_i = \frac{H_{hf}}{(N_A \mu_B)} \chi_i$$

to be $\sim 0.1\%$. Since the observed reduction 0.08–0.11% is comparable to K_i, the Knight shift due to quasi-particles becomes almost zero well below T_c for all directions. This result provides strong evidence that the superconductivity is due to singlet Cooper pairs in the clean limit.

Gap anisotropy   In the superconducting state, $1/T_1$ in zero field obeys the T^3 law very well down to 0.15 K, as seen in Fig. 5.7 [33]. This is the first case where the T^3 dependence of $1/T_1$ is observed down to such a low temperature as $0.08T_c$. Since the order parameter stays constant below $0.5T_c$, the T^3 law of $1/T_1$ in the low-T range of $0.5T_c$ (1 K)–$0.08T_c$ (0.15 K) is evidence of anisotropic superconductivity with vanishing gap along lines on the Fermi surface. It is noteworthy that at Al sites the T dependence of $^{27}(1/T_1)$ below T_c is determined only by quasi-particle excitations in the superconducting state. This might be partly because the magnetic form factor at the Al site filters away the fluctuating hyperfine field from the two adjacent uranium planes. However, the T^3 law of $1/T_1$ was also confirmed for the Pd sites, where the fluctuating hyperfine fields in the basal uranium plane are not filtered away [34].

In contrast, the specific heat $C(T)$ probes not only the quasi-particle excitations but also the magnetic excitations in the AF ordered state which coexists with the super-conductivity. In fact, $C(T)$ is not consistent with the line node [39]; it takes the form $\gamma_r T + \beta T^3$. It seems that the T-linear term with $\gamma_r = 24$ mJ/(mol K^2) well below T_c originates from the gapless magnetic excitations localized at U sites.

In anisotropic superconductors, the impurity effect gives an important clue to identify the symmetry of the order parameter. This is because even potential scattering causes the reduction of T_c, in contrast to the s-wave pairing. A slight inhomogeneity at the Al site, which shows up in a larger FWHM, causes a decrease of T_c from 2 to 1.75 K and yields a residual density of states. Actually, a $T_1 T = $ constant behaviour was reported on a sample of UPd$_2$Al$_3$ which has lower T_c ($= 1.75$ K) and larger FWHM($= 20$ kHz) than the standard sample with $T_c = 1.98$ K and FWHM $= 12$ kHz. Apparently, the T dependence of $^{27}(1/T_1)$ can sensitively probe an intrinsic quasi-particle excitation near the Fermi level. The rate R of T_c reduction was found to be $R = T_c(1.75\ K)/T_{co}(1.98\ K) = 0.88$ when the fraction of residual to normal density of states is $N_{res}/N_0 = [(T_1 T)_{res}^{-1}/(T_1 T)_n^{-1}]^{1/2} = 0.23$. This is consistent with a predicted value $N_{res}/N_0 = 0.36$ with $T_c/T_{co} = 0.88$ in the *unitarity limit* for anisotropic superconductivity with vanishing gap along a line [40,41].

The same d-wave model as the one used for CeCu$_2$Si$_2$ in the clean limit successfully interprets both the T dependence of $1/T_1$ and the Knight shift of ^{27}Al. The solid lines in

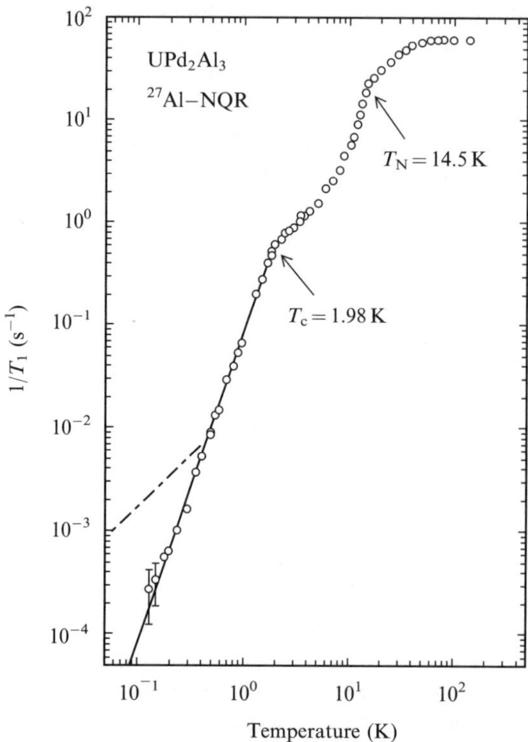

FIG. 5.7. Temperature dependence of $^{27}(1/T_1)$ in zero field Al-NQR for UPd$_2$Al$_3$. The solid line shows a T^3 dependence deduced from the d-wave model by using $\Delta(\theta) = \Delta\cos\theta$, with $2\Delta/T_c = 5.5$. Dash-dotted line shows the $T_1 T = \text{constant}$ law, which would be expected from the residual density of states by impurities [33].

Figs. 5.6 and 5.7 are the calculated results with the gap parameter $2\Delta_0 = 5.5\,T_c$. Here it is assumed that $\Delta(T)$ follows the same T dependence as that in the BCS case. Both experiments are in favour of the d-wave pairing model with a line node.

From the Knight shift result below T_c, it is shown that the large residual shift originates from the antiferromagnetic susceptibility, while the isotropic reduction of the spin shift below T_c is due to the formation of the singlet pairing among the quasi-particles near the Fermi level. Combining the results of both the Knight shift and $^{27}(1/T_1)$, which includes the impurity effect, it is concluded that UPd$_2$Al$_3$ is a d-wave superconductor characterized by a vanishing energy gap along the lines on the Fermi surface.

5.4.3　UPt$_3$

Superconducting phase diagram　　UPt$_3$ exhibits many indications of anisotropic super-conductivity [42–44]. Figure 5.8 shows the phase diagram in the magnetic field (B) vs temperature (T) plane, with the magnetic field perpendicular (a) and parallel (b) to the

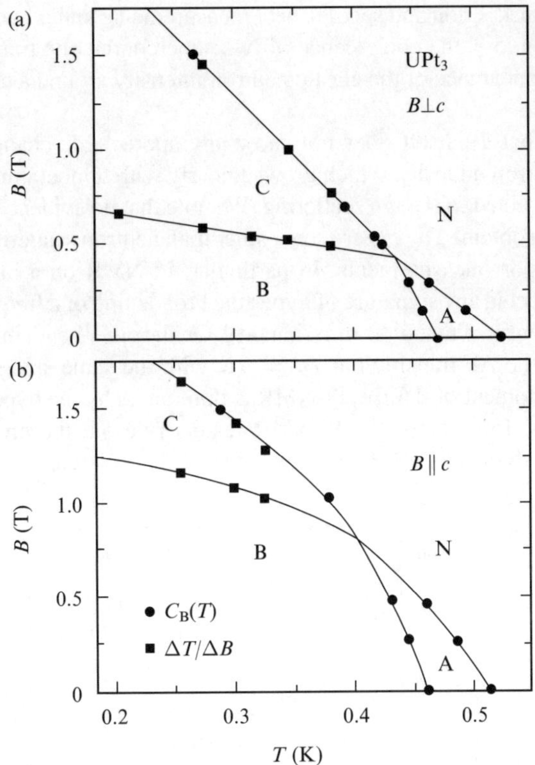

FIG. 5.8. The phase diagram of UPt$_3$ with the magnetic field B (a) perpendicular and (b) parallel to the c-axis determined by the specific heat, ultrasonic attenuation, and magnetocaloric effects [42–44].

hexagonal c-axis. These phases have been mapped out by a number of different measurements. The multiplicity of superconducting phases in UPt$_3$ is reminiscent of the complex phase diagram of superfluid ^{3}He. Various scenarios to interpret the complex phases of UPt$_3$ have been proposed with unconventional order parameters [45,46]. All the possible order parameter representations allowed for the hexagonal point group of UPt$_3$ have been enumerated under the assumption of strong spin–orbit coupling (see Table 5.2). In the presence of the inversion symmetry, the parity of order parameter is a good quantum number. A possible scenario is that the two phases at zero field belong to different order parameter representations which are nearly degenerate accidentally. The other scenario is that the two transitions arise from a splitting of an otherwise degenerate state within a single representation. In the latter case, a symmetry-breaking field is responsible for the splitting, in the same way as the magnetic field lifts the two-fold degeneracy of the A phase in superfluid ^{3}He. A candidate for the symmetry-breaking field in UPt$_3$ is the proposed weak AF order below $T_{\rm N} = 5\,{\rm K}$, with moments aligned in the basal plane. In fact, a direct coupling of the weak AF to the double superconducting transitions was found by neutron diffraction and specific heat measurements under hydrostatic pressure. As indicated in Fig. 5.9, the coalescence of two superconducting transitions correlates well with the disappearance of the elastic neutron intensity around a critical pressure of $P_c = 3.2\,\text{kbar}$ [47].

It is unusual that T_N itself does not show any appreciable change with pressure. Moreover, the neutron intensity, which grows linearly with temperature, behaves differently from that expected for Bragg scattering. We note that no evidence for the magnetic transition has been obtained by experiments other than neutron scattering. These include thermal and transport measurements. In particular, Pt-NMR on a high-quality single crystal has not detected any signature of hyperfine broadening or a frequency shift which should follow the onset of magnetic ordering [49]. On the other hand, in $U(Pt_{0.95}Pd_{0.05})_3$, which undergoes the AF transition at $T_N = 5\,\text{K}$ with the same spin structure as UPt_3 but with a larger moment of $0.6\,\mu_B$, Pt-NMR is dominated by the hyperfine field which amounts to $32\,\text{kOe}$ [50]. Thus, the Pt-NMR results rule out the presence of a static hyperfine field of about $1\,\text{kOe}$, which is expected if the ordered moment of $0.02\mu_B$ is

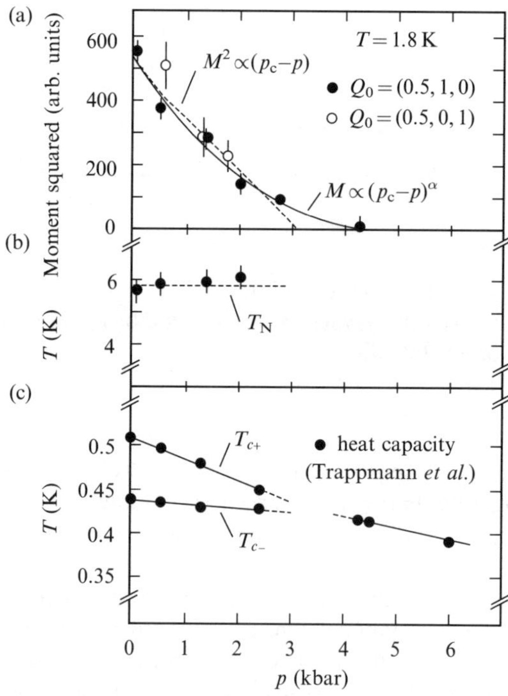

FIG. 5.9. (a) Variation of integrated neutron intensity of magnetic peak $(1/2, 1, 0)$ (closed circle) and $(1/2, 0, 1)$ (open circle) with hydrostatic pressure. This is a measure of the square of the sublattice magnetization, M_Q^2. Solid line is a fit of $M_Q^2 \propto (p_c - p)^\alpha$ with $\alpha = 2.6 \pm 1.9$ and $p_c = 5.4 \pm 2.9$. The dashed line shows $M_Q^2 \propto (p_c - p)$ which yields and $3.2 \pm 0.2\,\text{kbar}$. (b) Pressure dependence of T_N. (c) Pressure dependence of the temperatures of double transitions [47,48].

present. A possible way out is that uranium-derived moments fluctuate with a frequency larger than the Pt-NMR frequency. It seems unlikely that such fluctuating moment can be the symmetry-breaking field to lift the two-fold degeneracy of the order parameter. Most phenomenological approaches incorporate this as the symmetry-breaking field to interpret the multiplicity of the superconducting phase [45,46,51].

The parity of the order parameter in UPt$_3$ has been determined by precise Pt Knight shift measurements of a high-quality single crystal. As shown in Fig. 5.10 [52], the spin part of the Knight shift does not decrease below T_c at all in a field range of 4.4–15.6 kOe down to 25 mK. This result provides evidence that odd-parity superconductivity with parallel-spin pairing is realized. In contrast, the Knight shift decreases below T_c for a magnetic field parallel to the b-axis with $H < 5$ kOe, and for a magnetic field parallel to the c-axis with $H_c < 2.3$ kOe. Magnetic field along the a-axis does not lead to a change of Knight shift down to 1.7 kOe. This set of Pt Knight shift data seems consistent with odd-parity superconductivity including a nonunitary pairing [52,53].

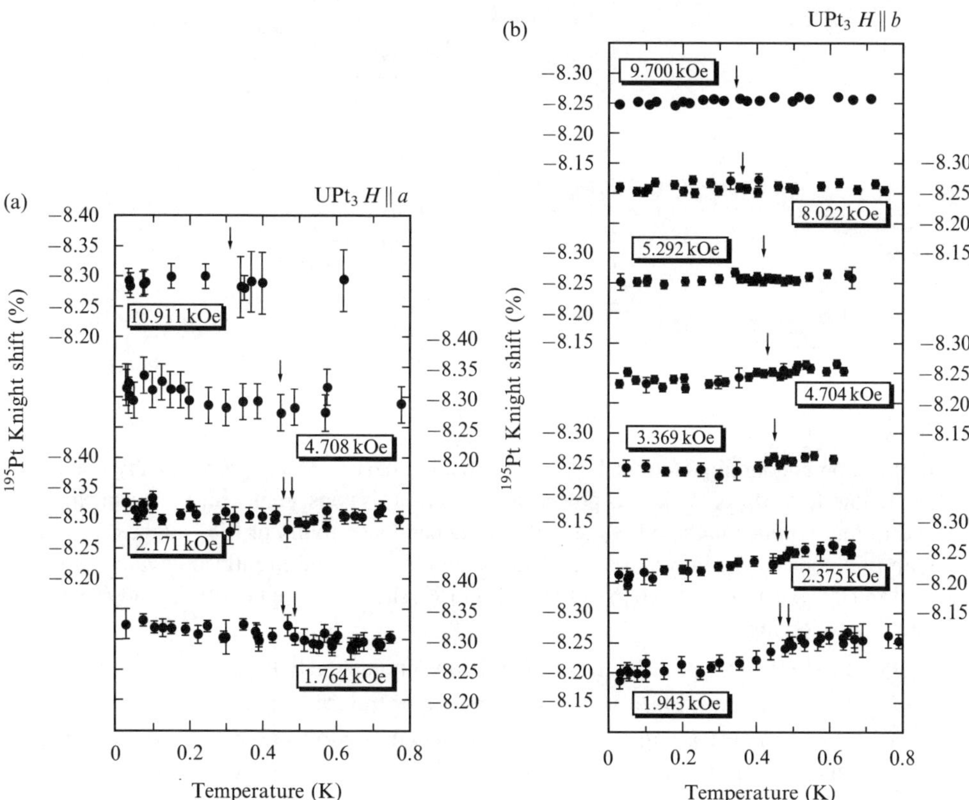

FIG. 5.10. T dependence of the ^{195}Pt Knight shift for various magnetic fields: (a) K_a for $H \parallel a$, (b) K_b for $H \parallel b$, and (c) K_c for $H \parallel c$. respectively [52]. Arrows ($\downarrow$) show the superconducting transition temperatures.

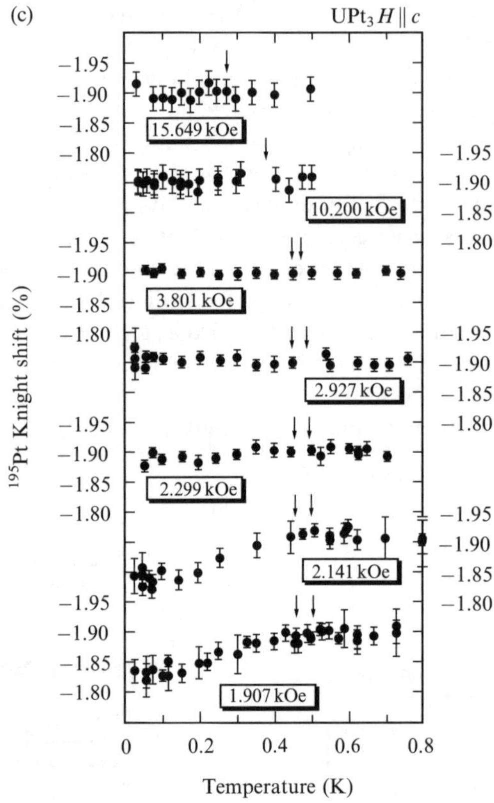

FIG. 5.10. (c)

5.4.4 UBe_{13}

In the superconducting state of UBe_{13}, the specific heat follows approximately the T^3 law below $T_c = 0.9\,\text{K}$ [54]. This power-law behaviour is consistent with a gap vanishing at different points. On the other hand, the T^3 dependence found in $1/T_1$ suggests a gap with line nodes [55]. Power-law behaviours are also reported in ultrasonic attenuation [56] and the penetration depth [57]. Figure 5.11 [58] shows the upper critical field $H_{c2}(T)$ with the following features:

(1) The slope $|dH_{c2}/dT|$ of the critical field is anomalously large;

(2) $H_{c2}(T)$ is almost linear in T over a large temperature range;

(3) $H_{c2}(0)$ at $T = 0$ is larger than the value calculated within the BCS theory assuming a magnetic moment comparable to μ_B.

The superconducting states of heavy electrons are very sensitive to impurities and defects. The critical temperature T_c can be strongly depressed in UPt$_3$ by grinding the sample. The substitution of a small amount of Th for U in UBe_{13} leads to a rapid decrease of T_c. For Th concentrations between $x = 0.017$ and 0.05 in $U_{1-x}Th_xBe_{13}$, two phase transitions were observed in specific heat measurements [59] with comparable

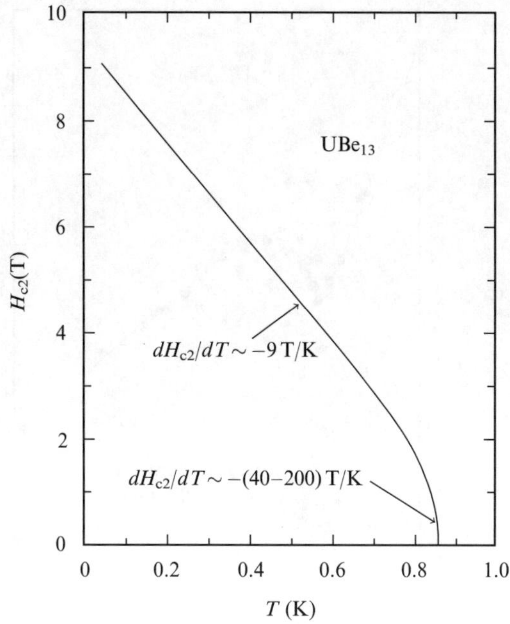

FIG. 5.11. Upper critical magnetic field H_{c2} v temperature T for UBe$_{13}$ [58].

discontinuities at T_c's. The phase diagram is shown in Fig. 5.12. It has been proposed [60] that the order parameter belongs to a one-dimensional representation Γ_1 for $x < x_c \sim 0.018$, and a three-dimensional one Γ_5 for $x > x_c$ and $T > T_{c2}$ (see Fig. 5.12). It is possible that Γ_1 and Γ_5 are interchanged. For $T < T_{c2}$ the time-reversal symmetry may be broken and a mixture of the two representations may be realized.

5.4.5   *Implication for the superconducting mechanism*

In the three compounds considered above, the dynamical magnetic properties are different in each case. Nevertheless, in all three the superconducting energy gap vanishes along lines on the Fermi surface. It is almost certain that anisotropic order parameters with spin singlet are realized in CeCu$_2$Si$_2$ and UPd$_2$Al$_3$, and that a different anisotropic order parameter with nonunitary spin-triplet pairing is realized in UPt$_3$, which is the first example of this pairing symmetry in charged many-body systems. These variations of the anisotropic order parameter could be due to the different characters of the magnetic fluctuations which lead to the Cooper pairing. We note that the integrated spectral weight of magnetic fluctuations is smaller for Ce compounds with the 4f^1 than for U compounds with 5f^2–5f^3. Although it is not certain to what extent the fluctuation frequency is relevant to the pairing, this difference in the number of f electrons may be a reason why the magnetism and superconductivity compete in CeCu$_2$Si$_2$ and can coexist in UPd$_2$Al$_3$.

In the phase A of CeCu$_2$Si$_2$, the AF spin fluctuations which have a large spectral weight at low energy are expelled by the onset of the superconducting transition. The application of pressure makes T_K shift to high temperature and hence transfers the spectral

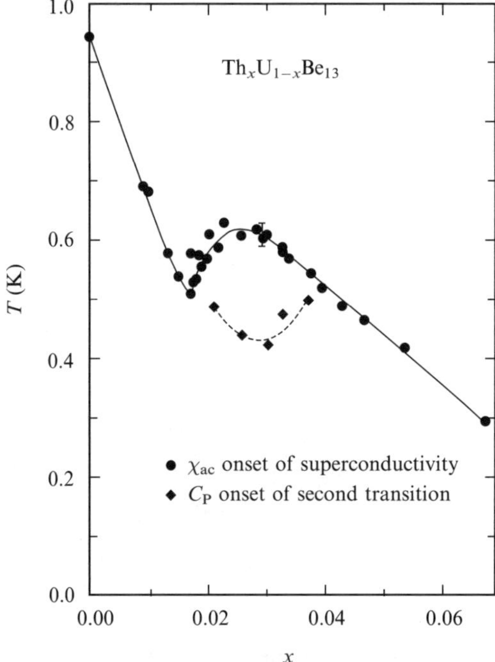

FIG. 5.12. The temperature vs Th concentration (x) phase diagram of $U_{1-x}Th_xBe_{13}$. Note the rapid change in $T_c(x)$ (solid line) as x increases up to $x_c \sim 0.018$, and the second phase transition at $T_{c2}(x)$ (dashed line) [54].

weight of the spin fluctuations to higher energy. As a result, T_c increases rapidly from 0.7 K at ambient pressure to around 2 K at 3 GPa. It is noteworthy that the superconductivity is enhanced when the spectral weight is transferred to a relatively higher-energy region. This transfer is derived from the law $T_1T = $ constant, which means that the Fermi-liquid description is valid for the lowest energy excitations. An important feature is that the ground state adjacent to the AF phase in $CeCu_2Si_2$ and $CeCu_2Ge_2$ is not the normal phase but the superconducting one.

In contrast, UPd_2Al_3 is characterized by the AF magnetic order with a large moment of $0.8\,\mu_B$ per U atom. At the same time, the Fermi-liquid description is valid well below T_N with a large density of states $\gamma = 150\,mJ/(mol\,K^2)$. The record high value $T_c = 2\,K$ is realized in the presence of the static AF order. These features seem to originate from the large degrees of freedom for magnetic fluctuations associated with the $5f^2$–$5f^3$ electrons.

UPt_3 has remarkable features of spin fluctuations. First, although any conventional AF order is not present, its spectral weight arises partly from a quasi-elastic AF contribution with the wave vector Q_b in the hexagonal plane. Secondly, there is also a sizable AF fluctuation with wave vector Q_c along the c-axis and with a rather high excitation energy of 5 meV [61]. Thirdly, the ferromagnetic fluctuation near $q = 0$ (the so-called paramagnon) is also observed in neutron scattering experiments [62]. The Fermi-liquid description applies below 1 K with the density of states $\gamma = 420\,mJ/(mol\,K^2)$. It seems possible that paramagnons play a role in producing the anisotropic p- or f-wave super-conductivity with parallel-spin pairing in UPt$_3$.

5.5 Toward a microscopic theory

The fundamental problem in the superconductivity of heavy electrons is why the strongly repulsive f electrons form pairs. There is no definite answer to this question yet. It is instructive to compare the situation with the triplet pairing in ^{3}He, where there is also a strongly repulsive core in the interaction [63]. At the same time, there is an attractive region in the pair potential which arises from the van der Waals force. The triplet pairing has zero amplitude at the origin, so that the repulsive core is not critical. We recall that the van der Waals force originates from the virtual polarization of valence electrons which has much higher characteristic energy than the Fermi energy of liquid ^{3}He. In the BCS theory, the electrons feel the attractive force mediated by the virtual lattice vibrations which have much lower energy than the Fermi energy. In the case of ^{3}He, there is no lattice in the background, but the role of phonons is played partly by nuclear spin fluctuations of ^{3}He. It has been shown that the ferromagnetic fluctuations favour triplet pairing rather than singlet (see e.g. Ref. [63]).

In heavy electrons, the relative importance of phonons and spin fluctuations is still uncertain. In any case, because of the strongly repulsive core, the s-wave pair seems to be difficult to be realized. The anisotropic pairing force due to the electron–phonon interaction is much weaker than the isotropic one.

We discuss the pairing force qualitatively using the Landau quasi-particle picture. In the usual strong coupling theory for superconductors [64], one derives the self-energy and the pairing force simultaneously. Then the solution of the resultant Eliashberg equation gives the transition temperature and the quasi-particle spectrum microscopically. This is unfortunately not possible for heavy electrons, because the renormalization leading to the Kondo state cannot be represented by any simple perturbation processes. Thus, one has to assume the spectrum of the quasi-particles, and then consider the interaction among them. A similar situation occurs in the superfluidity of ^{3}He [65].

We start with the linearized gap equation which is formally exact:

$$\Delta_{\alpha\beta}(\boldsymbol{k}, i\epsilon_n) = -T_c \sum_m \sum_{\boldsymbol{p}} \sum_{\mu\nu} \langle \alpha\beta | \Gamma(\boldsymbol{k}, \boldsymbol{p}; i\epsilon_n, i\epsilon_m) | \nu\mu \rangle$$

$$\times \, G_\mu(\boldsymbol{p}, i\epsilon_m) G_\nu(-\boldsymbol{p}, i\epsilon_m) \Delta_{\mu\nu}(\boldsymbol{p}, i\epsilon_m),$$

where $\langle \alpha\beta | \Gamma(\boldsymbol{k}, \boldsymbol{p}; i\epsilon_n, i\epsilon_m) | \nu\mu \rangle$ is the irreducible vertex part for the pairing interaction. Being irreducible, it means that the Feynman diagram cannot be separated into two pieces by cutting the two Green functions going parallel. We assume that T_c is much smaller than T_K, which plays the role of the Fermi energy for heavy electrons. Then we make an approximation, replacing the vertex part by its value at the Fermi surface, and imposing the cutoff energy ω_c in the summation over the Matsubara frequencies. The result is

$$\Delta_{\alpha\beta}(\boldsymbol{k}) = -\ln\left(\frac{\omega_c}{T_c}\right) \rho(0) a_f^2 \int \frac{d\Omega_p}{4\pi} \langle \alpha\beta | \Gamma(\boldsymbol{k}, \boldsymbol{p}) | \nu\mu \rangle \Delta_{\mu\nu}(\boldsymbol{p}), \qquad (5.43)$$

where the integral is over the solid angle of the Fermi surface. This equation takes almost the same form as the mean-field approximation for the pairing. What is different here is that the renormalization factor a_f is present. In other words, it is in the quasi-particle picture that we construct the mean-field theory in the spirit of the Landau theory.

In order to see the effect of the ferromagnetic and antiferromagnetic spin fluctuations in the simplest way, we assume, for the moment, spherical symmetry and neglect the spin–orbit interaction. Then the effective interaction is decomposed as

$$\langle \alpha\beta|\Gamma(\boldsymbol{k},\,\boldsymbol{p})|\mu\nu\rangle = U(\boldsymbol{k},\,\boldsymbol{p})\delta_{\alpha\nu}\delta_{\beta\mu} + J(\boldsymbol{k},\,\boldsymbol{p})\boldsymbol{\sigma}_{\alpha\nu}\cdot\boldsymbol{\sigma}_{\beta\mu}. \tag{5.44}$$

We note that the right-hand side gives $U + J$ for the triplet pair and $U - 3J$ for the singlet pair. Since both $\boldsymbol{k}$ and $\boldsymbol{p}$ are at the Fermi surface, i.e. with length k_{F}, the vertex part can be expanded in terms of spherical harmonics as

$$\langle \alpha\beta|\Gamma(\boldsymbol{k},\,\boldsymbol{p})|\mu\nu\rangle = \sum_{lm}\langle \alpha\beta|\Gamma_l|\mu\nu\rangle Y_{lm}(\hat{k})Y_{lm}(\hat{p})^*, \tag{5.45}$$

where $\hat{k}$ denotes the solid angle of $\boldsymbol{k}$. This equation, together with eqn (5.44), leads to the parameters U_l and J_l which characterize the strength of the potential and exchange scattering, respectively.

The pairing amplitude is also expanded as

$$\Delta_{\alpha\beta}(\hat{k}) = \sum_{lm}\Delta_{\alpha\beta}^{lm}Y_{lm}(\hat{k}).$$

Then the gap equation given by eqn (5.43) turns into

$$1 = -\ln(\omega_{\mathrm{c}}/T_{\mathrm{c}})\rho(0)a_{\mathrm{f}}^2(U_l + J_l) \quad \text{(triplet, } l\text{: odd)}, \tag{5.46}$$

$$1 = -\ln(\omega_{\mathrm{c}}/T_{\mathrm{c}})\rho(0)a_{\mathrm{f}}^2(U_l - 3J_l) \quad \text{(singlet } l\text{: even)}. \tag{5.47}$$

The actual pairing occurs for the type with the highest T_{c}, i.e. with a negative coupling constant with the largest absolute magnitude.

In order to clarify the role of spin fluctuations, we consider the so-called potential scattering model [65], where U and J depend only on the momentum transfer $\boldsymbol{k} - \boldsymbol{p}$. This model enables us to make a connection between $\Gamma(\boldsymbol{k},\,\boldsymbol{p})$ and the Landau parameters. The spin-independent scattering $U(\boldsymbol{k} - \boldsymbol{p})$ should be dominated by the strong on-site repulsion. Then its angle dependence should be weaker, and U_l is dominated by the component $l = 0$. Because of the large positive value of U_0, the s-wave pairing is strongly suppressed. On the other hand, the spin-dependent part $J(\boldsymbol{k} - \boldsymbol{p})$ is related to the magnetic susceptibility $\chi(\boldsymbol{q},\,\omega)$. If we assume a constant λ for the spin–fermion coupling constant, the relation reads

$$J(\boldsymbol{k} - \boldsymbol{p}) = \tfrac{1}{4}\lambda^2\chi(\boldsymbol{k} - \boldsymbol{p},\,0).$$

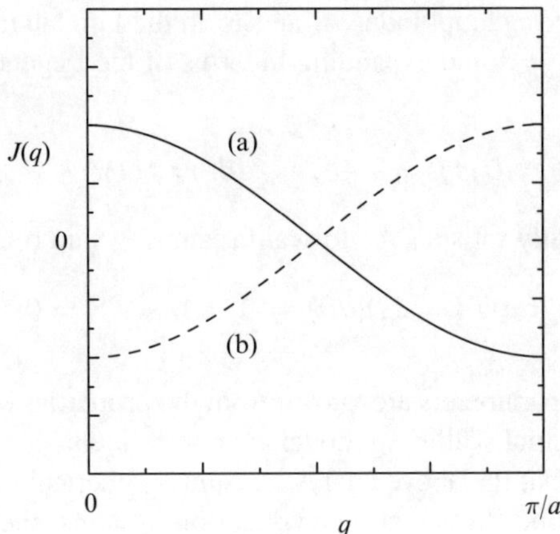

FIG. 5.13. Momentum dependence of the exchange interaction shown schematically:
(a) antiferromagnetic case, (b) ferromagnetic case.

If the spin fluctuation is dominantly ferromagnetic, $J(q)$ becomes negative near
the centre of the Brillouin zone, as shown schematically in Fig. 5.13(b). The angular
momentum component is given by the transformation which is the inverse of eqn (5.45):

$$J_l = \left(l + \frac{1}{2}\right) \int_{-1}^{1} dx\, P_l(x) J(q),$$

where $x = 1 - q^2/(2k_F^2)$. In this case, J_1 tends to be negative because $P_1(x)J(q)$ is
negative over most of the integration range. Thus, the triplet pairing is favourable with the
ferromagnetic spin fluctuations. On the other hand, if the spin fluctuation is dominantly
antiferromagnetic, $J(q)$ becomes negative near the boundary of the Brillouin zone. This
is illustrated in Fig. 5.13(a). In this case, J_1 tends to be positive and the triplet pairing
does not occur. With antiferromagnetic fluctuations, J_2 becomes positive because of the
dominant contribution from $q \sim 2k_F$. Thus, the d-wave singlet pairing is favoured [66].

The connection with the Landau parameters is given via a consideration of the general
form of the t-matrix for the quasi-particles. For incoming particles of momenta p_1 and
p_2, and outgoing particles with p_3 and p_4, the t-matrix T depends on two parameters
$x_3 = \hat{p}_1 \cdot \hat{p}_3$ and $x_4 = \hat{p}_1 \cdot \hat{p}_4$. Then the spin dependence can be parametrized as

$$\langle \alpha\beta | T(x_3, x_4) | \mu\nu \rangle = T^{\rm d}(x_3, x_4)\delta_{\alpha\nu}\delta_{\beta\mu} + T^x(x_3, x_4)\boldsymbol{\sigma}_{\alpha\nu} \cdot \boldsymbol{\sigma}_{\beta\mu}. \tag{5.48}$$

The t-matrix is in general different from the vertex part in eqn (5.43), since the latter
should exclude the pair propagator in the intermediate state. In the potential scattering
model, this intermediate state is not included even in the t-matrix and we obtain the
relations

$$T^{\rm d}(x_3, x_4) = U(x_3) - U(x_4), \qquad T^x(x_3, x_4) = J(x_3) - J(x_4),$$

where the second terms with minus signs take account of the exchange of the outgoing
momenta.

The forward scattering amplitudes A_l and B_l in the Landau theory (see § 1.3.4) are obtained by setting $x_3 = 1$ and expanding in terms of the Legendre polynomials. As a result, we obtain

$$A_l = U(1)\delta_{l,0} - U_l, \qquad B_l = J(1)\delta_{l,0} - J_l.$$

This model automatically satisfies the forward scattering sum rule

$$\langle \alpha\alpha | T(x_3, x_4) | \alpha\alpha \rangle = \sum_l (A_l + B_l) = 0.$$

Thus, if some Landau parameters are known from the properties of the normal state, the parameters in the potential scattering model are constrained.

We remark again that the above analysis assumes spherical symmetry and neglects the spin–orbit interaction. In actual heavy-electron systems, the quantitative analysis should respect the point group symmetry. One can deal with the symmetry in terms of appropriate basis functions instead of the spherical harmonics used above. Much less is known about the effect of strong anisotropy in the g-factor.

Bibliography

[1] J. Bardeen, J. N. Cooper, and J. R. Schrieffer, *Phys. Rev.* **108**, 1175 (1957).

[2] F. Steglich *et al.*, *Phys. Rev. Lett.* **43**, 1892 (1979).

[3] For review, see for example, Proc. Int. Conf. on Strongly Correlated Electron Systems, *Physica B* **259–261** (1999).

[4] N. Grewe and F. Steglich, *Handbook on the physics and chemistry of rare earths*, Eds, K. A. Gschneider and L. Eyring, North-Holland, Amsterdam, 1991, Vol. 14.

[5] A. J. Leggett, *Rev. Mod. Phys.* **47**, 331 (1975).

[6] P. W. Anderson and P. Morel, *Phys. Rev.* **123**, 1911 (1961).

[7] J. G. Bednorz and K. A. Müller, *Z. Phys. B* **64**, 189 (1986).

[8] For review, see for example, Proc. Int. Conf. on Material and Mechanism of Superconductivity: High-Temperature Superconductivity V, *Physica C* **282–287** (1997).

[9] T. Moriya and K. Ueda, *J. Phys. Soc. Jpn.* **63**, 1871 (1994).

[10] P. Monthoux and D. Pines, *Phys. Rev. B* **49**, 4261 (1994).

[11] R. Balian and N. R. Werthamer, *Phys. Rev.* **131**, 1553 (1963).

[12] P. W. Anderson and W. F. Brinkman, *Phys. Rev. Lett.* **30**, 1108 (1973).

[13] G. E. Volovik and L. P. Gorkov, *Sov. Phys. JETP* **61**, 843 (1985).

[14] M. Sigrist and K. Ueda, *Rev. Mod. Phys.* **63**, 239 (1991).

[15] S. Yip and A. Garg, *Phys. Rev. B* **48**, 3304 (1993).

[16] E. Blount, *Phys. Rev. B* **22**, 2935 (1985).

[17] K. Yosida, *Phys. Rev.* **110**, 769 (1958).

[18] P. W. Anderson, *Phys. Rev. Lett.* **3**, 325 (1959).

[19] L. C. Hebel and C. P. Slichter, *Phys. Rev.* **113**, 1504 (1959).

[20] S. Ohsugi, Y. Kitaoka, M. Kyogaku, K. Ishida, K. Asayama, and T. Ohtani, *J. Phys. Soc. Jpn.* **61**, 3054 (1992).

[21] S. Wada and K. Asayama, *J. Phys. Soc. Jpn.* **34**, 1168 (1973).

[22] D. Jaccard, K. Behnia, and J. Sierro, *Phys. Lett. A* **163**, 475 (1992).

[23] F. Thomas, J. Thomasson, C. Ayache, C. Geibel, and F. Steglich, *Physica B* **186–188**, 303 (1993).

[24] Y. Kitaoka, H. Tou, G.-q. Zheng, K. Ishida, K. Asayama, T. C. Kobayashi, A. Kohda, N. Takeshita. K. Amaya, G. Geibel, C. Schank, and F. Steglich, *Physica B* **206 & 207**, 55 (1995).

[25] F. M. Grosche, S. R. Julian, N. D. Mathur, and G. G. Lonzarich, *Physica B* **223 & 224**, 50 (1996).

[26] I. R. Walker, F. M. Grosche, D. M. Freye and G. G. Lonzarich, *Physica C* **282–287**, 303 (1997).

[27] Y. Kitaoka, H. Yamada, K. Ueda, Y. Kohori, T. Kohara, Y. Odd, and K. Asayama, *Jpn. Apply Phys. Suppl.* **26**, 1221 (1987).

[28] K. Ishida *et al.*, *Phys. Rev. Lett.* **82**, 5353 (1999).

[29] F. Steglich, *Springer Series in Solid-State Sciences* **62**, 23 (1985).

[30] C. Geibel, C. Shank, S. Thies, H. Kitazawa, C. D. Bredl, A. Bohm, M. Rau, A. Grauel, R. Caspary, R. Hefrich, U. Ahlheim, G. Weber, and F. Steglich, *Z. Phys. B* **84**, 1 (1991).

[31] N. Sato, N. Aso, G. H. Lander, B. Roessli, T. Komatsubara, and Y. Endoh, *J. Phys. Soc. Jpn.* **66**, 2981 (1997).

[32] N. Metoki, Y. Haga, Y. Koike, N. Aso, and Y. Onuki, *J. Phys. Soc. Jpn.* **66**, 2560 (1997).

[33] H. Tou, Y. Kitaoka, K. Asayama, C. Geibel, C. Schank and F. Steglich, *J. Phys. Soc. Jpn.* **64**, 725 (1995).

[34] K. Matsuda *et al.*, *Phys. Rev. B* **55**, 15 223 (1997).

[35] R. Caspary, P. Hellmann, M. Keller, G. Sparn, C. Wassilew, R. Kohler, C. Geibel, C. Schank, F. Steglich, and N. E. Phillips, *Phys. Rev. Lett.* **71**, 2146 (1993).

[36] N. Metoki *et al.*, *Phys. Rev. Lett.* **24**, 5417 (1998).

[37] M. Kyogaku, Y. Kitaoka, K. Asayama, N. Sato, T. Sakon, and T. Komatsubara, *Physica B* **186–188**, 285 (1993).

[38] R. Feyerherm *et al.*, *Phys. Rev. Lett.* **73**, 1849 (1994).

[39] F. Steglich *et al.*, *Physical phenomena at high magnetic fields-II*, (Eds, Z. Fisk *et al.*, World Scientific, Singapore, 1996) p. 185.

[40] S. Shimitt-Rink, K. Miyake, and C. M. Varma, *Phys. Rev. Lett.* **57**, 2575 (1986).

[41] P. Hirschfeld, D. Vollhardt, and P. Wölfle, *Solid State Commun.* **59**, 111 (1986).

[42] L. Taillefer, J. Flouquet, and G. G. Lonzarich, *Physica B* **169**, 257 (1991).

[43] H. v. Lohneysen, *Physica B* **197**, 551 (1994).

[44] K. Hasselbach, L. Taillefer, and J. Flouquet, *Phys. Rev. Lett.* **63**, 93 (1989).

[45] T. Ohmi and K. Machida, *Phys. Rev. Lett.* **71**, 625 (1993); K. Machida *et al.*, *J. Phys. Soc. Jpn.* **62**, 3216 (1993).

[46] J. A. Sauls, *J. Low Temp Phys.* **95**, 153 (1994).

[47] S. M. Hayden, L. Taillefer, C. Vettier, and J. Flouquet, *Phys. Rev. B* **46**, 8675 (1992).

[48] T. Trappmann, H. v. Lohneysen, and L. Teillefer, *Phys. Rev. B* **43**, 13714 (1991).

[49] H. Tou, Y. Kitaoka, K. Asayama, N. Kimura, Y. Ōnuki, E. Yamamoto, and K. Maezawa, *Phys. Rev. Lett.* **77**, 1374 (1996).

[50] Y. Kohori, M. Kyogaku, T. Kohara, K. Asayama, H. Amitsuka, and Y. Miyako, *J. Magn. Magn. Mater.* **90 & 91**, 510 (1990).

[51] D. W. Hess, T. A. Tokuyasu, and J. A. Sauls, *J. Phys. Condens. Matter* **1**, 8135 (1989).

[52] H. Tou *et al.*, *Phys. Rev. Lett.* **80**, 3129 (1998).

[53] T. Ohmi and K. Machida, *J. Phys. Soc. Jpn.* **65**, 4018 (1996).

[54] H. R. Ott, H. Rudigier, P. Delsing, and Z. Fisk, *Phys. Rev. Lett.* **50**, 1595 (1983).

[55] D. E. MacLaughlin, *J. Magn. Magn. Mater.* **47 & 48**, 121 (1985).

[56] B. Golding, D. J. Bishop, B. Batlogg, W. H. Haemmerle, Z. Fisk, J. L. Smith, and H. R. Ott, *Phys. Rev. Lett.* **55**, 2479 (1985).

[57] D. Einzel, P. J. Hirschfeld, F. Gross, B. S. Chandrasekhar, K. Andres, H. R. Ott, J. Beuers, Z. Fisk, and J. L. Smith, *Phys. Rev. Lett.* **56**, 2513 (1986).

[58] M. B. Maple, J. W. Chen, E. E. Lambert, Z. Fisk, J. L. Smith, H. R. Ott, J. S. Brooks, and M. J. Naughton, *Phys. Rev. Lett.* **54**, 477 (1985).

[59] H. R. Ott, H. Rudigier, E. Felder, Z. Fisk, and J. L. Fisk, *Phys. Rev. B* **33**, 126 (1986).

[60] M. Sigrist and T. M. Rice, *Phys. Rev. B* **39**, 2200 (1989).

[61] G. Aeppli *et al.*, *Phys. Rev. Lett.* **60**, 615 (1988); ibid., **63**, 676 (1989).

[62] N. R. Bernhoeft and G. C. Lonzarich, *J. Phys. C* **7**, 7325 (1995).

[63] D. Vollhardt and P. Wölfle, *The superfluid phases of helium 3* (Taylor and Francis, London, 1990).

[64] A. A. Abrikosov, L.'P. Gorkov, and I. E. Dzyaloshinskii, *Methods of quantum field theory in statistical physics* (Dover, New York, 1975).

[65] J. W. Serene and D. Rainer, *Phys. Rep.* **101**, 221 (1983).

[66] K. Miyake, *J. Magn. Magn. Mater.* **63 & 64**, 411 (1987).

6

COMPARISON WITH HIGH-TEMPERATURE SUPERCONDUCTORS

6.1 Characteristics of copper oxides

The discovery by Bednorz and Müller (1986) of superconductivity at $30\,\mathrm{K}$ in the ceramic copper oxides [1] $La_{2-x}Ba_xCuO_4$ (abbreviated as 2-1-4) has had great impact on solid state physics. This breakthrough was followed soon by the discovery of the $90\,\mathrm{K}$-class superconductor $YBa_2Cu_3O_{7-y}$ (1-2-3) and recently by that of $130\,\mathrm{K}$-class $HgBa_2Ca_2Cu_3O_{8+y}$ (Hg1223). These cuprates have the following characteristic properties [2]:

1. The CuO_2 planes are commonly present, and are separated by bridging blocks which act as charge reservoirs for the planes. Figure 6.1 illustrates the structure of the (2-1-4) system.

2. Undoped La_2CuO_4 and $YBa_2Cu_3O_6$ are antiferromagnetic (AF) insulators (Mott insulators) in which strong Coulomb correlations act to localize the Cu^{2+} electrons with spin 1/2, so that the system behaves like a two-dimensional (2D) Heisenberg antiferromagnet. The long-range order is caused by weak interlayer magnetic coupling between the CuO_2 planes. As holes are added into the plane through the substitution of Sr into La sites for the 2-1-4, or through oxygen doping into the CuO chain sites in the 1-2-3, the Néel temperature is reduced. The Cu^{2+} spins no longer exhibit long-range AF order beyond a critical hole content which corresponds to $La_{1.95}Sr_{0.05}CuO_4$ and $YBa_2Cu_3O_{6.4}$. At higher doping levels, the system enters an anomalous metallic phase which exhibits the superconducting transition with high T_c value. Figure 6.2 shows the phase diagram for the 2-1-4 including magnetic and structural phase transitions as a function of the doping level [3].

3. Properties in the normal state deviate from those of the Fermi liquid, in contrast to the conventional and heavy-electron superconductors. For example, the resistivity shows a T-linear behaviour over a wide temperature range for most cases. With increasing carrier content, however, it changes from T-linear to T^2 behaviour characteristic of the Fermi liquid. Eventually, the superconductivity disappears. Figure 6.3 shows the T dependence of the resistivity for $Tl_2Ba_2CuO_{6+y}$ (Tl2201) compounds with different T_c [4].

4. None of the superconducting properties resemble those of the singlet s wave, in which the energy gap is finite and isotropic. A growing list of experimental results are consistent with the $d_{x^2-y^2}$ pairing state, where the energy gap vanishes along lines. The NMR experiments, such as nuclear relaxation [5,6],

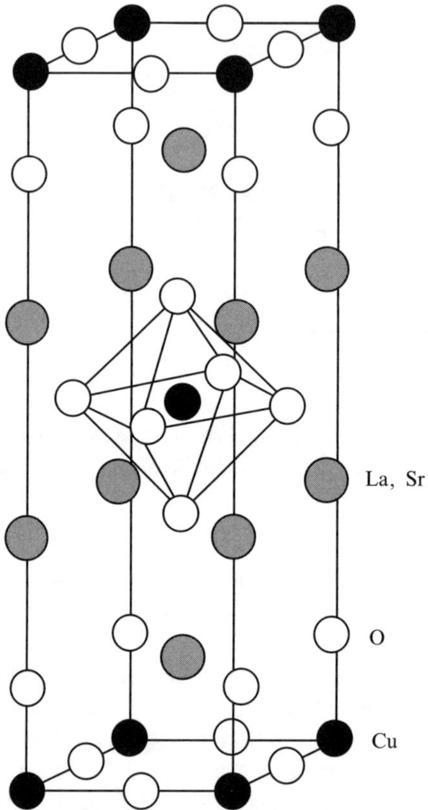

FIG. 6.1. Crystal structure of La_2CuO_4.

Knight shift, and impurity effect [7–10], have provided important clues to iden-
tifying the pairing symmetry of the high-T_c superconductivity as the d wave.
The NMR study was the first to find the unconventional impurity effect: T_c is
insensitive to the presence of impurities and other imperfections, whereas the
low-T properties in the superconducting state are extremely sensitive to the
presence of a minute concentration of impurities and imperfections. In contrast to the novel properties in the normal state, the superconducting properties share the anisotropic character with that in heavy-electron superconductors.
This commonality seems to originate from the strong repulsive interaction in
both systems; the node is favourable to reducing the energy cost of forming
the pair.

There is no consensus on the theoretical models to explain the anomalous normal
and superconducting properties over the entire doping level in a unified way. However,
a modified form of the BCS theory, which incorporates an unconventional pairing state
mediated by magnetic fluctuations, seems to be a valid phenomenology in describing
the anisotropic superconducting state with the d-wave pairing symmetry.

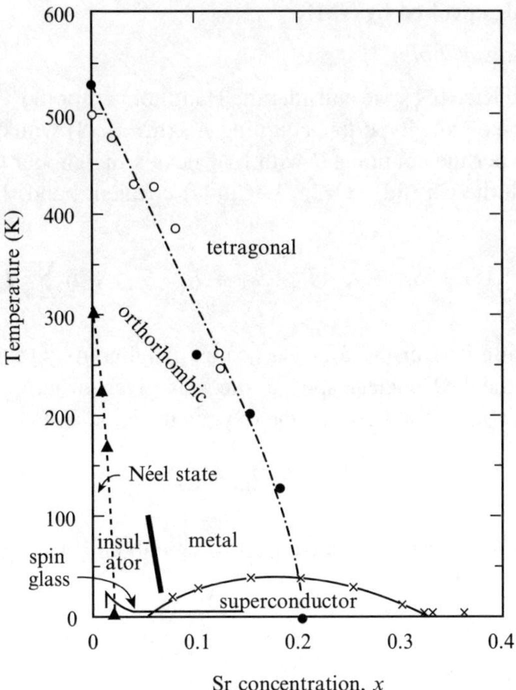

FIG. 6.2. Antiferromagnetic, superconducting and structural phase diagram of $La_{2-x}Sr_xCuO_4$ [3].

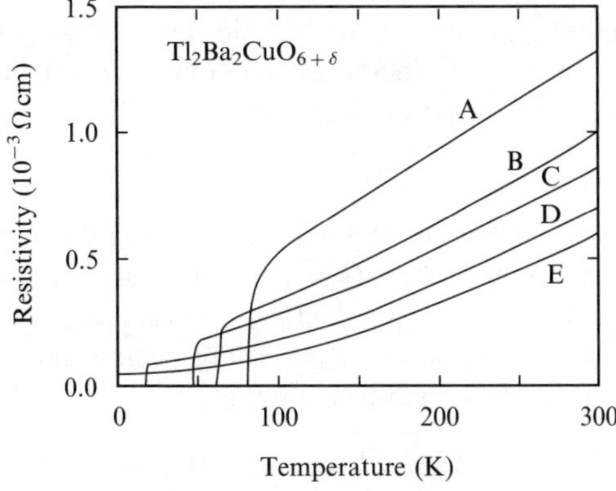

FIG. 6.3. Temperature dependence of resistivity in $Tl_2Ba_2CuO_{6+\delta}$ with various T_c values. All measurements were made using the same sample whose T_c was controlled [4].

6.2   Spin dynamics probed by NMR

6.2.1   Static spin susceptibility

Following Mila and Rice [11], we consider the Hamiltonian for the ^{63}Cu nuclear spin as given in terms of the on-site hyperfine coupling A_α ($\alpha = \perp, \parallel$) with d electrons, and the supertransferred hyperfine coupling B with four nearest-neighbour Cu sites. The latter coupling is through the Cu $(3d)-O(2p_\sigma)-Cu(4s)$ covalent bond. The Hamiltonian at site i is expressed as

$$^{63}H_i = A_\parallel I_{iz} \cdot S_{iz} + A_\perp (I_{ix} \cdot S_{ix} + I_{iy} \cdot S_{iy}) + B \sum_j I_i \cdot S_j, \tag{6.1}$$

where the summation is over the four Cu nearest neighbours. On the other hand, the Hamiltonian with the ^{17}O nuclear spin at site k is given in terms of the transferred hyperfine coupling constant C between the oxygen nuclei and the Cu spins as follows:

$$^{17}H_k = C \sum_l I_k \cdot S_l, \tag{6.2}$$

where the summation is over the two Cu nearest neighbours. The spin part $^{63}K_s$ of the Knight shift is expressed as

$$^{63}K_s(T) = (A_\alpha + 4B)\chi_s(T) \quad (\alpha = \parallel, \perp). \tag{6.3}$$

From the relation

$$\chi_s(T) = (K_{\perp \mathrm{obs}}(T) - K_{\mathrm{orb}})N\mu_B/(A_\perp + 4B), \tag{6.4}$$

the T dependence of the spin susceptibility $\chi_s(T)$ is deduced as indicated in Fig. 6.4 [33]. Here we use the results for the T-independent orbital shift of $K_{\mathrm{orb}} = 0.24\%$ and the on-site anisotropic and the isotropic hyperfine field, $A_\perp$ and B with 35 and 40 kOe/μ_B, respectively. Experimentally, the value of K_{orb} is found to be independent of the hole content over a wide doping level, while the value of B increases with increasing hole content beyond the optimum value.

In underdoped high-T_c cuprates where the hole content is smaller than the optimum value, the spin susceptibility $\chi_s(T)$ increases monotonically with increasing T, whereas in the overdoped cases, $\chi_s(T)$ stays constant.

6.2.2   Spin dynamics

The nuclear spin–lattice relaxation rate $1/T_1$ for ^{63}Cu is dominated by the AF spin fluctuations around the zone boundary $Q = (\pi/a, \pi/a)$ for the square lattice, whereas for ^{17}O it is not affected by the AF spin fluctuations since they are filtered away by the form factor. The form factors $F_\perp(q)$ and $F_\parallel(q)$ for respective directions of the magnetic field are written in terms of the hyperfine couplings as

$$F_\perp(q) = A_\perp + 2B[\cos(q_x a) + \cos(q_y a)], \tag{6.5}$$

$$F_\parallel(q) = A_\parallel + 2B[\cos(q_x a) + \cos(q_y a)],$$

$$G(q) = 2C \cos\left(\tfrac{1}{2}q_x a\right), \tag{6.6}$$

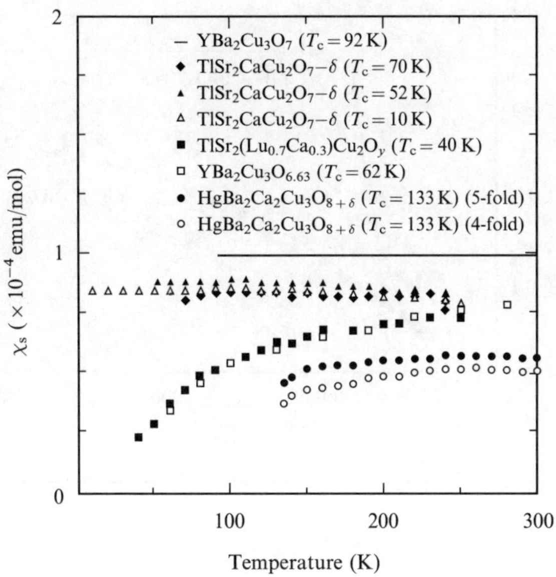

FIG. 6.4. Temperature dependence of spin susceptibility deduced from ^{63}Cu Knight shift for various high-T_c compounds [33].

where a is the distance between the Cu sites. Then, with the use of the dynamical susceptibility, $\chi(\boldsymbol{q}, \omega)$, $^{63}(1/T_1)$ and $^{17}(1/T_1)$ are given by

$$^{63}\left(\frac{1}{T_1}\right)_{\parallel} = \frac{3k_{\mathrm{B}}T}{2}\frac{1}{\mu_{\mathrm{B}}^2\hbar^2}\sum_{\boldsymbol{q}} F_\perp(\boldsymbol{q})^2 \frac{\mathrm{Im}\,\chi(\boldsymbol{q},\omega_{\mathrm{n}})}{\omega_{\mathrm{n}}}, \tag{6.7}$$

$$^{63}\left(\frac{1}{T_1}\right)_{\perp} = \frac{3k_{\mathrm{B}}T}{4}\frac{1}{\mu_{\mathrm{B}}^2\hbar^2}\sum_{\boldsymbol{q}} (F_\perp(\boldsymbol{q})^2 + F_\parallel(\boldsymbol{q})^2) \frac{\mathrm{Im}\,\chi(\boldsymbol{q},\omega_{\mathrm{n}})}{\omega_{\mathrm{n}}}, \tag{6.8}$$

$$^{17}\left(\frac{1}{T_1}\right) = \frac{3k_{\mathrm{B}}T}{4}\frac{1}{\mu_{\mathrm{B}}^2\hbar^2}\sum_{\boldsymbol{q}} G(\boldsymbol{q})^2 \frac{\mathrm{Im}\,\chi(\boldsymbol{q},\omega_{\mathrm{n}})}{\omega_{\mathrm{n}}}, \tag{6.9}$$

where ω_{n} is the NMR frequency.

In the case of AF spin fluctuations, the sums over $\boldsymbol{q}$ in eqns (6.7) and (6.8) are dominated by contributions around $\boldsymbol{q} = \boldsymbol{Q} = (\pi/a, \pi/a)$. However, the sum around $\boldsymbol{q} = 0$ is dominant in eqn (6.9) because $G(\boldsymbol{Q}) = 0$ at $\boldsymbol{Q} = (\pi/a, \pi/a)$. Then we have

$$^{63}\left(\frac{1}{T_1 T}\right) = {}^{63}\Delta(A - 4B)^2 \frac{\mathrm{Im}\,\chi(\boldsymbol{Q},\omega_{\mathrm{n}})}{\omega_{\mathrm{n}}}, \tag{6.10}$$

$$^{17}\left(\frac{1}{T_1 T}\right) = {}^{17}\Delta C^2 \frac{\mathrm{Im}\,\chi(0,\omega_{\mathrm{n}})}{\omega_{\mathrm{n}}}, \tag{6.11}$$

with $^{63}\Delta$ and $^{17}\Delta$ describing the respective width in the $\boldsymbol{q}$-space.

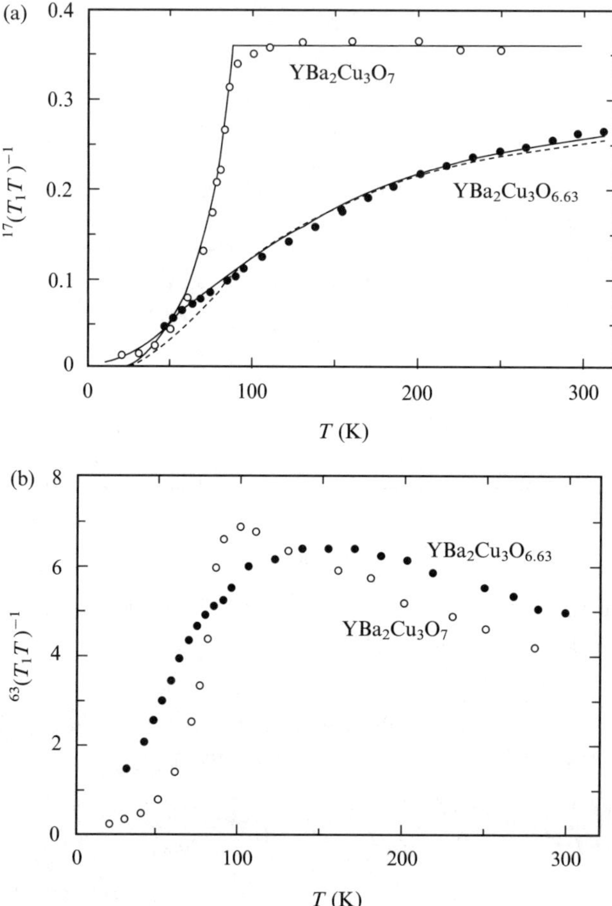

FIG. 6.5. Temperature dependence of (a) $^{17}(T_1T)^{-1}$ and (b) $^{63}(T_1T)^{-1}$ in YBa$_2$Cu$_3$O$_7$ with $T_c = 92$ K and YBa$_2$Cu$_3$O$_{6.65}$ with $T_c = 60$ K [13].

Figures 6.5(a) and (b) indicate the contrasting T dependence of $1/T_1T$ in ^{63}Cu and ^{17}O in the CuO$_2$ plane for both YBa$_2$Cu$_3$O$_7$ (YBCO$_7$) with $T_c = 90$ K and YBCO$_{6.65}$ with $T_c = 60$ K. Namely, $^{17}(1/T_1T)$ is found to be proportional to the static spin susceptibility $\chi_s(T)$, as indicated by the solid lines in Fig. 6.5(a). A comparison with Fig. 6.5 (b) shows that the dynamical susceptibility is significantly peaked around $\mathbf{Q} = (\pi/a, \pi/a)$ [13].

If we assume that Im $\chi(\mathbf{Q} + q, \omega)$ obeys the double-Lorentzian form in momentum (q) and energy (ω) distributions, eqns (1.84)–(1.89) lead to $^{63}(1/T_1T)$ in the form

$$^{63}\left(\frac{1}{T_1T}\right) \propto (A - 4B)^2 \frac{\chi_Q(T)}{\Gamma_0}. \tag{6.12}$$

Figures 6.6(a) and (b) show the T dependences of $^{63}(1/T_1T)$ and $^{63}(T_1T)$ in $La_{2-x}Sr_xCuO_4$ with various Sr contents, x [14,15]. This demonstrates that $\chi_Q(T)$ follows the Curie–Weiss law due to the substantial spin fluctuations around $\boldsymbol{Q}$. We remark that, in the case of uniform susceptibility, the Curie–Weiss law is not observed, in contrast to the situation for heavy-electron systems.

Since we have $\Gamma_Q(T)\chi_Q(T) = $ constant, according to eqn (1.87), the ensuing relation

$$^{63}(T_1T) = \alpha\Gamma_Q(T) \tag{6.13}$$

provides a direct measurement of the damping rate Γ_Q of the AF spin fluctuations around $\boldsymbol{Q}$. Since T_1T in $La_{2-x}Sr_xCuO_4$ varies linearly with temperature, as displayed in Fig. 6.6(b), we obtain

$$\Gamma_Q(T) = \beta[T + T^*(x)]. \tag{6.14}$$

From the experimental relation $T^*(x) \propto (x - 0.05)$, $\Gamma_Q(T)$ for $x = 0.05$ approaches zero near $T = 0$. This is consistent with the fact that $x = 0.05$ is just the phase boundary between the superconducting and the magnetic phases.

On the other hand, the nuclear spin–spin Gaussian decay rate, $1/T_{2G}$ provides a measure of the magnetic correlation length $\xi(T)$ [16–18]. In a two-dimensional system, the integration in eqn (1.146) leads to the relation $1/T_{2G} \propto \xi(T)$. In underdoped high-T_c compounds, such as $YBCO_{6.65}$ [17] and $YBa_2Cu_4O_8$ [19], the temperature dependence of T_{2G} follows a Curie–Weiss law. Hence, we have the relation $1/\xi(T) = a + bT$. From the experimental relation $(T_1T)/T_{2G} = $ constant, as shown in Fig. 6.7, a remarkable relation, namely,

$$\Gamma_Q(T) = \frac{c}{\xi(T)} \tag{6.15}$$

follows for spin fluctuations in underdoped high-T_c cuprates. A scaling argument for quantum critical fluctuations has been put forth [20,21] in describing the above relation between $\Gamma_Q(T)$ and $\xi(T)$ in the normal state for underdoped high-T_c cuprates.

In the underdoped region, the high-T_c cuprates deviate most strongly from the conventional materials. As seen in Fig. 6.5(b), for the underdoped $YBCO_{6.65}$ $^{63}(1/T_1T)$ has a shallow peak around $150\,K$ followed by the significant decrease down to $T_c = 60\,K$ [13,22]. This behaviour suggests the presence of a pseudo-spin-gap in spin excitations, and has been confirmed by subsequent neutron experiments as well [23]. In addition to these anomalies in magnetic excitations, the pseudo-gap of the same size and q dependence as the d-wave superconducting gap has been revealed by the photoemission experiments [24]. The persistence of the gap above T_c leads to an argument that phase fluctuation may be the determining factor for T_c. At present, it is controversial whether the pseudo-gap is interpreted as some superconducting fluctuation [25] or the spin excitation gap from some resonating-valence-bond (RVB)-type singlet state [26,27].

In contrast, in optimum and overdoped high-T_c compounds such as $YBCO_7$ [28], Hg1223 [29], and $TlSr_2CaCu_2O_{8+\delta}$ (Tl1212) [30], the experimental relation

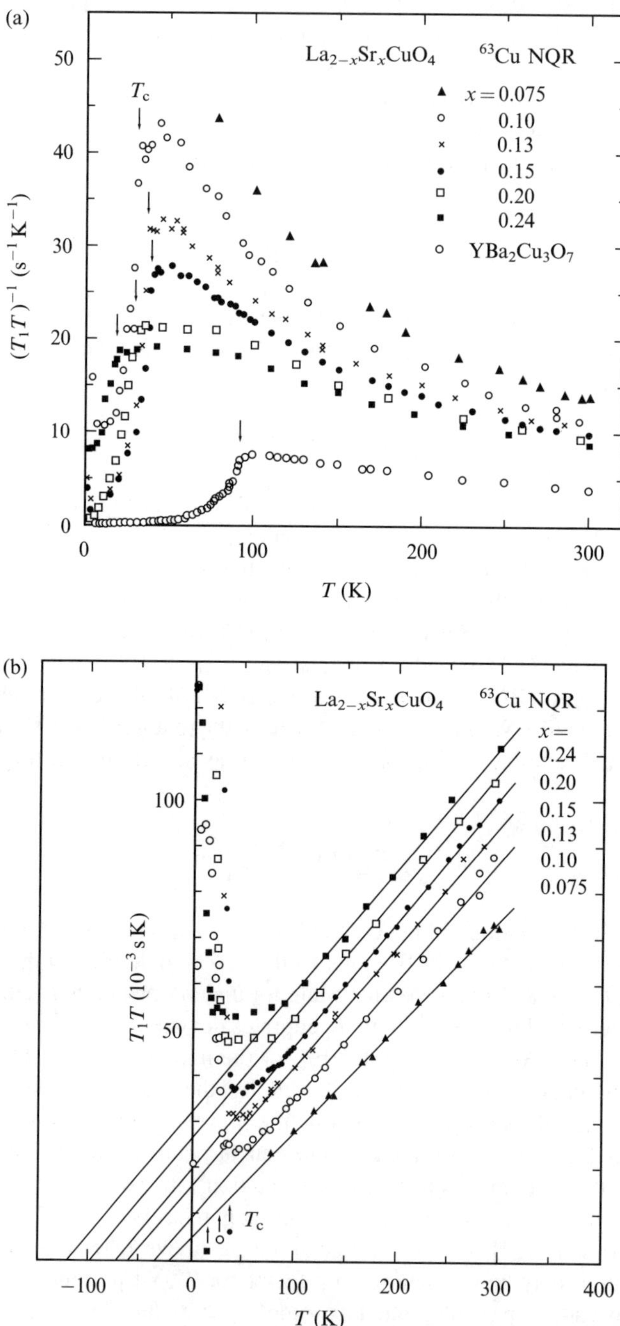

FIG. 6.6. Temperature dependence of $^{63}(T_1T)^{-1}$ (a) and $^{63}(T_1T)$ (b) in La$_{2-x}$Sr$_x$CuO$_4$ with various x values [14,15].

$(T_1 T)/(T_{2G})^2$ = constant holds, as shown in Fig. 6.8. This leads to the relation

$$\Gamma_Q(T) = \frac{c'}{\xi^2(T)}, \tag{6.16}$$

which is consistent with phenomenological RPA-type approaches [31,32].

By combining both NMR and neutron experimental results, it is suggested that the quantum critical aspect of spin correlations in underdoped systems is relevant to the onset

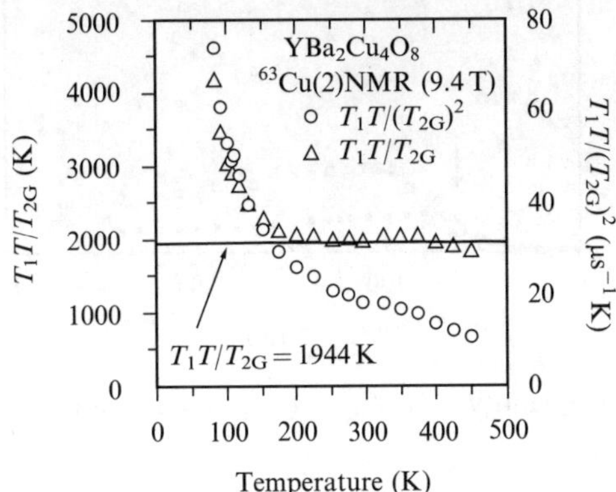

FIG. 6.7. T dependence of the $(T_1 T)/T_{2G}$ for the underdoped YBa$_2$Cu$_4$O$_8$ [19].

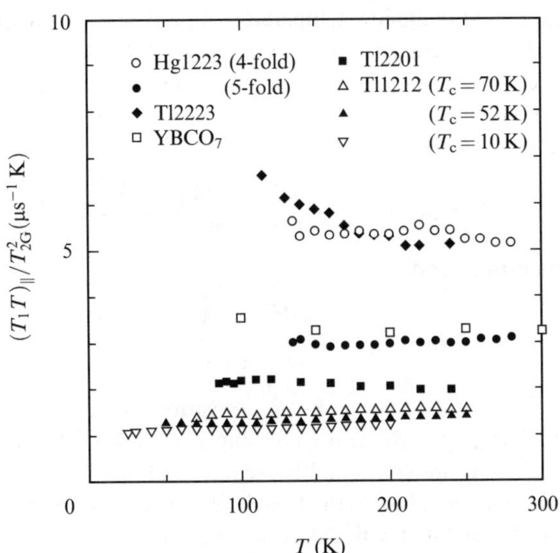

FIG. 6.8. Temperature dependence of $(T_1 T_\parallel)/(T_{2G})^2$ for the optimum and overdoped high-T_c cuprates [29,30]. Differences in the hyperfine coupling constant have been corrected.

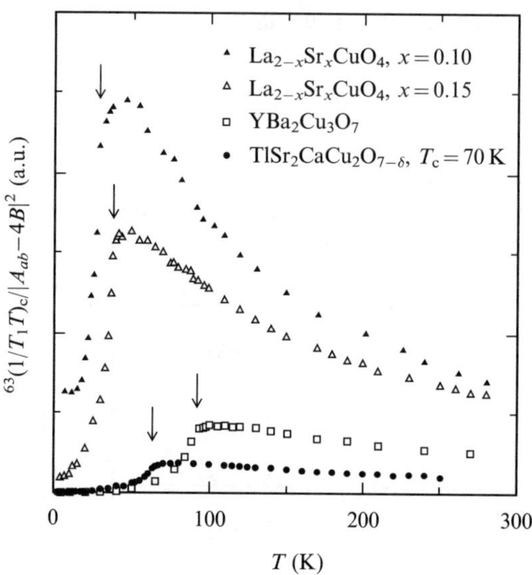

FIG. 6.9. Change of the quantity $^{63}(T_1T)_c^{-1}/(A_{ab}-4B)^2 = (\alpha\Gamma_{\mathbf{Q}})^{-1}$ with increasing hole content [29,30].

of the high-T_c superconductivity. Increasing the number of holes causes spin correlations to be overdamped and leads to an enhancement of T_c. With further increase of the hole content, a dramatic change in the spin dynamics occurs as shown in Fig. 6.9 [33]. Thus, Im $\chi(\mathbf{Q}, \omega)$ is suppressed in the low-temperature region, suggesting that $\Gamma_{\mathbf{Q}}$ increases. As a result, $(1/T_1T)$ stays constant in a wide range of T. Doping of holes above an optimum value makes the AF correlation length shorter and decreases the T_c. This change was extracted from the anisotropy of $R = (T_1T)_\perp/(T_1T)_\parallel$. For the case where $\chi(\mathbf{q})$ is strongly peaked at $\mathbf{q} = \mathbf{Q}$, we obtain

$$^{63}R_{\mathrm{AF}} = \frac{(A_\perp - 4B)^2 + (A_\parallel - 4B)^2}{2(A_\perp - 4B)^2}, \tag{6.17}$$

whereas, if $\chi(\mathbf{q})$ is $\mathbf{q}$-independent,

$$^{63}R_{\mathrm{r}} = \frac{(A_\perp^2 + 4B^2) + (A_\parallel^2 + 4B^2)}{2(A_\perp^2 + 4B^2)}, \tag{6.18}$$

follows. Figure 6.10 shows the values of ^{63}R's in various high-T_c cuprates. It decreases progressively from $^{63}R_{\mathrm{AF}}$ to $^{63}R_{\mathrm{r}}$ with increasing B or holes. The AF spin correlation in Tl1212 with $T_c = 10$ K are considerably weaker, and that in Tl2201 with $T_c = 0$ is almost absent. Both cases are close to the limit $^{63}R_{\mathrm{r}}$. This shows that the presence of the AF spin fluctuation is correlated with the occurrence of superconductivity.

We note that the spin fluctuation spectrum becomes broader in the q-space as the characteristic energy $\Gamma_{\mathbf{Q}}$ increases with increasing hole content. This variation of q dependence of spin dynamics suggests that T_c has a maximum at an optimum characteristic energy $\Gamma_{\mathbf{Q}}$ in the range 20–40 meV. The q dependence of spin fluctuations is a

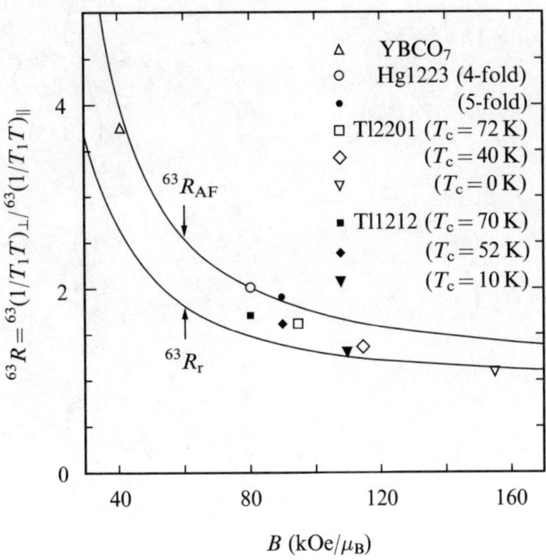

$^{63}R = {}^{63}(1/T_1 T)_\perp / {}^{63}(1/T_1 T)_\parallel$

B (kOe/μ_B)

FIG. 6.10. The anisotropy ratio $^{63}R_{ex} = {}^{63}(1/T_1 T)_{ab}/{}^{63}(1/T_1 T)_c$ in high-T_c cuprates with various T_c's and hole contents. Lines indicated by $^{63}R_{AF}$ and $^{63}R_r$ describe the limiting cases for strongly $\boldsymbol{q}$-dependent AF spin correlations and $\boldsymbol{q}$-independent ones, respectively [30].

necessary condition for the occurrence of superconductivity [30,34]. These properties are consistent with some phenomenological models [32,35,36] for spin-fluctuation-induced superconductivity. They also provide an explanation for the d-wave superconductivity established experimentally. However, any microscopic theory is not yet presented.

It is noteworthy that the nature of low-energy spin fluctuations in high-T_c cuprates is *nonlocal*, dominated by the superexchange interaction $J_{ex} \sim 1500$ K, which is clearly present in the parent AF insulator. On the other hand, the spin fluctuation in heavy-electron systems is nearly *local* due to the weak hybridization between the f and conduction electrons. In the latter systems, the q-dependent magnetic response emerges only well below the effective Fermi temperature in the range 10–100 K.

6.3 The p–d hybridization model

As the simplest model to investigate the quasi two-dimensional cuprates, we consider a square lattice where each copper site at a lattice point $\boldsymbol{R}_j = (n, m)$ has the $3d_{x^2-y^2}$ orbital. The model is shown schematically in Fig. 6.11. The annihilation operator of the corresponding hole is represented by $d_{j\sigma}$ with spin σ. The d-hole has Coulomb repulsion U_d. For each oxygen at $\boldsymbol{r}_i = (n + 1/2, m)$, we take the $2p_x$ antibonding orbital which interacts most strongly with the 3d hole. The annihilation operator of the hole is written as $c_{i\sigma}$. Similarly, for each oxygen site at $\boldsymbol{r}_l = (n, m + 1/2)$, we take the $2p_y$ orbital with

FIG. 6.11. The geometry of the p–d hybridization model. Each Cu site with the $3d_{x^2-y^2}$ orbital is surrounded by four O sites with antibonding $2p_x$ or $2p_y$ orbitals.

the annihilation operator written as $c_{l\sigma}$. The Hamiltonian is given by

$$H = H_1 + H_2, \tag{6.19}$$

$$H_1 = \sum_{ij}\sum_{\sigma} t_{ij} c_{i\sigma}^{\dagger} c_{j\sigma} + \sum_{i}\sum_{\sigma} \epsilon_d d_{i\sigma}^{\dagger} d_{i\sigma}$$

$$+ V \sum_{<ij>}\sum_{\sigma} (c_{i\sigma}^{\dagger} d_{j\sigma} + d_{j\sigma}^{\dagger} c_{i\sigma}), \tag{6.20}$$

$$H_2 = U_d \sum_{j} n_{d\uparrow} n_{d\downarrow}, \tag{6.21}$$

where $t_{ii} = \epsilon_p$ is the level of a 2p hole and the nearest-neighbour 2p–2p transfer is given by $t_{ij} = -t_p$. The strong covalency between 2p and 3d holes is taken into account by V. The sum is over nearest neighbours of Cu and O sites.

In order to simplify the notation, we have chosen the phase of the operator $d_{j\sigma}$ at $\boldsymbol{R}_j = (n, m)$, so that the odd parity of the p state is compensated. Namely, if $n + m$ is odd, the phase is chosen opposite to that at a site with $n + m$ even. Then V is taken to be a constant. The change of phase corresponds to the shift $(\pi/a, \pi/a)$ of the Brillouin zone. This p–d hybridization model has an obvious similarity with the Anderson lattice. However, the relative magnitudes of parameters are much different from those for heavy electrons. The typical values are:

$$U_d = 10.5, \qquad t = 1.3, \qquad \epsilon_p - \epsilon_d \equiv \Delta = 3.6,$$

in units of eV [37].

Since U_d is much larger than any other parameter in the model, it is reasonable to work with another effective Hamiltonian which eliminates the double occupation of a

3d orbital which corresponds to a $3d^8$ configuration. One can also assume that ϵ_d is deep enough below the Fermi level and can neglect the $3d^{10}$ configuration. Then the spin operator S_j is sufficient to describe the dynamics of the 3d hole. For 2p holes we introduce the notation:

$$x_{i\sigma} = \frac{1}{\sqrt{2}}(c_{i-\hat{x}/2,\sigma} + c_{i+\hat{x}/2,\sigma}), \qquad (6.22)$$

$$y_{i\sigma} = \frac{1}{\sqrt{2}}(c_{i-\hat{y}/2,\sigma} + c_{i+\hat{y}/2,\sigma}), \qquad (6.23)$$

$$a_{i\sigma} = \frac{1}{\sqrt{2}}(x_{i\sigma} + y_{i\sigma}), \qquad (6.24)$$

where $\hat{x}$ and $\hat{y}$ denote unit vectors along each axis. Note that these operators use a nonorthogonal basis, so that the fermion anticommutator $\{x_{i\sigma}, y_{j\sigma}^{\dagger}\}$ does not vanish if $x_{i\sigma}$ and $y_{j\sigma}^{\dagger}$ have a common oxygen orbital. The effective Hamiltonian is given by

$$H_{\text{eff}} = \sum_i H_i + J \sum_{<ij>} S_i \cdot S_j - \frac{4NV^2}{\Delta}, \qquad (6.25)$$

$$H_i = -2t_{\text{p}} \sum_\sigma (x_{i\sigma}^{\dagger} y_{i\sigma} + y_{i\sigma}^{\dagger} x_{i\sigma}) + 2t_{\text{s}} \sum_\sigma a_{i\sigma}^{\dagger} a_{i\sigma}$$

$$+ 4J_{\text{s}} \sum_{\alpha\beta} S_i \cdot a_{i\alpha}^{\dagger} \sigma_{\alpha\beta} a_{i\beta}, \qquad (6.26)$$

where the origin of energy is taken to be the 2p hole level ϵ_{p}. Here the new parameters J, J_{s} and t_{s} are given by

$$J = \frac{4V^4}{\Delta^2}\left(\frac{1}{U_{\text{d}}} + \frac{1}{\Delta}\right), \quad J_{\text{s}} = V^2\left(\frac{1}{\Delta} + \frac{1}{U_{\text{d}} - \Delta}\right), \quad t_{\text{s}} = V^2\left(\frac{1}{\Delta} - \frac{1}{U_{\text{d}} - \Delta}\right). \qquad (6.27)$$

The parameter t_{s} describes the effective transfer between 2p orbitals induced by the p–d hybridization. The transfer occurs not only between $2p_x$ and $2p_y$, but also between $2p_x$ and $2p_x$ sandwiching the copper site. The spectra of the 2p bands without including J_{s} is given by

$$E(\boldsymbol{k}) = t_{\text{s}}(\cos k_x + \cos k_y)$$
$$\pm (t_{\text{s}}(\cos k_x - \cos k_y)^2 + 16(t_{\text{p}} - t_{\text{s}})^2 \cos^2 \tfrac{1}{2}k_x \cos^2 \tfrac{1}{2}k_y)^{1/2}. \qquad (6.28)$$

It is instructive to compare eqn (6.25) with the Kondo lattice model discussed in Chapter 3. The two models are essentially the same except for the difference in the relative magnitudes of parameters, and the filling of the conduction band. The large J_{s} is favourable to the formation of a singlet which is composed of 3d and 2p holes sitting at nearest neighbour sites. This state is often called the Zhang–Rice singlet [38]. The Zhang–Rice singlet in the p–d hybridization model evolves continuously into the Kondo singlet as the ratio $J_{\text{s}}/t_{\text{s}}$ becomes smaller, as long as there are enough 2p holes to screen the 3d spin. Accordingly, the extension of the singlet wave function grows as the singlet energy scale decreases. Let us consider the case where there is precisely one 2p hole per unit cell of the lattice. In the limit of large J_s, all 2p holes form the Zhang–Rice singlet, and there are no mobile carriers. Then the system should become an insulator. This state corresponds to the Kondo insulator in the opposite limit of small J_s. The important feature common to both insulating states is that the ground state connects continuously to the band insulator. We recall that for a description of the hydrogen molecule, both the Heitler–London picture and the molecular orbital picture are useful. The Zhang–Rice singlet is close to the Heitler–London picture.

In contrast to heavy-electron metals, however, the 2p holes are absent unless the material is doped. Without the 2p holes, the effect of J_s does not appear explicitly in physical properties, and the effective model to describe the dynamics of the 3d holes reduces to the Heisenberg model. Actually, from numerical work [39], some important corrections to the Heisenberg model have been suggested. Namely, taking a small cluster Cu_4O_8 with a free boundary condition, one computes the low-lying levels by an exact numerical diagonalization of the Hamiltonian given by eqn (6.25). Then, fitting of these levels by an effective model requires not only the Cu–Cu nearest-neighbour exchange but a large four-spin interaction. However, it is not clear whether this result remains the same as the system size increases.

Let us consider a situation where a single 2p hole is doped in the insulating state. Because of the large coupling J_s, the 2p hole forms a singlet with any of the 3d holes, and the entity can move due to the nonorthogonality of 2p orbitals. The resultant moving unit can alternatively be viewed as a defect in 3d spins. The mathematical expression of the latter view is called the t–J model, as discussed in the next section.

6.4   Separation of spin and charge in dynamics

The characteristic energy of spin fluctuations in heavy-electron systems is much smaller than that of the f-charge excitation which is $|\epsilon_f|$ or U. This is seen in the dynamical susceptibilities of spin and charge, as discussed in Chapter 3. In this sense, the spin and charge have different dynamical scales. The velocity of the excitations, however, is the same in the low-energy limit. This is because the Fermi-liquid ground state has the single Fermi velocity which sets the energy of the quasi-particles in the metallic state. Having the same velocity does not depend on the strength of the correlation. A similar situation holds in the insulating case called the Kondo insulator (except in one dimension; this is discussed below). Namely, the energy gap is common for both spin and charge excitations, as in usual semiconductors. The strong correlation manifests in the spectral intensity which is much smaller for the charge excitation near the threshold. If a bound state is present, such as excitons, the spin excitation has a lower threshold than the charge excitation because of the exchange interaction between the excited pair. This difference of the threshold, however, is also seen for semiconductors.

In one dimension, on the other hand, the spin and charge excitations can really have different velocities. This is in strong contrast with the Fermi-liquid state, and is called spin–charge separation. We take the t–J model as the simplest model to display the characteristics in one dimension. The model under finite magnetic field h is given by

$$\mathcal{H} = \mathcal{P} \sum_{i \neq j} \left[-t_{ij} \sum_{\sigma} c_{i\sigma}^{\dagger} c_{j\sigma} + \frac{1}{2} J_{ij} \left(\mathbf{S}_i \cdot \mathbf{S}_j - \frac{1}{4} n_i n_j \right) \right] \mathcal{P} - 2h \sum_i S_{iz}, \quad (6.29)$$

where $\mathcal{P}$ is the projection operator to exclude double occupation of each site. The t–J model can be derived from the Hubbard model as an effective model in the limit of large U. It can also be obtained from the p–d hybridization model in the limit of large binding energy of the Zhang–Rice singlet. Instead, here we regard the transfer $-t_{ij}$ and the exchange J_{ij} between the sites i, j as free parameters in the model. In some special cases, the model can be solved exactly. The first case is that the parameters satisfy the condition: $t_{ij} = J_{ij}/2 = J/2$ for nearest neighbours and zero otherwise. In this case, the model can be solved analytically by using the Bethe Ansatz [40,41]. Another solvable case is when the parameters take a long-range form

$$J_{ij} = 2t_{ij} = JD(x_i - x_j)^{-2}, \quad D(x_i - x_j) = (N/\pi) \sin\left[\frac{\pi}{N} (x_i - x_j) \right]. \quad (6.30)$$

The quantity $D(x_i - x_j)$ corresponds to the chord distance between the two sites in the ring-shaped system. In the second case, it has been shown [42] that the ground state can be obtained explicitly. Let us take the completely up-polarized state as the reference state, and let x_α denote the site occupied by a down spin, and y_l a vacant (hole) site. Then the eigenfunction of the ground state at $h = 0$ takes the following form:

$$\Psi_G(\{x\}, \{y\}) = \exp\left[-i\pi \left(\sum_{\alpha=1}^{M} x_\alpha + \sum_{l=1}^{Q} y_l \right) \right]$$
$$\times \prod_{\alpha > \beta} D(x_\alpha - x_\beta)^2 \prod_{l > m} D(y_l - y_m) \prod_{\alpha,l} D(x_\alpha - y_l). \quad (6.31)$$

This wave function has the same form as the Gutzwiller wave function, if the reference state is taken to be the vacant state instead of the fully polarized one.

One can also derive the low-lying excited states starting from this wave function. The procedure of derivation is rather involved, and we give here only the results [43]. The spin susceptibility χ_s at zero temperature is given by

$$\chi_s = \frac{\partial(n_\uparrow - n_\downarrow)}{\partial h} = \frac{4}{\pi^2 J(1 - |m|)} \equiv \frac{2}{\pi v_s}, \quad (6.32)$$

which depends only on the magnetization $m = \langle n_{i\uparrow} - n_{i\downarrow} \rangle$ per site and is independent of the filling $n = \langle n_{i\uparrow} + n_{i\downarrow} \rangle$. This is a manifestation of the spin–charge separation. The spin velocity v_s is defined in analogy with the Fermi velocity which would have arisen in the absence of interaction. Similarly, we introduce the charge velocity v_c through the charge susceptibility χ_c at $T = 0$, which is given by

$$\chi_c = \frac{\partial(n_\uparrow + n_\downarrow)}{\partial \mu} = \frac{4}{\pi^2 J(1 - n)} \equiv \frac{2}{\pi v_c}. \quad (6.33)$$

This is now independent of the magnetization m. It is remarkable that spin and charge susceptibilities have exactly the same form as functions of m and n, respectively.

The specific heat is linear in temperature, as in the case of the Fermi liquid. The coefficient γ is given by

$$\gamma = \chi_s + \chi_c, \tag{6.34}$$

which shows that both spin and charge excitations contribute to the linear specific heat. It is clear that the charge velocity vanishes in the high-density limit. As a result, the specific heat diverges due to the soft charge excitation. On the other hand, the spin velocity remains finite for all densities without the magnetization, but diverges in the limit $m \to 1$. Figure 6.12 shows χ_s at finite temperatures and with $m = 0$ [44]. It is seen that χ_s increases linearly from zero temperature, which is to be contrasted with the T^2 dependence in the Fermi liquid. The slope is independent of the filling and is given by $\partial \chi_s / \partial T = 16/(\pi J)^2$. However, in the presence of magnetization, χ_s has the T^2 dependence as given by

$$\chi_s = \frac{4}{\pi^2 (1-m) J} \left[1 + \frac{8T^2}{3\pi^2 (1-m)^4 J^2} \right] + O(T^3). \tag{6.35}$$

Remarkably, the charge susceptibility again has the same form as given by eqn (6.35) with m replaced by n. This nice symmetry is specific to the present supersymmetric model. In other one-dimensional fermion systems, the spin–charge separation appears as the different spin and charge velocities only. The spin susceptibility of the Heisenberg chain is given by the same formula as eqn (6.32) with $m = 0$ at $T = 0$, but the temperature dependence involves $\ln T$ [45]. This complication can be understood by perturbation theory from the high-density (spin-chain) limit of the present t–J model, which is called the Haldane–Shastry model [46,47].

Numerical studies in one dimension have also been made for the Anderson lattice model or the Kondo lattice model. This includes the Monte Carlo calculation and the numerical diagonalization of the Hamiltonian. At present, it is still not possible for

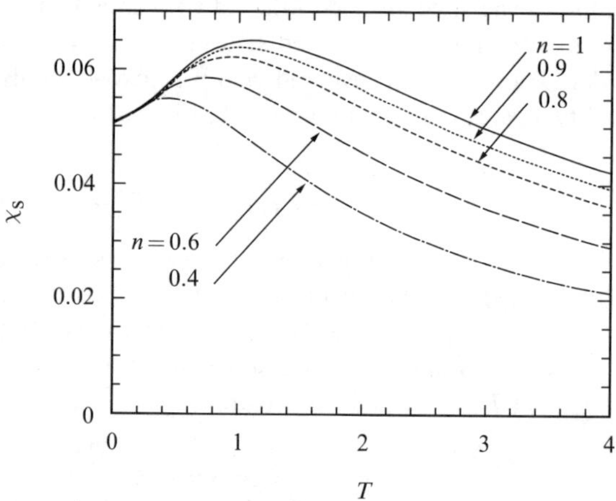

FIG. 6.12. The spin susceptibility of the supersymmetric t–J model with the long-range interaction.

these numerical studies to deal with the Kondo energy T_K if it is much smaller than the bare parameters in the model. Only the numerical renormalization group for the single impurity has been treated on the minute scale. Recently, a new numerical method, called the density matrix renormalization group, has been introduced [48] and is becoming an area of very active investigation.

6.5   Epilogue—to be or not to be a Fermi liquid

The strong correlation among electrons appears in two qualitatively different ways: perturbative and nonperturbative. This distinction persists even though we restrict the discussion to the paramagnetic state. In the case of the Fermi liquid, the ground and low-lying excited states are adiabatically connected to the corresponding states in the noninteracting Fermi gas. On the other hand, the electronic states in one dimension are not accessible by perturbation theory with respect to the interaction among the electrons. In this sense, at zero temperature, the distinction between the Fermi liquid and the non-Fermi liquid is clear. However, at finite temperature the distinction is not always possible; even in the Fermi liquid, the sharp discontinuity of the momentum distribution at the Fermi surface is absent. This ambiguity causes considerable dispute about the nature of electronic states in high-temperature superconductors in the normal state.

The terminology of 'spin–charge separation' is popular in the area of high-T_c superconductivity. As we have discussed, the phenomenon is well established in one-dimensional electron systems. Since there are no ideal one-dimensional systems in the real world, the question is how important is the three-dimensional perturbation. In two dimensions, on the other hand, there is still no established answer about the relevance of the non-Fermi-liquid state. The importance of forward scattering is emphasized, which might break the Fermi-liquid nature of excitations [49]. If one uses the renormalization group to decide the relevance or irrelevance of the Coulomb interaction, the critical dimension turns out to be one [50]. Hence, the two-dimensional system belongs to the universality class of higher dimensions. However, if the fixed-point behaviour is limited only to very narrow region, this behaviour is less relevant experimentally. Instead, some different behaviour in the crossover region may dominate the most interesting parameter regions. For example, in the quasi-one-dimensional system, where the interchain coupling is very weak, the one-dimensional behaviour will dominate although the system is driven eventually to the three-dimensional fixed point.

As explained in the present chapter, superconducting oxides probed by the NMR [13,22] and neutron scattering [23] show an apparent gap for spin excitations. The charge excitation as observed by optical absorption is gapless. Irrespective of the connection with the problem of Fermi liquid, it is very important whether the spin and charge excitations have really different thresholds of excitations, or if they have simply different scales for dynamics. In a spectroscopic experiment, it is obviously impossible to distinguish between zero and small but finite spectral intensity. In fact, even in three-dimensional systems, the optical experiment sometimes claims an energy gap apparently larger than the one measured by a magnetic probe such as neutron scattering and NMR. We have seen in many places in this book that the strong correlation brings about very different spectral weights between the spin and charge excitations even in the Fermi-liquid state. Thus, in order to decide the Fermi liquid nature, one must examine a more detailed dependence of the spectrum upon temperature and frequency. It seems that the experimental results so far are not sufficient by themselves to resolve the issue.

In heavy-electron systems, the non-Fermi-liquid ground state can appear even without the spin–charge separation. The first case is realized near the critical boundary between the paramagnetic ground state and a magnetic one. An example is an alloy system $CeCu_{6-x}Au_x$ [51]. With increasing x, there appears a region where the magnetic susceptibility and the specific heat show a logarithmic dependence on temperature. In this case, the characteristic energies on both sides of the phase boundary are reduced near the critical region. As a result, even a 'low temperature' in the usual experimental sense can already be above the characteristic energy.

The second instance for the non-Fermi liquid is closely related to the electronic structure of the U ions. For example, the quadrupolar Kondo effect discussed in Chapter 2 is a candidate to realize the non-Fermi liquid in impurity systems. In the dilute alloy $U_{1-x}Th_xRu_2Si_2$, the non-Fermi liquid seems to be realized as a consequence of an isolated U impurity [52]. Another example where the interplay between the strong correlation and the disorder effect is also important is $U_{1-x}Y_xPd_3$ [53,54]. In the latter case, the local Fermi liquid is realized in the dilute limit of U ions [55]. A deviation from the Fermi-liquid behaviour is also observed near the quantum critical point, where the ground state of the system changes from paramagnetic to antiferromagnetic or to spin-glass [56]. A distinction between these cases should be made possible by varying external parameters and noting the dependence of physical quantities on these parameters.

Although the present book focused mostly on heavy electrons in rare-earth and actinide compounds, comparative study together with 3d-electron systems should be rewarding [57]. The recent interest common in f- and d-electron systems is the coupling between the spin and orbital degrees of freedom. This appears in f-electron systems as the quadrupole ordering in CeB_6, as discussed in Chapter 4 and TmTe [58]. Essentially, the same ordering is called orbital order in 3d-electron systems such as $La_{1-x}Sr_xMnO_3$, where a hole is doped by addition of Sr [59] and the magnetic anisotropy changes drastically [60].

Another interesting phenomenon to be clarified in the near future is the occurrence of heavy electrons in systems with very small number of carriers. Typical examples such as CeP and Yb_4As_3 are discussed in Chapter 4. The antiferromagnetic correlation between localized pairs of electrons tend to form a singlet. If there are many fluctuating singlets around each spin, the correlation looks similar to the Kondo screening. Namely, instead of the mobile conduction electrons, surrounding spins may also provide a singlet cloud by slow fluctuation of spins. If the spin fluctuations are gapless, they lead to a large T-linear specific heat. In the Fermi liquid, the slow spin fluctuation appears as an enhanced effective mass. Even though the number of carriers is small, the slow spin fluctuation can appear as a large specific heat. Because of these new developments, the field of strongly correlated electrons including heavy electrons still remains one of the most challenging fields in condensed-matter physics for both theory and experiment.

Bibliography

[1] J. G. Bednorz and K. A. Müller, *Z. Phys.* B **64**, 189 (1986).

[2] For review, see for example, *Proc. Int. Conf. Material and Mechanism of Superconductivity High-Temperature Superconductivity V, Physica C* **282–287** (1997).

[3] B. Keimer, N. Belk, R. J. Birgeneau, A. Cassanho, C. Y. Chen, M. Greven, M. A. Kastner, A. Aharony, Y. Endoh, R. W. Erwin, and G. Shirane, *Phys. Rev. B* **46**, 14 034 (1992).

[4] Y. Kubo, Y. Shimakawa, T. Manako, and H. Igarashi, *Phys. Rev. B* **43**, 7875 (1991).

[5] Y. Kitaoka, *et al.*, *Physica C* **153–155**, 83 (1988).

[6] T. Imai, T. Shimizu, H. Yasuoka, Y. Ueda, and K. Kosuge, *J. Phys. Soc. Jpn.* **57**, 2280 (1988).

[7] K. Ishida *et al.*, *Physica C* **179**, 29 (1991).

[8] K. Ishida *et al.*, *J. Phys. Soc. Jpn.* **63**, 2803 (1993).

[9] K. Ishida, Y. Kitaoka, K. Asayama, K. Kadowaki, and T. Mochiku, *J. Phys. Soc. Jpn.* **63**, 1104 (1994).

[10] Y. Kitaoka, K. Ishida, and K. Asayama, *J. Phys. Soc. Jpn.* **63**, 2052 (1994).

[11] F. Mila and T.M. Rice, *Phys. Rev. B* **40**, 11 382 (1989).

[12] Y. Kitaoka *et al.*, *J. Phys. Chem. Solid* **56**, 1931 (1995).

[13] M. Takigawa, *Electronic properties and mechanism of high-T_c superconductors* (Eds, T. Oguchi *et al.*, North-Holland, 1992) 123.

[14] Y. Kitaoka, S. Ohsugi, K. Ishida, and K. Asayama, *Physica C* **170**, 189 (1990).

[15] S. Ohsugi, Y. Kitaoka, K. Ishida, and K. Asayama, *J. Phys. Soc. Jpn.* **60**, 2351 (1991).

[16] C. H. Pennington and C.P. Slichter, *Phys. Rev. Lett.* **66**, 381 (1991).

[17] M. Takigawa, *Phys. Rev. B* **49**, 4158 (1994).

[18] D. Thelen and D. Pines, *Phys. Rev. B* **49**, 3528 (1994).

[19] R. L. Corey, N. J. Curro, K. O'Hara, T. Imai, and C. P. Slichter, *Phys. Rev.* **53**, 5907 (1996).

[20] A. V. Chubukov, S. Sachdev, and J. Ye, *Phys. Rev. B* **49**, 11919 (1994).

[21] V. Barzykin and D. Pines, *Phys. Rev. B* **52**, 13585 (1995).

[22] H. Yasuoka, T. Imai, and T. Shimizu, Springer Series in Solid State Science 89, *Strong correlation and superconductivity* (Eds, H. Fukuyama *et al.*, Springer-Verlag, New York, 1989) p. 254.

[23] J. Rossat-Mignod *et al.*, *Physica B* **169**, 58 (1991).

[24] A. G. Loeser *et al.*, *Science* **273**, 325 (1996); H. Ding *et al.*, *Nature* (London) **382**, 511 (1996).

[25] V. Emery and S. Kivelson, *Nature* (London) **374**, 434 (1995).

[26] P. W. Anderson, *Science* **235**, 1196 (1987).

[27] H. Fukuyama, *Prog. Theor. Phys. Suppl.* **108**, 287 (1992).

[28] T. Imai, C. P. Slichter, A. P. Paulikas, and B. Veal, *Phys. Rev. B* **47**, 9158 (1993).

[29] K. Magishi *et al.*, *J. Phys. Soc. Jpn.* **64**, 4561 (1995).

[30] K. Magishi *et al.*, *Phys. Rev. B* **54**, 10 131 (1996).

[31] A. Millis, H. Monien, and D. Pines, *Phys. Rev. B* **42**, 167 (1990).

[32] T. Moriya, Y. Takahashi, and K. Ueda, *J. Phys. Soc. Jpn.* **59**, 2905 (1990).

[33] Y. Kitaoka, K. Ishida, G.-q. Zheng, S. Ohsugi, and K. Asayama, *J. Phys. Chem. Solid* **54**, 1358 (1993).

[34] Y. Kitaoka, K. Fujiwara, K. Ishida, K. Asayama, Y. Shimakawa, T. Manako, and Y. Kubo, *Physica C* **179**, 107 (1991).

[35] K. Ueda, T. Moriya, and Y. Takahashi, *J. Phys. Chem. Solid* **53**, 1515 (1992).

[36] P. Monthoux and D. Pines, *Phys. Rev. B* **50**, 16015 (1994).

[37] M. S. Hybertsen, M. Schlüter, and N. E. Chrestensen, *Phys. Rev. B* **39**, 9028 (1989).

[38] F. C. Zhang and T. M. Rice, *Phys. Rev. B* **37**, 3759 (1988).

[39] H.-J. Schmidt and Y. Kuramoto, *Physica C* **167**, 263 (1990).

[40] B. Sutherland, *Phys. Rev. B* **12**, 3795 (1975).

[41] P.-A. Bares, G. Blatter, and M. Ogata, *Phys. Rev. B* **44**, 130 (1991).

[42] Y. Kuramoto and H. Yokoyama, *Phys. Rev. Lett.* **67**, 1338 (1991).

[43] Y. Kuramoto and Y. Kato, *J. Phys. Soc. Jpn.* **64**, 4518 (1995).

[44] Y. Kato and Y. Kuramoto, *J. Phys. Soc. Jpn.* **65**, 1622 (1996).

[45] S. Eggert, I. Affleck, and M. Takahashi, *Phys. Rev. Lett.* **73**, 332 (1994).

[46] F. D. M. Haldane, *Phys. Rev. Lett.* **90**, 635 (1988)

[47] S. Shastry, *Phys. Rev. Lett.* **90**, 639 (1988)

[48] S. R. White, *Phys. Rev. B* **48**, 10345 (1993).

[49] P. W. Anderson, *Phys. Rev. Lett.* **64**, 1839 (1990).

[50] W. Metzner, *Physica B* **197**, 457 (1994).

[51] H. v. Lohneysen, *J. Phys. Cond. Matter* **8**, 9689 (1996).

[52] H. Amitsuka and T. Sakakibara, *J. Phys. Soc. Jpn.* **63**, 736 (1994).

[53] C. L. Seaman *et al.*, *Phys. Rev. Lett.* **67**, 2882 (1991).

[54] B. Andraka and A. M. Tsvelick, *Phys. Rev. Lett.* **67**, 2886 (1991).

[55] Y. Aoki *et al.*, *Physica B* **206 & 207**, 451 (1995).

[56] P. Gegenwart *et al.*, *Phys. Rev. Lett.* **81**, 1501 (1998).

[57] M. Imada, A. Fujimori, and Y. Tokura, *Rev. Mod. Phys.* **70**, 1039 (1998).

[58] T. Matsumura, S. Nakamura, T. Goto, H. Shida, and T. Suzuki, *Physica B* **223 & 224**, 385 (1996).

[59] Y. Tokura, A. Urushibara, Y. Moritomo, T. Arima, A. Asamitsu, and G. Kido, *J. Phys. Soc. Jpn.* **63**, 3931 (1994).

[60] K. Hirota, N. Kaneko, A. Nishizawa, and Y. Endoh, *J. Phys. Soc. Jpn.* **65**, 3736 (1996).

APPENDIX A

LINEAR RESPONSE THEORY

The linear response theory is based upon two basic assumptions [1]:

(a) the system is in thermal equilibrium before the external perturbation $H_{ex}(t)$, such as a magnetic field, is switched on;

(b) the system develops adiabatically in the course of infinitesimally slow increase of $H_{ex}(t)$;

Based on these assumptions, let us first consider the expectation value of an operator A for the ground state $|\Psi_g\rangle$. In the Schrödinger picture, the time dependence of the expectation value arises from that of the state vector $|\Psi_g(t)\rangle$. The latter is given by

$$i\frac{\partial}{\partial t}|\Psi_g(t)\rangle = [H + H_{ex}(t)]|\Psi_g(t)\rangle. \tag{A.1}$$

In order to deal with the external field perturbationally, we introduce a unitary operator $U(t)$ such that

$$|\Psi_g(t)\rangle = \exp(-iHt)U(t)|\Psi_g(-\infty)\rangle, \tag{A.2}$$

where $|\Psi_g(-\infty)\rangle$ is the ground state without H_{ex}. Substituting this into the Schrödinger equation, we obtain

$$i\frac{\partial}{\partial t}U(t) = \exp(iHt)H_{ex}(t)\exp(-iHt)U(t) \equiv H_{ex}^{H}(t)U(t), \tag{A.3}$$

where the Heisenberg representation $H_{ex}^{H}(t)$ has been introduced. In solving eqn (A.3), we use the assumption (a) as the boundary condition $U(-\infty) = 1$. The iterative solution starts with replacing $U(t)$ on the right-hand side by 1 and integrating with respect to t, giving

$$U(t) = 1 - i\int_{-\infty}^{t} dt'\, H_{ex}^{H}(t') + O(H_{ex}^2). \tag{A.4}$$

Then we get

$$\begin{aligned}
\langle\Psi_g(t)|A|\Psi_g(t)\rangle &= \langle\Psi_g(-\infty)|U(t)^{\dagger}A^{H}(t)U(t)|\Psi_g(-\infty)\rangle \\
&= \langle\Psi_g(-\infty)|A^{H}(t)|\Psi_g(-\infty)\rangle \\
&\quad -i\int_{-\infty}^{t} dt'\,\langle\Psi_g(-\infty)|[A^{H}(t), H_{ex}^{H}(t')]|\Psi_g(-\infty)\rangle + O(H_{ex}^2).
\end{aligned}$$
$$\tag{A.5}$$
$$\tag{A.6}$$

At finite temperature $T = \beta^{-1}$, the system is in thermal equilibrium at $t = -\infty$ characterized by the density operator $\exp(-\beta H)/Z$. By the assumption (b), the distribution function at any later time is given by the same density operator. Then the quantity $\delta\langle A^H(t)\rangle$, which describes the deviation of $\langle A^H(t)\rangle$ from equilibrium, is given by

$$\delta\langle A^H(t)\rangle = -i \int_{-\infty}^{t} dt' \, \langle [A^H(t), H_{ex}^H(t')] \rangle, \tag{A.7}$$

within linear order in H_{ex}. Equation (A.7) is called the Kubo formula of the linear response theory.

Let us write $H_{ex}^H(t')$ in terms of an external field $\phi \exp(-i\omega t')$ coupled with an operator $B(t')$ as $H_{ex}^H(t') = -B(t')\phi \exp(-i\omega t')$. Then eqn (A.7) gives

$$\delta\langle A^H(t)\rangle = \chi_{AB}(\omega) \exp(-i\omega t)\phi, \tag{A.8}$$

where the dynamical susceptibility $\chi_{AB}(\omega)$ is given by

$$\chi_{AB}(\omega) = i \int_{-\infty}^{t} dt' \, \langle [A^H(t), B(t')] \rangle \exp[i\omega(t - t')]. \tag{A.9}$$

In the special case of the magnetic response against the external magnetic field with frequency ω and wavenumber q, we obtain eqn (1.43). The superscript H has been omitted there.

If the external field is static, the susceptibility is obtained simply by putting $\omega = 0$. The resultant static response corresponds to the adiabatic susceptibility, as is clear by the derivation. If the system is always in contact with the heat bath, on the other hand, the static response should be isothermal. In order to obtain the isothermal susceptibility, we write

$$\exp[-\beta(H + H_{ex})] = \exp(-\beta H)U_T(\beta) \tag{A.10}$$

and derive $U_T(\beta)$ up to first order in the static perturbation H_{ex}. This is analogous to eqn (A.3), except for the boundary condition $U_T(0) = 1$ and the use of the imaginary time $i\lambda$ with $0 < \lambda < \beta$ in the integral. The result for the magnetic susceptibility is given by eqn (1.57) in Chapter 1.

APPENDIX B

SPECTRAL REPRESENTATION AND FLUCTUATION–DISSIPATION THEOREM

The retarded Green function with two physical observables A and B is defined as follows:

$$\langle [A, B] \rangle (z) \equiv -i \int_0^\infty dt \, \langle [A(t), B] \rangle e^{izt}, \tag{B.1}$$

where z lies in the upper half plane. As a special case, one obtains the dynamical susceptibility as

$$\chi_{\mu\nu}(\boldsymbol{q}, \omega) = -\langle [M_\mu(\boldsymbol{q}), M_\nu(-\boldsymbol{q})] \rangle (\omega + i\delta). \tag{B.2}$$

It is clear that $\langle [A, B] \rangle (z)$ is analytic in the upper half plane of z since it is finite and is derivable any number of times. This analyticity comes from the positive integration range and it means the causality principle that the response cannot precede the external perturbation is satisfied. In terms of the many-body eigenstates $|n\rangle$ and $|m\rangle$ with energies E_n and E_m, one can evaluate eqn (B.1) as

$$\langle [A, B] \rangle (z) = \mathrm{Av}_n \sum_m \frac{A_{nm} B_{mn}}{z - \omega_{mn}} [1 - \exp(-\beta \omega_{mn})], \tag{B.3}$$

where Av_n is the thermal average over n and $\omega_{mn} \equiv E_m - E_n$. Note that $\langle [A, B] \rangle (z)$ can be extended also to the lower half plane as an analytic function of z by eqn (B.3). This piece is in fact the advanced Green function defined by

$$\langle [A, B] \rangle (z) \equiv i \int_{-\infty}^0 dt \, \langle [A(t), B] \rangle e^{izt}, \tag{B.4}$$

the integral of which converges for z in the lower half plane. Finally we see that singularities of $\langle [A, B] \rangle (z)$ as defined by eqn (B.3) lie only on the real axis.

We introduce the spectral function $I_{AB}(\omega)$ by

$$I_{AB}(\omega) = \mathrm{Av}_n \sum_m A_{nm} B_{mn} [1 - \exp(-\beta\omega)] \delta(\omega - \omega_{mn})$$

$$= \int_{-\infty}^\infty \frac{dt}{2\pi} e^{i\omega t} \langle [A(t), B] \rangle. \tag{B.5}$$

In terms of $I_{AB}(\omega)$, the retarded Green function is expressed as

$$\langle [A, B] \rangle (z) = \int_{-\infty}^\infty d\omega \frac{I_{AB}(\omega)}{z - \omega}, \tag{B.6}$$

which is called the spectral representation or the Lehmann representation.

One then introduces the relaxation function (or the canonical correlation function) by

$$\langle A(t); B \rangle \equiv \frac{1}{\beta} \int_0^\beta d\lambda \, \langle A(t - i\lambda)B \rangle, \tag{B.7}$$

and its Laplace transform

$$\langle A; B \rangle(z) \equiv \int_0^\infty dt \, \langle A(t); B \rangle e^{izt}. \tag{B.8}$$

This quantity gives the time development $\langle A(t) \rangle_{\mathrm{ex}}$ when a constant external field coupled with B is suddenly switched off at $t = 0$, hence the name 'relaxation function'. Note that $\langle A(t = 0); B \rangle$ is the static susceptibility, and that the statistical operator $\exp[-\beta(H + H_{\mathrm{ex}})]/Z$ as occurring in eqn (A.10) remains the same for $\langle A(t > 0) \rangle_{\mathrm{ex}}$. In terms of the spectral function, the relaxation function is given by

$$\langle A(t); B \rangle = \int_{-\infty}^\infty d\omega \frac{I_{AB}(\omega)}{\beta\omega} e^{-i\omega t}, \tag{B.9}$$

as can be checked by taking the matrix elements with respect to the eigenstates. Hence, we obtain a relation to the Green function:

$$\langle A; B \rangle(z) = (iz\beta)^{-1} \{ \langle [A, B] \rangle(z) - \langle [A, B] \rangle(0) \}. \tag{B.10}$$

Let $\dot{B}$ denote the time derivative of the operator B. The spectral function $I_{A\dot{B}}(\omega)$ is related to $I_{AB}(\omega)$ by $I_{A\dot{B}}(\omega) = i\omega I_{AB}(\omega)$, which is easily checked using eqn (B.5). Then the following identities can be proved:

$$\beta \langle A; \dot{B} \rangle(z) = -\langle [A, B] \rangle(z), \tag{B.11}$$

$$\beta \langle A(t); \dot{B} \rangle = i \langle [A(t), B] \rangle. \tag{B.12}$$

Finally, one introduces the symmetrized correlation function by

$$\langle \{A(t), B\} \rangle \equiv \langle A(t)B + BA(t) \rangle. \tag{B.13}$$

By taking matrix elements as in eqn (B.1), we obtain

$$\langle \{A(t), B\} \rangle = \int_{-\infty}^\infty d\omega \, \coth\left(\frac{\beta\omega}{2}\right) I_{AB}(\omega) \exp(-i\omega t), \tag{B.14}$$

where eqn (B.5) has been used. The case $B = A^\dagger$ is the most interesting physically. The spectral function is then a real positive quantity for positive ω and satisfies

$$I_{AA^\dagger}(\omega) = -I_{AA^\dagger}(-\omega) = -\frac{1}{\pi} \mathrm{Im} \, \langle [A, A^\dagger] \rangle(\omega + i\delta). \tag{B.15}$$

In the case when $t = 0$ and $B = A^\dagger$, the left-hand side of eqn (B.14) describes the fluctuation of the quantity A, while $I_{AA^\dagger}$ on the right-hand side describes the dissipation of energy by the linear response. Hence, eqn (B.14) is called the fluctuation–dissipation theorem.

APPENDIX C

RAYLEIGH–SCHRÖDINGER PERTURBATION THEORY AND HIGHER-ORDER RENORMALIZATION

C.1 Expansion of the effective Hamiltonian

As discussed in Chapter 1, the effective Hamiltonian H_{eff} in the Brillouin–Wigner perturbation theory depends on the eigenenergy to be derived. This dependence becomes inconvenient for direct perturbation theory in higher orders. The alternative scheme, called the Rayleigh–Schrödinger (RS) perturbation theory, is designed to give Ω, and hence H_{eff}, in a form which involves only the eigenenergies of H_0. Thus, the RS perturbation theory is free from the unknown energy E_i. In order to derive H_{eff} in the RS perturbation theory, we write the Schrödinger equation in the form

$$(E - H_0)\psi = E\psi - H_0\Omega P\psi = V\Omega P\psi, \tag{C.1}$$

where the index i specifying an eigenstate is omitted, and we have used the property $\Omega P\psi = \psi$ for an eigenstate of $H_0 + V$. Applying the projection operator P on both sides of eqn (C.1), and further applying Ω, we obtain the alternative form

$$\Omega(E - H_0)P\psi = E\psi - \Omega H_0 P\psi = \Omega P V\Omega P\psi, \tag{C.2}$$

where we used the property $PH_0 = H_0 P$. Subtracting eqn (C.2) from eqn (C.1), we obtain

$$[\Omega, H_0]P\psi = (1 - \Omega P)V\Omega P\psi. \tag{C.3}$$

We make the power series expansion:

$$\Omega = \Omega_0 + \Omega_1 + \Omega_2 + \cdots, \tag{C.4}$$

where Ω_n denotes the $O(V^n)$ contribution, with $\Omega_0 = 1$. Comparing terms with the same order of magnitudes on both sides of eqn (C.3), we obtain

$$[\Omega_n, H_0] = QV\Omega_{n-1} - \sum_{j=1}^{n-1} \Omega_j P V\Omega_{n-j-1}. \tag{C.5}$$

Then the effective Hamiltonian is also expanded as

$$H_{\mathrm{eff}} = P(H_0 + V)P + H_2 + H_3 + \cdots, \tag{C.6}$$

with $H_n = PV\Omega_{n-1}P$ for $n \geq 2$. The matrix elements of some lower-order parts are given explicitly by

$$\langle a|H_2|b\rangle = \langle a|V(\epsilon_b - H_0)^{-1}QV|b\rangle, \tag{C.7}$$

and

$$\langle a|H_3|b\rangle = \left\langle a \left| V \frac{1}{\epsilon_b - H_0} QV \frac{1}{\epsilon_b - H_0} QV \right| b \right\rangle$$
$$- \sum_c \left\langle a \left| V \frac{1}{\epsilon_b - H_0} \frac{1}{\epsilon_c - H_0} QV \right| c \right\rangle \langle c|V|b\rangle, \tag{C.8}$$

where $Q = 1 - P$ and all states represented by $|a\rangle$, $|b\rangle$, and $|c\rangle$ belong to the model space.

In the RS perturbation theory, there are two kinds of terms in H_n with $n \geq 3$; the first one is of the same form as the one in the Brillouin–Wigner perturbation theory (except for the eigenenergy in the denominator), and the second constitutes the correction terms which originate from the second term in eqn (C.5) or in eqn (C.8). The latter terms play an important role in accomplishing the linked-cluster expansion. Thus, by removing the restriction that the number of local electron is unity at any stage of the intermediate states, one can apply a variant of Wick's theorem to prove the cancellation of unlinked parts in the RS perturbation theory [2]. The final form of the effective Hamiltonian conserves the number of local electrons in each order of perturbation. Hence, in the model space with one and only one local electron which obeys either the Fermi or the Bose statistics, one can reproduce the actual situation in the Kondo-like impurity model with the constraint on the local number.

In order to deal with the energy denominator in the second term of eqn (C.8), the concept of the 'folded diagram' is useful. Thus, one draws a local electron line propagating from left to right in order to represent the presence of P in the intermediate state. An example is shown in Fig. C.1. Then the energy denominator is associated with the difference of the left-going energies and the right-going ones as in the other diagrams, e.g. Fig. C.2. By comparing the energy denominators given by eqn (C.8), and those given by the rule mentioned above, one can confirm that both give identical results.

C.2   Renormalization of the Kondo model

In Chapter 2, we described the lowest-order renormalization of the Kondo model, where the fixed point of the model corresponds to infinity of the exchange coupling. Because of the infinite coupling constant, all higher-order terms become large and there is no analytical way to control the renormalization. However, in the multi-channel Kondo model given by eqn (2.102) with a large number of the orbitals n, the third-order renormalization describes the nontrivial fixed point correctly [3]. The reason for this is explained below.

In order to perform renormalization explicitly, we first introduce the approximation to set $\epsilon_b = 0$ in eqn (C.8). This approximation is equivalent to neglecting the energy dependence of the effective interaction. Then, in the third-order, we deal with effective Hamiltonian given by

$$H_3^L = V(H_0^{-1}QV)^2 + \sum_c VH_0^{-1}(\epsilon_c - H_0)^{-1}QVPV. \tag{C.9}$$

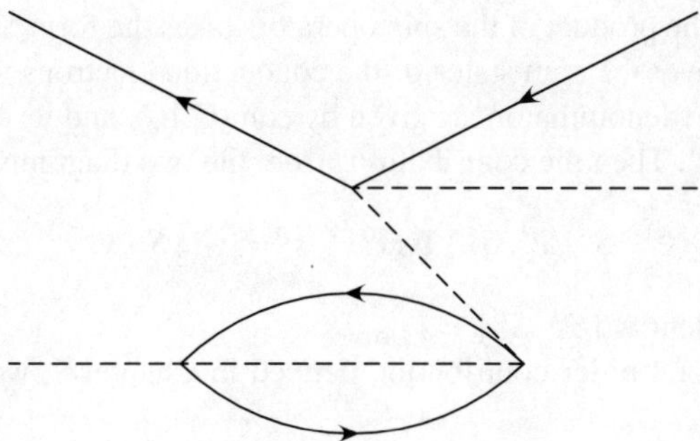

FIG. C.1. The third-order folded diagram. The assignment of energy denominators are explained in the text.

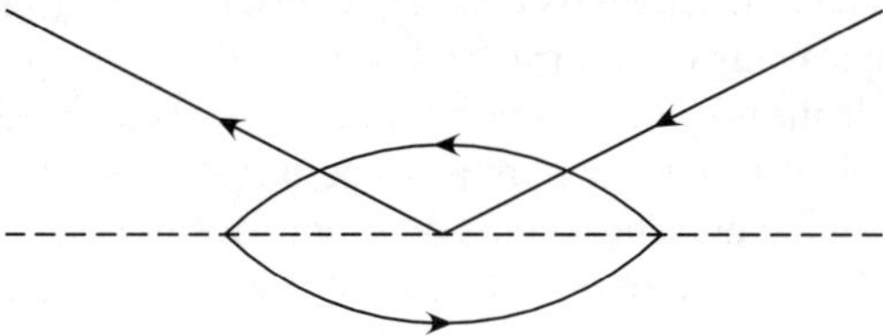

FIG. C.2. An exchange scattering diagram in the third order. The solid line shows a conduction-electron state, while the dashed line the local electron state making up the impurity spin. The projection operator Q requires one of the two conduction-electron states resulting in the loop to have energies near the band edges.

Let us assume a constant density of states for n conduction bands each of which, per spin, is given by

$$\rho_c(\epsilon) = (2D)^{-1}$$

for $|\epsilon| < D$ and zero otherwise. The model space consists of the local spin and such conduction electrons whose energy ϵ is within the range $[-D + |dD|,\ D - |dD|]$, where $dD\ (< 0)$ is the infinitesimal change of the cut off energy. In order to minimize the number of electrons to deal with, the Fermi sea is chosen as the vacuum for the conduction electrons, and the excitations are described as particles and holes generated from the vacuum.

In the third-order effective Hamiltonian, we have to consider only two diagrams shown in Figs. C.1 and C.2. It can be shown by using eqn (C.9) that other diagrams without a loop of conduction electrons describe a shift of the ground-state energy, and do not contribute to renormalization [4]. In Fig. C.2, the energy denominators are obtained by associating the excitation energy D with one of the two lines in the electron loop. The result of energy integration is given by

$$\int_{-D}^{0} d\epsilon' \frac{1}{(-D + \epsilon')^2} + \int_{0}^{D} d\epsilon \frac{1}{(-D - \epsilon)^2} = \frac{1}{D}. \tag{C.10}$$

On the other hand, the product of the spin operators takes the form $S^\alpha S^\beta S^\gamma \operatorname{Tr}(s^\alpha s^\gamma)s^\beta$, where the trace is over the spin states of the conduction electrons. The folded diagram has the same energy denominators as given by eqn (C.10), and its spin part is given by $S^\alpha S^\gamma S^\beta \operatorname{Tr}(s^\alpha s^\gamma)s^\beta$. Then the contributions from the two diagrams combine to give

$$S^\alpha[S^\beta, S^\gamma]\operatorname{Tr}(s^\alpha s^\gamma)s^\beta = -\tfrac{1}{2}\mathbf{S}\cdot\mathbf{s}, \tag{C.11}$$

where we use the identity $[S^\alpha, S^\beta] = i\epsilon_{\alpha\beta\gamma}S^\gamma$.

Adding the second-order contribution derived in Chapter 2, we obtain the scaling equation:

$$\frac{dg}{dl} = -g^2 + \tfrac{1}{2}g^3, \tag{C.12}$$

where $l = \ln D$ and the dimensionless coupling constant $g = J\rho_c$ has been introduced. Note that the present scheme of renormalization extends the the poor man's scaling [5] with a different idea. In the poor man's scaling, one considers the t-matrix of the exchange scattering which involves an infinite number of perturbation terms. On the contrary, the effective interaction in the present scheme does not have intermediate states belonging to the model space. In the case of an n-fold degenerate conduction band, each loop of conduction-electron lines acquires the factor n by summing over the degenerate orbitals. This loop appears from the third-order diagrams as shown in Figs C.1 and C.2. Instead of eqn (C.12), we then obtain

$$\frac{dg}{dl} = -g^2 + \frac{n}{2}g^3. \tag{C.13}$$

The fixed point of the renormalization group is given by $g = g_c = 2/n$, which is small in the case of large n. Thus, the fixed point is within the reach of the perturbative renormalization. If one considers still higher-order diagrams, one would have the expansion [6,7]:

$$\frac{dg}{dl} = -g^2 + \frac{n}{2}g^3 + (an + b)g^4 + (cn^2 + dn + e)g^5 + \cdots, \tag{C.14}$$

where a–e are numerical constants independent of n. Since the total contribution from the third- and higher-order terms on the right-hand side is of $O(1/n^3)$ with $g = O(1/n)$, the first two terms of $O(1/n^2)$ are sufficient in deriving the fixed point in the leading order of $1/n$.

APPENDIX D

SPECTRAL SHAPE AND RELAXATION RATE

We introduce an inner product $\langle A|B\rangle$ between the two operators A and B by

$$\langle A|B\rangle = \int_0^\beta d\tau \, \langle \exp(\tau H)A \exp(-\tau H)B\rangle \equiv \beta\langle A; B\rangle, \qquad (\text{D.1})$$

where the canonical correlation function $\langle A; B\rangle$ has been defined by eqn (B.7). The inner product corresponds to the static susceptibility χ_{AB}. By introducing the Liouville operator $\mathcal{L}$ as $\mathcal{L}A \equiv [H, A]$, a simple calculation shows that

$$\langle A|\mathcal{L}|B\rangle \equiv \langle A|\mathcal{L}B\rangle = \langle \mathcal{L}A|B\rangle. \qquad (\text{D.2})$$

Hence, $\mathcal{L}$ is Hermitian with respect to this inner product. The dynamical susceptibility defined by eqn (A.9) is written as

$$\chi_{AB}(z) = \left\langle A \left| \frac{\mathcal{L}}{\mathcal{L} - z} \right| B \right\rangle. \qquad (\text{D.3})$$

The relaxation function defined by eqn (B.8) is represented by

$$\langle A; B\rangle(z) = \left\langle A \left| \frac{i}{z - \mathcal{L}} \right| B \right\rangle \equiv C_{AB}(z). \qquad (\text{D.4})$$

We now consider the dynamical magnetic susceptibility $\chi_M(z)$ by setting $A = B = M$ in eqn (D.3), with M being the z-component of magnetic moment. The projection operator $\mathcal{P}$ is defined by

$$\mathcal{P} \equiv |M\rangle \chi_M^{-1} \langle M|, \qquad (\text{D.5})$$

where χ_M is the static susceptibility given by $\chi_M = \langle M|M\rangle$. It is obvious that $\mathcal{P}^2 = \mathcal{P} = \mathcal{P}^\dagger$, which is a necessary condition for a projection operator. We also introduce a projection operator $\mathcal{Q}$ complementary to $\mathcal{P}$ by $\mathcal{Q} = 1 - \mathcal{P}$. In order to obtain the relaxation function $C_M(z)$, we have to invert the operator $z - \mathcal{L}$. We note the following identity for the 2×2 matrix:

$$\left[\begin{pmatrix} a & b^* \\ b & c \end{pmatrix}^{-1} \right]_{11} = \frac{1}{a - b^* c^{-1} b}, \qquad (\text{D.6})$$

which can be generalized to the case where the entries c and b are a matrix and a column vector, respectively. Thus, we identify

$$a = \mathcal{P}(z - \mathcal{L})\mathcal{P}, \qquad b = -\mathcal{Q}\mathcal{L}\mathcal{P}, \qquad c = \mathcal{Q}(z - \mathcal{L})\mathcal{Q}. \qquad (\text{D.7})$$

Then we obtain the following result [8]:

$$C_M(z) = i\chi_M[z - \Omega + i\Gamma(z)]^{-1}, \tag{D.8}$$

where

$$\Omega = \chi_M^{-1}\langle M|\mathcal{L}|M\rangle \tag{D.9}$$

describes the systematic motion corresponding to precession. The effect of the random force is described by $\Gamma(z)$, which is given by

$$\Gamma(z) = \chi_M^{-1}\left\langle Q\dot{M}\left|\frac{i}{z - QLQ}\right|Q\dot{M}\right\rangle, \tag{D.10}$$

where $\dot{M} = i\mathcal{L}M$.

This exact expression is very useful for evaluating the relaxation rates in the lowest-order perturbation theory. Let us assume that the zeroth-order Hamiltonian H_0 of the total system $H = H_0 + H_1$ conserves M. Then, up to $O(H_1^2)$, we can approximate

$$\Gamma(z) = \chi_M^{-1}\langle [H_1, M]|i(z - \mathcal{L}_0)^{-1}|[H_1, M]\rangle + O(H_1^2), \tag{D.11}$$

where $\mathcal{L}_0$ is the Liouville operator corresponding to H_0. This result can be derived by noting that $\langle M|\dot{M}\rangle = \langle [M, M]\rangle = 0$ and $Q\dot{M} = \dot{M}$. As a physical relaxation function, we put $z = \omega + i\delta$. The imaginary part of $\Gamma(\omega)$ describes a shift in the resonance frequency. The real part, on the other hand, corresponds to the magnetic relaxation rate. The lowest-order result for it is given by

$$\Gamma(\omega) = \pi\chi_M^{-1}\langle [H_1, M]|\delta(\omega - \mathcal{L}_0)|[H_1, M]\rangle. \tag{D.12}$$

Usually, we have $\mathcal{L}_0\dot{M} \neq 0$; in this case, $\Gamma(z)$ is a smooth function of $z = \omega + i\delta$ around $\omega = 0$ in the upper half plane. As long as ω is much smaller than the characteristic frequency of $\mathcal{L}_0\dot{M}$, we may put $\omega = 0$ and neglect $\mathrm{Im}\,\Gamma$. In this case, the spectral shape takes the Lorentzian form

$$\frac{1}{\omega}\mathrm{Im}\,\chi_M(\omega) = \frac{\chi_M\Gamma}{\omega^2 + \Gamma^2}. \tag{D.13}$$

It is straightforward to generalize the above formalism to the case where a set of operators A_i ($i = 1, 2, \ldots$) constitute slow variables. The projection operator to this set is defined by

$$\mathcal{P} \equiv \sum_{ij} |A_i\rangle\chi^{-1}\langle A_j|, \tag{D.14}$$

where χ is the susceptibility matrix with elements $\chi_{ij} = \langle A_i|A_j\rangle$. Then we obtain the following identity for the relaxation function matrix $C(z)$:

$$C(z) = i\chi[z - \Omega + i\Gamma(z)]^{-1}, \tag{D.15}$$

where the elements of the matrix Ω are given by

$$\Omega_{ij} = \sum_l (\chi^{-1})_{il}\langle A_l|\mathcal{L}|A_j\rangle, \tag{D.16}$$

and those of $\Gamma(z)$ by

$$\Gamma_{ij}(z) = \sum_l (\chi^{-1})_{il} \left\langle Q\dot{A}_l \left| \frac{i}{z - Q\mathcal{L}Q} \right| Q\dot{A}_j \right\rangle. \tag{D.17}$$

We now turn to a general spectral shape of $\chi_M(z)$ including non-Lorentzian cases. The relaxation function in the real time domain is written as

$$\langle M(t)|M\rangle = \langle M|\exp(-i\mathcal{L}t)|M\rangle \equiv \langle\langle\exp(-i\mathcal{L}t)\rangle\rangle\langle M|M\rangle, \tag{D.18}$$

where we have introduced the average $\langle\langle\cdots\rangle\rangle$. The spectral moments $\langle\omega^n\rangle$ ($n = 0, 1, 2, \ldots$) are given by

$$\langle\omega^n\rangle \equiv \int_{-\infty}^{\infty} \frac{d\omega}{\pi} \omega^{n-1}\mathrm{Im}\,\frac{\chi_M(\omega)}{\chi_M} = \langle\langle\mathcal{L}^n\rangle\rangle. \tag{D.19}$$

Thus, the full set of moments completely determines the spectrum. Corresponding to the average $\langle\langle\cdots\rangle\rangle$ we can define the cumulant average $\langle\langle\cdots\rangle\rangle_c$ [9]:

$$\langle\langle\exp(-i\mathcal{L}t)\rangle\rangle = \exp[\langle\langle\exp(-i\mathcal{L}t)\rangle\rangle_c - 1] \equiv \exp[-i\langle\omega\rangle t + X(t)]. \tag{D.20}$$

In eqn (D.20), $X(t)$ is determined by the set of cumulants $\langle\omega^n\rangle_c \equiv \langle\langle\mathcal{L}^n\rangle\rangle_c$ as

$$X(t) = \sum_{n=2}^{\infty} \frac{(-it)^n}{n!}\langle\omega^n\rangle_c. \tag{D.21}$$

Thus, the full set of cumulants also determines the spectrum completely. The particularly important quantity is the second-order cumulant

$$\langle\omega^2\rangle_c = \langle\langle\mathcal{L}^2\rangle\rangle - \langle\langle\mathcal{L}\rangle\rangle^2. \tag{D.22}$$

The function $X(t)$ starts from zero at $t = 0$ and should decrease with increasing t. We introduce a characteristic time τ_R in $X(t)$ such that $\mathrm{Re}\,X(\tau_R) \sim -1$. Then the relaxation function becomes negligible for $t \gg \tau_R$. For large enough t, on the other hand, $X(t)$ should behave as

$$X(t) \to -i(\delta\omega - i\Gamma)t, \tag{D.23}$$

with $\delta\omega$ being the shift of the resonance frequency. The long-time behaviour is nonanalytic in t because of the dissipation in the system. Hence, this behaviour does not follow from the first-order term in eqn (D.21). A characteristic time τ_C is defined as the one after which the short-time behaviour crosses over to the long-time one given by eqn (D.23). We consider the extreme cases: (i) $\tau_C \gg \tau_R$ and (ii) $\tau_C \ll \tau_R$, assuming $\langle\omega\rangle = \delta\omega = 0$ for simplicity. In the case (i), the short-time behaviour of $X(t)$ determines the spectrum. Then we can rely on the expansion given by eqn (D.21). The resultant spectrum is of the Gaussian shape given by

$$\mathrm{Im}\,\frac{\chi_M(\omega)}{\omega} = \sqrt{\frac{\pi}{2}}\frac{\chi_M}{D}\exp\left[-\frac{1}{2}\left(\frac{\omega}{D}\right)^2\right],\tag{D.24}$$

where

$$D^2 = \chi_M^{-1}\langle \dot{M}|\dot{M}\rangle,\tag{D.25}$$

and we have $D \sim \tau_R^{-1}$. It is possible to apply the perturbation theory to derive D. The lowest-order result is used in estimating T_{2G} in Chapter 1. In the case (ii), on the contrary, the long-time behaviour dominates the spectrum, which gives rise to the Lorentzian shape. The relaxation rate in the lowest-order perturbation theory is given by eqn (D.12) with $\omega = 0$, which is given approximately by $\Gamma \sim D^2\tau_C \sim \tau_C/\tau_R^2$. The relaxation rate is smaller than in the case (i) with the same magnitude of D. According to specific mechanism which controls τ_C, this reduction of the relaxation rate is called motional narrowing or exchange narrowing.

APPENDIX E

GREEN FUNCTION IN THE IMAGINARY TIME

We introduce the Matsubara representation by

$$e^{\tau H} A e^{-\tau H} \equiv A^{(M)}(\tau),$$

(E.1)

with $-\beta < \tau < \beta$. This corresponds to the Heisenberg representation in the imaginary time τ. Then the Matsubara Green function $D_{AB}[\tau]$ is defined by

$$D_{AB}[\tau] = -\langle T_\tau A^{(M)}(\tau) B \rangle = \begin{cases} -\langle A^{(M)}(\tau) B \rangle & (\tau > 0), \\ -\langle B A^{(M)}(\tau) \rangle & (\tau < 0), \end{cases}$$

(E.2)

where T_τ is the time-ordering symbol. If A and B are generalized to fermion operators, the Green function becomes $+\langle B A^{(M)}(\tau) \rangle$ for $\tau < 0$, as discussed later.

By taking matrix elements, one can show that the Green function is periodic in τ with period β, i.e.

$$D_{AB}[\tau] = -\int_{-\infty}^{\infty} d\omega \, I_{AB}(\omega) \frac{e^{-(\beta+\tau)\omega}}{1 - e^{\beta\omega}} = D_{AB}[\tau + \beta],$$

(E.3)

with $-\beta < \tau < 0$. Therefore, one can carry out a Fourier transform

$$D_{AB}[\tau] = T \sum_n D_{AB}(i\nu_n) \exp(-i\nu_n \tau),$$

(E.4)

where $i\nu_n = 2\pi i n T$, with n integer, is the (even) Matsubara frequency. The Fourier component $D_{AB}(i\nu_n)$ is calculated as

$$D_{AB}(i\nu_n) = \int_0^\beta d\tau \, D_{AB}[\tau] \exp(i\nu_n \tau) = \int_{-\infty}^{\infty} d\omega \frac{I_{AB}(\omega)}{i\nu_n - \omega}.$$

(E.5)

Thus, we obtain a remarkable relation

$$D_{AB}(i\nu_n) = \langle [A, B] \rangle (i\nu_n).$$

(E.6)

This means that the Matsubara Green function with $\nu_n > 0$ is continued analytically to the retarded Green function, and with $\nu_n < 0$ to the advanced Green function [10].

In the case of fermionic operators ψ_A and $\psi_B^\dagger$, we define the Matsubara Green function $G_{AB}[\tau]$ as follows:

$$G_{AB}[\tau] = -\langle T_\tau \psi_A^{(M)}(\tau) \psi_B^\dagger \rangle = \begin{cases} -\langle \psi_A^{(M)}(\tau) \psi_B^\dagger \rangle & (\tau > 0), \\ \langle \psi_B^\dagger \psi_A^{(M)}(\tau) \rangle & (\tau < 0). \end{cases}$$

(E.7)

The Green function now has the antiperiodic property, with $-\beta < \tau < 0$,

$$G_{AB}[\tau] = -\int_{-\infty}^{\infty} d\epsilon \, \rho_{AB}(\epsilon) \frac{e^{-(\beta+\tau)\omega}}{1 + e^{\beta\omega}} = -G_{AB}[\tau + \beta], \qquad (E.8)$$

where the spectral function $\rho_{AB}(\epsilon)$ is defined by

$$\rho_{AB}(\epsilon) = \mathrm{Av}_n \sum_m (\psi_A)_{nm}(\psi_B^\dagger)_{mn}[1 + \exp(-\beta\epsilon)]\delta(\epsilon - \epsilon_{mn}). \qquad (E.9)$$

The Fourier transform $G_{AB}(i\epsilon_n)$ is now with odd Matsubara frequency $i\epsilon_n = \pi i (2n + 1)T$ with n integer, and has the spectral representation

$$G_{AB}(i\epsilon_n) = \int_0^\beta G_{AB}[\tau]\exp(i\epsilon_n\tau) = \int_{-\infty}^{\infty} d\epsilon \frac{\rho_{AB}(\epsilon)}{i\epsilon_n - \epsilon}. \qquad (E.10)$$

The analytic continuation $G_{AB}(z)$, with z in the upper half-plane, corresponds to the retarded Green function, i.e.

$$G_{AB}(z) = -i\int_0^\infty dt \, e^{izt} \langle\{\psi_A(t), \psi_B^\dagger\}\rangle, \qquad (E.11)$$

with the anticommutator $\{\cdots\}$. If z is in the lower half plane, $G_{AB}(z)$ gives the advanced Green function

$$G_{AB}(z) = +i\int_{-\infty}^0 dt \, e^{izt} \langle\{\psi_A(t), \psi_B^\dagger\}\rangle. \qquad (E.12)$$

The spectral intensity $\rho_{AB}(\epsilon)$ can be given also in terms of an integral,

$$\rho_{AB}(\epsilon) = \int_{-\infty}^\infty dt \, e^{i\epsilon t} \langle\{\psi_A(t), \psi_B^\dagger\}\rangle. \qquad (E.13)$$

APPENDIX F

PATH INTEGRAL REPRESENTATION OF THE PARTITION FUNCTION

F.1 Grassmann numbers and coherent states

The basic property of the Grassmann numbers is that they anticommute, in contrast to the c-numbers. Let η_1 and η_2 be Grassmann numbers. Then we require

$$\eta_1\eta_2 = -\eta_2\eta_1. \tag{F.1}$$

We also require that the Grassmann numbers commute with any c-number and anticommute with fermion creation and annihilation operators. The complex conjugate has the following property:

$$(\eta_1\eta_2)^* = \eta_2^*\eta_1^* = -\eta_1^*\eta_2^*, \tag{F.2}$$

which may appear to be strange if Grassmann numbers are real, but assures the reality condition $(\eta_1^*\eta_1)^* = \eta_1^*\eta_1$.

The derivative of Grassmann numbers is defined by

$$\frac{\partial}{\partial\eta_i}\eta_j = \delta_{ij}, \tag{F.3}$$

which is analogous to the usual derivative. We note, however, that the differential $d\eta_i$ is also a Grassmann number. Then we introduce the definite integrals $\int d\eta\, 1$ and $\int d\eta\, \eta$ which are sufficient for the single variable because higher powers of η vanish. The first integral is the Grassmann number, while the second commutes with all other Grassmann numbers and, hence, is a c-number. In order to make the two integrals invariant against the linear shift $\eta \to \eta + \xi$, with ξ a constant Grassmann number, we require

$$\int d\eta\, \eta = 1, \qquad \int d\eta\, 1 = 0. \tag{F.4}$$

The first equality sets the normalization of the integral. The second one, where 0 is a Grassmann number, follows from the invariance.

Let us take a fermionic single-particle state $|1\rangle = f^\dagger|0\rangle$, where $f^\dagger$ denotes the creation operator. The coherent state $|\eta\rangle$ associated with a complex Grassmann number η is constructed by

$$|\eta\rangle = \exp(f^\dagger\eta)|0\rangle = (1 + f^\dagger\eta)|0\rangle = |0\rangle - \eta|1\rangle. \tag{F.5}$$

The coherent state is a superposition of zero- and single-particle states. It is the eigenstate of the annihilation operator f. In fact, the action of f upon $|\eta\rangle$ gives

$$f|\eta\rangle = \eta|\eta\rangle. \tag{F.6}$$

The Hermitian conjugate of this equation is $\langle\eta| = \langle\eta|f^\dagger$. Furthermore, the inner product of the coherent states is given by

$$\langle\eta_1|\eta_2\rangle = \exp(\eta_1^*\eta_2). \tag{F.7}$$

Thus, we obtain for an operator function $F(f^\dagger, f)$ which is f or $f^\dagger$ or $f^\dagger f$:

$$\langle\eta_1|F(f^\dagger, f)|\eta_2\rangle = F(\eta_1^*, \eta_2)\exp(\eta_1^*\eta_2). \tag{F.8}$$

The following relation is the most important one:

$$\int d\eta^* \, d\eta \, |\eta\rangle\langle\eta| \exp(-\eta^*\eta) = |0\rangle\langle0| + |1\rangle\langle1| = \mathbf{1}, \tag{F.9}$$

which is the identity operator. It is used in the next section to set up the path integral representation of the partition function. It is clear that the coherent states span the overcomplete set for $|0\rangle$ and $|1\rangle$.

F.2   Partition function

As an illustrative example of using the Grassmann integration technique, we begin with a trivial Hamiltonian $H(f^\dagger, f) = \epsilon f^\dagger f$ for which the (grand) partition function Z is given by $Z = 1 + \exp(-\beta\epsilon)$. Here the chemical potential is chosen as the origin of energy.

We consider the partition function for a Hamiltonian H and represent the trace by the integral over the coherent states

$$Z = \int d\eta_1^* \, d\eta_1 \, \langle\eta_1|e^{-\beta H}| - \eta_1\rangle \exp(-\eta_1^*\eta_1), \tag{F.10}$$

where we have used eqn (F.9) and $\langle n|\eta\rangle\langle\eta|e^{-\beta H}|n\rangle = \langle\eta|e^{-\beta H}|n\rangle\langle n| - \eta\rangle$, with $n = 0$ or 1. We decompose $\exp(-\beta H) \sim (1 - \delta\tau H/M)^M$, with $\delta\tau = \beta/M$ for a large integer M $(\gg 1)$ and insert the identity given by eqn (F.9) $(M - 1)$ times between each decomposed factor. This is often called the Suzuki–Trotter decomposition. Then we obtain, using eqn (F.8),

$$Z = \int \prod_{i=1}^{M} d\eta_i^* \, d\eta_i \, R(\eta_1^*, \eta_2)\exp[-\eta_1^*(\eta_1 - \eta_2)]R(\eta_2^*, \eta_3) \times \cdots$$
$$\times \exp[-\eta_M^*(\eta_M - \eta_{M+1})]R(\eta_M^*, \eta_{M+1}), \tag{F.11}$$

where $R(\eta_1^*, \eta_2) = 1 - \delta\tau H(\eta_1^*, \eta_2) \sim \exp[-\delta\tau H(\eta_1^*, \eta_2)]$ and $\eta_{M+1} \equiv -\eta_1$. In the limit of large M, one may write η_i as $\eta(\tau)$ for the range $0 < \tau < \beta$ with the boundary condition $\eta(\beta) = -\eta(0)$. Then one can make a Fourier expansion of $\eta(\tau)$

using the odd Matsubara frequency. Provided the highest frequency appearing in the summation is much smaller than $(\delta\tau)^{-1}$, one may replace the difference $\eta(\tau_i) - \eta(\tau_{i+1})$ by $\delta\tau\,\partial\eta(\tau)/\partial\tau$.

It is also possible to represent the partition function with many degrees of freedom and with interparticle interactions as an integral over the Grassmann numbers. Instead of single $\eta(\tau)$, we have to introduce for each single-particle state α the corresponding Grassmann variable η_α. The multiple integrals over Grassmann numbers is called the functional integral or the path integral and is written as $\mathcal{D}\eta^*\,\mathcal{D}\eta$. The latter arises from the analogy with the path integral in single-particle quantum mechanics.

As is clear in the derivation, the Hamiltonian must have the normal-ordered form where the creation operators stand to the left of annihilation operators. Then one may extend eqn (F.8) to many variables as

$$\langle\{\eta_\alpha\}|F(\{f_\alpha^\dagger\},\{f_\alpha\})|\{\eta_\alpha'\}\rangle = F(\{\eta_\alpha^*\},\{\eta_\alpha'\})\exp\left(\sum_\alpha \eta_\alpha^*\eta_\alpha'\right). \tag{F.12}$$

The final formula for a general normal-ordered Hamiltonian turns out to be

$$Z = \int \mathcal{D}\eta^*\,\mathcal{D}\eta\,\exp(-S), \tag{F.13}$$

$$S = \int_0^\beta d\tau\left[\sum_\alpha \eta_\alpha^*\frac{\partial}{\partial\tau}\eta_\alpha + H(\tau)\right]. \tag{F.14}$$

The quantity S is often called the action and the integrand in S the Lagrangian, in analogy with the real-time formulation of the path integral.

APPENDIX G

MANY-BODY PERTURBATION THEORY

G.1 Gaussian integral over Grassmann numbers

By noting the identity $\exp(-\eta^* A\eta) = 1 - \eta^* A\eta$, with A a c-number, we obtain the simplest Gaussian integral over the Grassmann numbers η and η^* as

$$\int d\eta^* \, d\eta \, \exp(-\eta^* A\eta) = A. \tag{G.1}$$

Similarly, with ξ a Grassmann number, one can prove that

$$\int d\eta^* \, d\eta \, \exp(-\eta^* A\eta + \xi^* \eta + \eta^* \xi) = A \exp(\xi^* A^{-1}\xi), \tag{G.2}$$

either by expanding the exponential or by the change $\eta \to \eta - A^{-1}\xi$, $\eta^* \to \eta - \xi^* A^{-1}$ of integration variables. These results can be extended to the case of many variables. Namely, for a Hermitian c-number matrix A, we obtain

$$\int \prod_i d\eta_i^* \, d\eta_i \, \exp\left(-\sum_{ij} \eta_i^* A_{ij}\eta_j \right) = \det A, \tag{G.3}$$

which can be checked by diagonalizing the matrix A by a unitary transformation and recognizing that $\det A$ is the product of eigenvalues. Equation (G.3) should be compared with the usual Gaussian integration over N c-numbers. The latter gives $\pi^N \det A^{-1}$. Similarly, as the multivariable extension of eqn (G.2), we get

$$\int \prod_i d\eta_i^* \, d\eta_i \, \exp\left[-\sum_{ij} \eta_i^* A_{ij}\eta_j + \sum_i (\xi_i^* \eta_i + \eta_i^* \xi_i) \right]$$

$$= \det A \, \exp\left(-\sum_{ij} \xi_i^* G_{ij}\xi_j \right) \equiv Z(\xi^*, \xi), \tag{G.4}$$

with $G = -A^{-1}$.

We introduce the average of Grassmann numbers by the relation

$$\langle \eta_i^* \eta_j \rangle_0 \equiv \int \prod_m d\eta_m^* \, d\eta_m \, \eta_i^* \, \eta_j \, \exp\left(-\sum_{ij} \eta_i^* A_{ij}\eta_j \right) \Big/ \det A. \tag{G.5}$$

This average is easily evaluated by taking the derivative of $Z(\xi^*, \xi)$:

$$\langle \eta_i^* \eta_j \rangle_0 \equiv -\lim_{\xi^*, \xi \to 0} \frac{1}{Z} \frac{\partial^2}{\partial \xi_i \partial \xi_j^*} Z(\xi^*, \xi) = G_{ji}. \tag{G.6}$$

Thus, there is a correspondence $\eta_j \to \partial/\partial\xi_j^*$, $\eta_i^* \to -\partial/\partial\xi_i$ in taking the average. We can generalize this result to averages involving $2n$ Grassmann variables as follows:

$$\langle \eta_1^* \eta_{n+1} \eta_2^* \eta_{n+2}^* \cdots \eta_n^* \eta_{2n} \rangle_0 = \sum_P \text{sgn}\, P\, G_{n+1,P(1)} G_{n+2,P(2)} G_{2n,P(n)}, \qquad \text{(G.7)}$$

where P denotes a permutation of n variables, and the sign factor sgn P accounts for the anticommuting property. This relation is equivalent to Wick's theorem, as will be explained shortly.

We interpret the set of variables η_i as representing not only the set of N single-particle states but also the set of M imaginary times between 0 and β generated by the Suzuki–Trotter decomposition . For example, we consider the trivial case of $H = \epsilon f^\dagger f$ with $N = 1$. Then the integral for Z is evaluated as

$$Z = \det_\tau \left(\frac{\partial}{\partial \tau} + \epsilon \right), \qquad \text{(G.8)}$$

where the determinant is for the $M \times M$ matrix with $M \to \infty$. The meaning of the derivative becomes clear if we first rewrite

$$Z = \exp\left[\text{Tr}_\tau \ln \beta \left(\frac{\partial}{\partial \tau} + \epsilon \right) \right], \qquad \text{(G.9)}$$

and use the complete set of antiperiodic functions $\exp(-i\epsilon_n \tau)/\sqrt{\beta}$ in taking the trace. Multiplication of β for the argument of the logarithm is only to make the argument dimensionless. The exponent corresponds to $\ln Z = -\beta\Omega$, which becomes

$$-\beta\Omega = \sum_n \ln \beta(-i\epsilon_n + \epsilon). \qquad \text{(G.10)}$$

This sum is actually divergent and does not reproduce the correct answer $\ln[1 + \exp(-\beta\epsilon)]$. The divergence originates from the replacement of the difference in eqn (F.11) by the differential. In a practical application explained below, this difficulty is not a serious one because we can use a reference state for which the thermodynamic potential is derived without using the path integral. Then the deviation $\Delta\Omega$ from this reference state is convergent. Alternatively, one can absorb the divergence by requiring that both Ω and $\partial\Omega/\partial\epsilon = \langle f^\dagger f \rangle$ go to zero as $\epsilon \to \infty$. Then, using

$$\sum_n (i\epsilon_n - \epsilon)^{-2} = \beta^2 e^{\beta\epsilon}/(e^{\beta\epsilon} + 1)^2, \qquad \text{(G.11)}$$

and integrating twice with respect to ϵ, we obtain the correct result.

G.2   Wick's theorem

The path integral representation applies also to Green functions. By using the Heisenberg representation $f_\alpha(\tau) = \exp(\tau H) f \exp(-\tau H)$ in the imaginary time (Matsubara representation) with a general normal-ordered Hamiltonian H, one can show the equivalence of averages:

$$\langle T_\tau f_\alpha^\dagger(\tau_1) f_\beta(\tau_2)\rangle = \int \mathcal{D}\eta^* \mathcal{D}\eta \, \eta_\alpha^*(\tau_1)\eta_\beta(\tau_2) \exp(-S)/Z \equiv \langle \eta_\alpha^*(\tau_1)\eta_\beta(\tau_2)\rangle,$$

$$(G.12)$$

the left-hand side of which is the Green function. This equivalence can be proved for the case of $\tau_1 > \tau_2$ by the Trotter decomposition of

$$\exp[-(\beta - \tau_1)H]f_\alpha^\dagger \exp[-(\tau_1 - \tau_2)H]f_\beta \exp[-\tau_2 H]. \qquad (G.13)$$

The other case, $\tau_1 < \tau_2$, can be treated similarly. Note that T_τ does not appear in the average of η variables.

It is convenient to introduce the partition function $Z(\xi^*, \xi)$ in the presence of the Grassmann external fields $\xi_\alpha(\tau)^*$, $\xi_\alpha(\tau)$. They appear in the Lagrangian as $\xi^\dagger \eta + \eta^\dagger \xi$, where the variables are regarded as vectors with state and time indices. In a particular case of a noninteracting Hamiltonian $H = \sum_{\alpha\beta} h_{\alpha\beta} f_\alpha^\dagger f_\beta$, the matrix element of $-A$ is given by $i\epsilon_n \delta_{\alpha\beta} - h_{\alpha\beta}$, where we have made the Fourier transform to Matsubara frequencies. Then we can use the result of Gaussian integration explained above. For example, the inverse matrix $-A^{-1} = G$ is just the Green function matrix. In analogy with eqn (G.6), one can show that

$$\langle \eta_\alpha^*(\tau_1)\eta_\beta(\tau_2)\rangle = - \lim_{\xi^*,\xi \to 0} \frac{1}{Z} \frac{\delta^2}{\delta\xi_\alpha(\tau_1)\delta\xi_\beta^*(\tau_2)} Z(\xi^*, \xi)$$

$$= G_{\beta\alpha}(\tau_2 - \tau_1), \qquad (G.14)$$

where the last quantity denotes the Green function.

There is now a correspondence,

$$\eta_\alpha(\tau) \to \delta/\delta\xi_\alpha^*(\tau), \qquad \eta\alpha^*(\tau) \to -\delta/\delta\xi_\alpha(\tau), \qquad (G.15)$$

in taking the average. By using this relation and eqn (G.4) for $Z(\xi^*, \xi)$, one can derive the Wick's theorem as follows:

$$\langle T_\tau f^\dagger(\tau_1) f(\tau_{n+1}) \cdots f^\dagger(\tau_n) f(\tau_{2n})\rangle = \sum_P \text{sgn} \, P \, G(\tau_{n+1} - \tau_{P(1)}) \cdots G(\tau_{2n} - \tau_{P(n)}),$$

$$(G.16)$$

where the index α for single-particle states has been omitted for notational simplicity. The result at finite temperature is also called the Bloch–de Dominicis theorem. Use of the correspondence given by eqn (G.15) for the Hamiltonian with the two-body interaction $V(f^\dagger, f)$ leads to a formal representation of the partition function Z in terms of $Z_0(\xi^*, \xi)$ without V but with the external fields as [11]

$$Z = - \lim_{\xi^*,\xi \to 0} \exp\left[-\int_0^\beta d\tau \, V\left(\frac{\delta}{\delta\xi}, -\frac{\delta}{\delta\xi^*} \right) \right] Z_0(\xi^*, \xi). \qquad (G.17)$$

By expanding this expression in terms of V and using Wick's theorem, one can derive the perturbation theory in terms of the Feynman diagrams.

Alternatively, we consider the partition function without the external fields and write the action as $S_0 + S_1$, where $S_1 = \int_0^\beta d\tau\, V(\tau)$. The partition function is written as

$$Z = Z_0 \langle \exp(-S_1) \rangle_0, \qquad (G.18)$$

where Z_0 corresponds to S_0, and $\langle \cdots \rangle_0$ denotes the average in terms of the unperturbed action. Taking the logarithm of eqn (G.18), we obtain the shift Ω_1 of the thermodynamic potential as

$$\Omega_1 = -T[\langle \exp(-S_1) \rangle_c - 1], \qquad (G.19)$$

where $\langle \cdots \rangle_c$ is the unperturbed cumulant average as explained in Appendix D. In terms of the Feynman diagrams, the cumulant average eliminates all unlinked parts in each order of S_1. Hence, eqn (G.19) gives the linked-cluster expansion.

G.3 Variational property and the Luttinger–Friedel sum rule

We consider a situation where a coupling

$$\int_0^\beta d\tau \int_0^\beta d\tau'\, f_\alpha^\dagger(\tau) U_{\alpha\beta}(\tau - \tau') f_\beta(\tau') \qquad (G.20)$$

with the fictitious external field $U_{\alpha\beta}(\tau - \tau')$ is present in the system. $U_{\alpha\beta}(\tau - \tau')$ may be viewed as a Gaussian average of the product of two external fields. In the following, we use a simplified notation where Tr now means the trace over both the spatial and imaginary-time degrees of freedom.

With slight change δU in U, the thermodynamic potential Ω varies by the amount

$$\beta \delta \Omega(U) = \mathrm{Tr}(G \delta U), \qquad (G.21)$$

where G is the Green function, which is a matrix in space and time indices. We now eliminate δU in favour of the changes δG of the Green function and $\delta \Sigma$ of the self-energy matrix Σ. Using the Dyson equation $G^{-1} = g^{-1} - U - \Sigma$, where g is the bare Green function, we obtain

$$G \delta U = -G \delta(G^{-1} + \Sigma) = -\delta(\ln G^{-1} + G\Sigma) + (\delta G)\Sigma. \qquad (G.22)$$

The trace of the last term on the right-hand side is given as the change $\delta \Phi$ of certain quantity Φ. In the diagrammatic language, Φ is given by the sum of skeleton diagrams for Ω. A skeleton diagram is defined such that it does not include any self-energy parts as subdiagrams. This identification makes it possible to integrate the differential relation (G.21). We obtain

$$\beta(\Omega\{G\} - \Omega_0) = \Phi\{G\} - \mathrm{Tr}(\Sigma G) - \mathrm{Tr}\ln(G^{-1}g), \qquad (G.23)$$

where Ω_0 is the thermodynamic potential for the noninteracting system. Note that there is no explicit appearance of U in eqn (G.23). As a functional of G, Ω is stationary against the change of G, as can be checked easily [12,13]. This variational property is very analogous to that in the Helmholtz free energy F against the change in the chemical potential, or the Gibbs energy G against the change in the number of particles [14]. In all cases, the role of the Legendre transformation of natural variables is significant in the variational property.

Let us consider the zero-temperature limit of eqn (G.23) in the imaginary frequency domain. The contribution of a given diagram to Φ remains the same if the internal frequency $i\epsilon$ of a Green function is increased. In other words, for this change we have

$$\delta\Phi = \delta\epsilon \, \mathrm{Tr}\left(\Sigma\frac{\partial G}{\partial\epsilon}\right) = 0. \tag{G.24}$$

By partial integration, the above equality leads to

$$\mathrm{Tr}\left(G\frac{\partial\Sigma}{\partial\epsilon}\right) = 0. \tag{G.25}$$

With this setup, we represent the total number of electrons N in the system as

$$N = \mathrm{Tr}\exp(i\epsilon 0_+)G = -\mathrm{Tr}\exp(i\epsilon 0_+)\frac{\partial\ln G}{i\partial\epsilon}, \tag{G.26}$$

where we have used eqn (G.25) in the second equality. For definiteness, we first consider the homogeneous system and replace the trace by

$$\mathrm{Tr} \rightarrow \sum_k \int_{-\infty}^{\infty}\frac{d\epsilon}{2\pi} = \sum_k \int_C \frac{d\omega}{2\pi i}, \tag{G.27}$$

with $i\epsilon = \omega$ and the integration contour C as shown in Fig. G.1. Here we have utilized the convergence factor $\exp(i\epsilon 0_+)$ in deforming the integration contour. Then we obtain

$$N = \sum_k \int_{-\infty}^{0}\frac{d\omega}{\pi}\,\mathrm{Im}\,\frac{\partial\ln G(k,\omega+i\delta)}{\partial\omega} = \frac{1}{\pi}\mathrm{Im}\sum_k \ln[-G(k,i0_+)]. \tag{G.28}$$

The minus sign with $G(k,i0_+)$ arises from the contribution at $\omega \rightarrow -\infty$. In the Fermi liquid, the self-energy $\Sigma(k,i0_+)$ becomes real because of the restriction in the available damping processes; hence, $G(k,i0_+)$ is also real. If $G(k,i0_+)$ is positive, which is the case for small $|k|$, the phase $\mathrm{Im}\,\ln[-G(k,i0_+)]$ of the Green function is π. On the other hand, $G(k,i0_+)$ becomes negative as $|k| \rightarrow \infty$. The surface of k where $G(k,i0_+)$ changes sign is just the Fermi surface. Thus, eqn (G.28) tells us that the volume inside the Fermi surface of the interacting system remains the same as that of the noninteracting one with the same number of particles, N [12]. This property is called the Luttinger sum rule, which distinguishes the itinerant f electrons from the localized ones with no participation in the Fermi volume.

We remark that the sum rule can in fact be applied to systems with finite magnetization or without translational symmetry. By shifting the internal energy for up-spin Green functions only, we obtain eqn (G.24) for the up spin. This is because the Coulomb interaction does not mix different spin components. Then we obtain eqn (G.25) for each

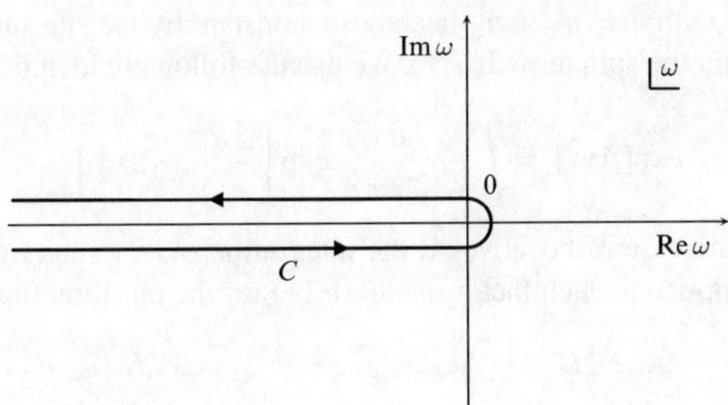

FIG. G.1. Deformation of the integration contour from along the imaginary axis to the one encircling the negative real axis.

spin component. In terms of the Green function $G_\sigma(\boldsymbol{k}, \omega + i\delta)$ with z-component of spin $\sigma/2$, the total magnetization $M = N_\uparrow - N_\downarrow$ is given by

$$M = \frac{1}{\pi}\operatorname{Im}\sum_{\boldsymbol{k}\sigma}\sigma\,\ln[-G_\sigma(\boldsymbol{k}, i0_+)]. \tag{G.29}$$

For systems without the translational invariance, the most successful application is the proof of the Friedel sum rule for the Anderson model [15]. If the perturbation theory converges, in the latter case we have the relation

$$n_{\mathrm{f}} = \frac{2}{\pi}\operatorname{Im}\ln[-G_{\mathrm{f}}(i0_+)], \tag{G.30}$$

where n_{f} is the number of f electrons and $G_{\mathrm{f}}(i0_+)$ is the f-electron Green function at the Fermi surface. The above relation is called the Friedel sum rule. The number n_{f} is not necessarily an integer in this case.

G.4 Path integral over auxiliary fields

Instead of dealing with the electron–electron interaction as it stands, it is often useful to introduce auxiliary fields so that the electrons interact via these fields. Consider the operator

$$\exp(-\delta\tau\, U n_\uparrow n_\downarrow), \tag{G.31}$$

which appears in eqn (F.11) for the partition function. Here we concentrate on a particular site and, for simplicity, discard the site index. The operator $n_\uparrow n_\downarrow$ can be written in alternative forms as

$$n_\uparrow n_\downarrow = \tfrac{1}{2}n - 2(S^z)^2 = \tfrac{1}{2}n - \tfrac{2}{3}\boldsymbol{S}\cdot\boldsymbol{S}. \tag{G.32}$$

The density term with $n = n_\uparrow + n_\downarrow$ becomes a constant by the site summation and we concentrate first on the spin term $2(S^z)^2$. We use the following identity:

$$\exp[Ay^2] = \int_{-\infty}^{\infty} \frac{dx}{\sqrt{2\pi A}} \exp\left[-\frac{x^2}{A} - 2xy\right], \qquad (G.33)$$

with $A > 0$. In the case of negative A, the integration over x runs from $-i\infty$ to $i\infty$. This identity is applied to each factor in eqn (F.11) for the partition function with

$$A = 2U, \qquad y = \sqrt{\delta\tau}S^z, \qquad x = \sqrt{\delta\tau}\phi. \qquad (G.34)$$

We then obtain the following expression for the Hubbard model:

$$Z = \int \mathcal{D}\eta^* \,\mathcal{D}\eta \,\mathcal{D}\phi \exp\left[-\int_0^\beta d\tau \,\mathcal{L}(\tau)\right], \qquad (G.35)$$

$$\mathcal{L}(\tau) = \mathcal{L}_0(\tau) + \frac{1}{2U}\sum_i \phi_i(\tau)^2 + 2\sum_i \phi_i(\tau)S_i^z(\tau)$$

$$= \mathcal{L}_0(\tau) + \mathcal{L}_\phi(\tau), \qquad (G.36)$$

where $\mathcal{L}_0$ is the Lagrangian without the interaction, and we have discarded the constant term in the Lagrangian. The procedure of introducing the auxiliary fields ϕ_i is called the Hubbard–Stratonovich transformation.

One can perform the integration over the Grassmann numbers exactly because the action is bilinear with respect to them. The result is given by

$$Z = \int \mathcal{D}\phi \exp\left[-\int_0^\beta d\tau \sum_i \frac{1}{2U}\phi_i(\tau)^2 + \mathrm{Tr}\ln(g^{-1} - \phi\sigma_z)\right], \qquad (G.37)$$

where Tr is the trace over imaginary time and space, and g in the logarithm is the unperturbed green function matrix. Thus, the original Fermion problem is reduced to another one with fluctuating magnetic field ϕ.

The spin–spin correlation function in the imaginary time is given by

$$\langle T_\tau S^z(\tau)S^z(\tau')\rangle = Z^{-1}\int \mathcal{D}\eta^*\,\mathcal{D}\eta\,\mathcal{D}\phi \exp\left[-\int_0^\beta d\tau\,\mathcal{L}_0(\tau)\right]$$

$$\times \frac{\partial^2}{4\partial\phi_i(\tau)\partial\phi_j(\tau')}\exp\left[-\int_0^\beta d\tau\,\mathcal{L}_\phi(\tau)\right]. \qquad (G.38)$$

Thus, $\partial/\partial\phi_i$ acts like the spin operator, which is natural since ϕ_i means a fluctuating magnetic field conjugate to S_i^z. In the present case of the Gaussian distribution of ϕ_i, one can carry out a partial integration in eqn (G.38) and obtain, for the magnetic susceptibility,

$$\chi(\boldsymbol{q}, i\nu) = \frac{1}{U}\left[\frac{1}{U}\langle|\phi(\boldsymbol{q}, i\nu)|^2\rangle - 1\right], \qquad (G.39)$$

where we have introduced the Fourier transform $\phi(\boldsymbol{q}, i\nu)$ of $\phi_i(\tau)$ and $\langle\cdots\rangle$ means the statistical average. This equation shows that the field $\phi_i(\tau)$ also implies a fluctuating magnetization upon proper scaling and subtraction. The subtraction removes the part of $\langle |\phi(q, i\nu)|^2 \rangle$ which remains in the high-frequency limit.

As the simplest application of the above formalism, we derive the RPA susceptibility. We expand the logarithm in eqn (G.37) up to second order in ϕ. Then, after a Fourier transformation, we obtain

$$\chi(q, i\nu) = \frac{1}{U} \left(\frac{1}{1 - U\chi_0(q, i\nu)/2} - 1 \right) = \frac{\chi_0(q, i\nu)}{1 - U\chi_0(q, i\nu)/2}, \tag{G.40}$$

which corresponds to the RPA discussed in Chapter 1.

We note that the result obtained above in fact depends on the particular choice of the decomposition in eqn (G.32). If we chose the second decomposition, we would have obtained the susceptibility where U in eqn (G.40) replaced by $U/3$. It is evident that the lowest-order expansion of the logarithm caused the difference. In order to obtain the result correct to $O(U)$ for $\chi(q, i\nu)$, one has to expand the logarithm up to $O(\phi^4)$, which gives the correct result for $\langle |\phi(q, i\nu)|^2 \rangle$ to $O(U^2)$. The first factor $1/U$ in eqn (G.39) reduces the order of U by one, so that the lowest-order expansion of the logarithm is not sufficient to obtain the $O(U)$ result for the susceptibility. Thus, the RPA result obtained above is fortuitous as the lowest-order theory.

It is more convenient for numerical calculation, such as the Monte Carlo simulation, to use the Ising-type discrete auxiliary field $\sigma = \pm 1$ instead of using the continuous one [16]. The identity we use now is the following:

$$\exp[A(S^z)^2] = \frac{1}{2} \sum_\sigma \exp(-B\sigma S^z), \tag{G.41}$$

where B satisfies the relation $\exp(A/4) = \cosh(B/2)$ with $A, B > 0$. This relation is easily confirmed by applying both sides of eqn (G.41) to each eigenstate of S_z. We can solve for B as

$$B = 2\ln[\exp(A/4) + \sqrt{\exp(A/2) - 1}]. \tag{G.42}$$

Thus, the partition function is written, after introducing the Ising variable σ_i to each site and discretizing τ, as

$$Z = \prod_{i,j} \int d\eta_i(\tau_j)^* \, d\eta_i(\tau_j) \sum_{\sigma_i(\tau_j)} \exp[-\delta\tau \sum_j \mathcal{L}(\tau_j)], \tag{G.43}$$

$$\mathcal{L}(\tau_j) = \mathcal{L}_0(\tau_j) + I \sum_i S_i^z(\tau_j)\sigma_i(\tau_j), \tag{G.44}$$

where $\cosh(\delta\tau I/2) = \exp(\delta\tau U/2)$ with $\delta\tau = \tau_j - \tau_{j-1}$. Note that one has $I \gg U$ as $\delta\tau$ approaches 0. The path integral over the fermions can be carried out exactly as before. Thus, one is left with the partition function of a fictitious Ising model with interactions dependent on the space and time difference. The summation over the Ising variables is conveniently dealt with by the Monte Carlo sampling.

Bibliography

[1] R. Kubo, *J. Phys. Soc. Jpn.* **12**, 570 (1957).

[2] I. Lindgren and J. Morrison, *Atomic many-body theory* (Springer Verlag, Berlin, 1986).

[3] P. Nozieres and A. Blandin, *J. Physique* **41**, 193 (1980).

[4] Y. Kuramoto, *European J. Phys. B* **5**, 457 (1998).

[5] P. W. Anderson, *J. Phys. C* **3**, 2439 (1970).

[6] A. A. Abrikosov and A. A. Migdal, *J. Low Temp. Phys.* **3**, 519 (1970).

[7] J. Gan, N. Andrei, and P. Coleman, *Phys. Rev. Lett.* **70**, 686 (1993).

[8] H. Mori, *Prog. Theor. Phys.* **34**, 399 (1965).

[9] R. Kubo, *J. Phys. Soc. Jpn.* **17**, 206 (1962).

[10] A. A. Abrikosov, L. P. Gorkov, and I. E. Dzyaloshinskii, *Methods of quantum field theory in statistical physics* (Dover, New York, 1975).

[11] J. Zinn-Justin, *Quantum field theory and critical phenomena* (Oxford University Press, Oxford, 1993).

[12] J. Luttinger and J. Ward, *Phys. Rev.* **118**, 1417 (1960).

[13] G. Baym, *Phys. Rev.* **127**, 835 (1962).

[14] C. De Dominicis and P. C. Martin, *J. Math. Phys.* **5**, 14 (1964).

[15] D. C. Langreth, *Phys. Rev.* **150**, 516 (1966).

[16] J. E. Hirsch, *Phys. Rev. B* **28**, 4059 (1983).

INDEX